Master FreeCAD v1.1.0 for CNC

From First Sketch to G-Code with a Certification Project

Monte North
BEd.(Sec.) BSc. MInfoTech

Northward House Publishing
Brisbane, Australia

Northward House

First published 2026

Published by Northward House Publishing
Brisbane, Queensland, Australia
northwardhouse.com.au

No Liability / Accuracy Disclaimer

While every effort has been made to ensure the accuracy of the information contained in this publication, the author and publisher make no representations or warranties of any kind, express or implied, with respect to the accuracy, completeness, or fitness for purpose of the contents. Machine settings, feed rates, safety procedures, software interface descriptions, and GRBL parameters are provided as starting-point guidance only. CNC machining involves inherent risks. The author and publisher accept no responsibility or liability for any loss, injury, damage, or expense — whether direct, indirect, incidental, or consequential — arising from use of or reliance on information in this publication.

Users are responsible for verifying all settings, procedures, and safety measures against the specifications of their own equipment, software version, and working environment. Always follow the safety instructions provided by your machine manufacturer.

Software and Version Notice

All content in this textbook is written for FreeCAD version 1.1.0. FreeCAD is an open-source project under active development. Menu locations, tool names, and workflow steps may differ in other versions. FreeCAD is published under the LGPL v2+ licence. Universal G-code Sender (UGS) is published under the Apache 2.0 licence. GRBL is published under the MIT licence. None of these projects are affiliated with or endorsed by Northward House Publishing.

ISBN 978-1-7640698-8-5

Student Edition
Printed in Australia

For all lifelong learners.

"The fastest way to learn is to make something you can hold in your hands."

Contents

Preface

This textbook is founded on a simple principle: the most effective way to learn design software is through the creation of tangible outcomes. When learners move beyond screen-based exercises and produce physical artefacts, the feedback becomes immediate and meaningful. Errors are no longer abstract—they are observed in material, finish, and function. This direct connection between design decisions and physical results is central to skill development in CAD/CAM and digital fabrication.

Many introductory CAD courses emphasise procedural knowledge: identifying tools, following demonstrations, and reproducing guided examples. While these approaches build familiarity, they do not always develop independent design capability. The transition from replication to creation requires learners to apply judgement, adapt processes, and respond to real-world constraints. This textbook is structured to support that transition.

FreeCAD has been selected as the primary software platform due to its accessibility, capability, and longevity. As an open-source application, it provides a comprehensive environment for parametric modelling, multi-body design, and integrated CAM workflows without reliance on proprietary licensing. This ensures that the skills developed through its use remain accessible to learners beyond the duration of the course.

The CNC platform used throughout the course has been selected for similar reasons. Desktop-scale machines provide a practical and cost-effective entry point into digital fabrication while maintaining sufficient precision to produce high-quality outcomes. Together, FreeCAD and a GRBL-based CNC system form a complete and replicable workflow from concept through to manufactured object.

The structure of this textbook is informed by Bloom's Revised Taxonomy (Anderson & Krathwohl, 2001), which describes the progression from foundational knowledge to independent creation. The course is sequenced to develop capability incrementally: beginning with identification and understanding of tools and concepts, progressing through application and analysis, and culminating in evaluation and independent design. Each stage builds deliberately on prior learning.

The certification project represents the culmination of this progression. Learners are provided with a design brief and a defined set of tools, but no prescribed solution. The outcome is an original, self-directed design that demonstrates the integration of modelling, toolpath generation, and fabrication processes. The emphasis is not on a single correct result, but on the learner's ability to synthesise knowledge and produce a complete, functional artefact.

This textbook has been developed with reference to the broader FreeCAD community, whose contributions have supported the ongoing evolution of the software and its application in education and industry. Community resources, including documentation and forum discussions, have informed aspects of the workflows presented throughout this text.

Monte North
Everton Hills, Queensland, 2025

About the Author

Monte North

BEd.(Sec.) · BSc. (Mathematics & Computer Science) · MIT. (Information Technology). ·

Monte North is a secondary school teacher with Education Queensland, teaching Digital Technology, Design & Technology, Mathematics, and Science since 2007 across multiple schools in Queensland. This course is based on his classroom practice, focusing on hands-on, technology-integrated learning.

Before teaching, he worked as a Senior Engineer and Cybersecurity Consultant with organisations including CITEC, AAPT Networks, and NCS Professionals, building expertise in hardware, firmware, and systems integration. He holds a Master of Information Technology from Queensland University of Technology and degrees in Science and Education from The University of Queensland.

The course is structured using Bloom's Taxonomy to support progressive skill development. Tools such as FreeCAD and the CNC 4040 Pro Max are used to provide accessible, practical design and manufacturing experiences.

Education

Masters of Information Technology (Distinction) — QUT, Network Management & Cybersecurity
Bachelor of Education (Distinction) — UQ, Secondary Teaching
Bachelor of Science — UQ, Mathematics and Computer Science

info@northwardhousepublishing.com.au

Northward House Publishing

www.northwardhousepublishing.com.au

CHAPTER
1
of 24

PART ONE — BEFORE YOU BEGIN

Course Introduction & Philosophy

Why making things is the best way to learn — and what you are about to build

Chapter Learning Objectives

By the end of this chapter, you will be able to:

- Describe what FreeCAD is and why it is used in this course
- Explain the concept of parametric CAD in plain English
- Identify the key components of the CNC 4040 Pro Max
- Outline the structure of the 10-week course
- Explain Bloom's Taxonomy and identify which level applies to each week
- Navigate the textbook's layout, icons, and features confidently

A Letter to the Student

You are about to begin a journey that starts at a computer screen and ends with a physical object in your hands.

That object — whether it is a simple name badge, a precision finger joint box, or an interlocking gear pair — will be something you designed yourself, from scratch, using one of the world's most capable free software tools: FreeCAD. You will not just watch someone else build it, and you will not copy a template. You will design it, set up the machine instructions, load the file, press go, and hold the finished result in your hands.

This course was built on a simple idea: that the fastest way to learn design software is to make things you can actually touch. When a sketch constraint is wrong, your piece comes out the wrong size — and you feel it immediately. When a toolpath is set up incorrectly, you hear it in the first few seconds of cutting. That direct, physical feedback is something no amount of screen-based tutorials can replicate, and it is the reason this course is structured the way it is.

You do not need any prior experience with CAD software, engineering drawing, or CNC machines to begin this course. You need curiosity, a willingness to make mistakes and learn from them, and enough patience to follow a process carefully. Everything else will come.

Each week builds on the last. Each project is a little more complex than the one before it. By the time you reach the certification project at the end, you will be doing things in Week 11 that would have seemed impossibly advanced on Day 1. That progression is not an accident — it is exactly how the course was designed, using a well-established educational framework called Bloom's Taxonomy, which you will meet properly a little later in this chapter.

Welcome. Let's get started.

Why Learning Through Making Works

There is a long tradition in education of separating theory from practice — students learn concepts in a classroom, and then, separately, apply them in a workshop or laboratory. This course does not do that. In FreeCAD for CNC Makers, theory and practice happen at the same time, on the same screen, with the same file.

Consider what happens when you constrain a sketch incorrectly in FreeCAD. The sketch turns red, and the software refuses to let you proceed. The error message points directly to the problem. You fix it, the sketch turns green, and you move on — having learned precisely what that constraint does and why it matters. That feedback loop — design, attempt, error, correction, understanding — happens dozens of times in a single lesson, and each repetition deepens your knowledge in a way that reading about constraints never could.

Now consider what happens when your design reaches the CNC machine. If your Z zero is set incorrectly by even a millimetre, the engraving is too shallow or too deep. If your feed rate is too high, the machine chatters and the surface looks rough. If you forgot to add tabs to a profile cut, the piece breaks free and is thrown across the workshop. Each of these outcomes is a lesson — immediate, memorable, and permanent.

Research in educational psychology has consistently shown that learners retain information far more effectively when they encounter it in a context that is meaningful and emotionally engaging. Holding something you made yourself, understanding every decision that went into it, and seeing where it could be improved — that is deeply meaningful. It engages you in a way that passive learning simply does not.

> 🤔 **Stop and Think**
> Think about something practical you have learned well — cooking a dish, fixing something mechanical, playing a musical instrument. How did you actually learn it? Was it from reading about it, or from doing it repeatedly and adjusting as you went? Keep that experience in mind as you work through this course.

This course also acknowledges that making mistakes is not a sign of failure — it is the mechanism of learning. You will make mistakes in the Sketcher. You will produce G-code that does not cut what you expected. You will assemble a joint that is too tight. These are not setbacks; they are data. The course is designed to give you the tools to diagnose and fix every problem you encounter, and each time you do, your understanding of FreeCAD and CNC machining deepens.

By the end of the course, you will not just know how to use FreeCAD — you will understand why it works the way it does, and you will have the confidence to figure out what to do when something unexpected happens. That is the difference between knowing a software package and being competent in it.

What is FreeCAD?

FreeCAD is a free, open-source, parametric 3D CAD (Computer-Aided Design) application. That sentence contains a lot of important words, so let's unpack them one at a time.

Free and Open-Source

FreeCAD costs nothing to download, install, and use — not now, and not in the future. It is developed and maintained by a global community of volunteers and professionals who contribute their time and expertise because they believe that powerful design tools should be accessible to everyone, not just those who can afford expensive software licences.

Being open-source means that the source code — the actual instructions that make the software run — is publicly available. Anyone can inspect it, suggest improvements, or build extensions for it. This is why FreeCAD has a rich ecosystem of community-developed add-ons, including the FCGear workbench you will use in Week 7 to generate mathematically precise gear geometry.

> 📖 **Key Term: Open-source**
> Software whose source code is publicly available and free to use, modify, and distribute. FreeCAD is open-source under the LGPL v2+ licence.

3D CAD

CAD stands for Computer-Aided Design. It is the use of computer software to create precise technical drawings and 3D models. Before CAD, designers and engineers produced everything by hand on drawing boards — an enormously time-consuming process with no easy way to update a design once it was drawn.

FreeCAD allows you to create fully three-dimensional models of objects — not just flat pictures of them. You can rotate your design, view it from any angle, measure every dimension, and immediately see how a change to one part affects the whole. When your model is complete, it can be handed directly to a manufacturing process — in our case, the CNC 4040 Pro Max — with mathematical precision.

Version 1.1.0

All content in this textbook is written specifically for FreeCAD version 1.1.0. This is an important detail because FreeCAD is actively developed, and menu locations, tool names, and workflow steps can change between versions. If you are using a different version, most of the content will still apply, but some specific steps may differ slightly. The FreeCAD forum (forum.freecad.org) is an excellent resource for version-specific questions.

> 📋 **Note**
> All menu paths, screenshots, and step-by-step instructions in this textbook have been verified against FreeCAD v1.1.0. If your installation shows a different layout, check that you are using the correct version before assuming the textbook is wrong.

What is Parametric CAD? A Plain-English Explanation

The word 'parametric' is the single most important concept in this course. Understanding it early will change the way you think about design. Let's explain it with an everyday example.

Imagine you are designing a rectangular shelf bracket. You draw the bracket, set the width to 200 millimetres, the height to 150 millimetres, and the thickness to 10 millimetres. In a traditional ('dead') drawing — the kind produced by basic drawing software — those numbers are locked in. If your client changes the brief and needs the bracket to be 250 millimetres wide, you have to redraw it from scratch.

In a parametric CAD model, those dimensions are not locked. They are variables — called parameters — that drive the model's geometry. The model does not contain a fixed shape; it contains a set of rules that describe the shape. Change the width parameter from 200 to 250, press Refresh, and the entire model updates automatically. The height proportions shift to match, the mounting holes reposition themselves, and all the associated manufacturing instructions update in seconds.

In FreeCAD, the primary tool for managing parameters is the Spreadsheet workbench, which you will learn in Week 3. A parameter spreadsheet is simply a list of named values — Width = 200, Height = 150, Thickness = 10 — that your sketches and features reference instead of using fixed numbers. It sounds simple, and the principle is. But the practical power it gives you is enormous.

> 📖 **Key Term: Parametric design**
> A design approach in which the geometry of a model is controlled by variables (parameters) rather than fixed dimensions. Changing a parameter value automatically updates the entire model.

Here is a concrete example of why this matters in a CNC workshop. Suppose you design a set of 10 identical shelf brackets for a client. Three months later, they ask for the same brackets but 20% larger. In a traditional workflow, you redraw all 10 brackets. In a parametric workflow, you open your spreadsheet, change one value — say, a Scale parameter from 1.0 to 1.2 — press Refresh, and all 10 brackets update. You re-export the G-code files and you are done. What might have been a two-hour redrawing task becomes a two-minute update.

This is not an exotic, advanced concept reserved for professional engineers. It is a foundational way of working that you will begin using in Week 3 of this course and apply in every project from that point forward. By the time you reach the certification project, parametric design will be as natural as saving a file.

> **Stop and Think**
> Can you think of a situation in everyday life — a cooking recipe, a construction plan, a clothing pattern — that would benefit from this kind of 'change one value, update everything' approach? What would the equivalent of the spreadsheet be in that context?

The CNC 4040 Pro Max — Your Output Device

FreeCAD is where you design. The CNC 4040 Pro Max is where your design becomes a physical object. Understanding both — and the connection between them — is essential to making good decisions at every stage of the process.

A CNC (Computer Numerical Control) router is a machine that moves a spinning cutting bit across a flat material in precisely controlled patterns. The patterns are defined by G-code — a set of movement instructions that your computer sends to the machine. Every line of G-code says something like: 'Move to X=50, Y=30, Z=-2 at a speed of 1000 millimetres per minute.' The machine follows those instructions exactly, without judgment or interpretation.

This last point is worth dwelling on. The CNC machine does exactly what it is told. It has no way of knowing whether your instructions are sensible or dangerous. If you tell it to plunge 30 millimetres deep into a 12-millimetre piece of MDF, it will attempt to do precisely that. This is why learning to produce correct, safe G-code — through FreeCAD's CAM workbench and verification in UGS — is not just a technical skill. It is a safety responsibility.

> ⚠ **Important**
> The CNC machine does exactly what its G-code instructions say — nothing more, nothing less.
> Always verify your toolpath in the CAM Simulator and in UGS before sending any job to the machine.
> Never send a job to the machine that you have not personally inspected in the visualiser.

Key Specifications

The CNC 4040 Pro Max is a desktop-format CNC router well-suited to timber, MDF, plastics, and soft metals. Its key specifications are worth knowing, because they define what is and is not possible with your designs.

Work area	400 × 400 mm (X × Y axes)
Z-axis travel	Approximately 70 mm
Controller	GRBL — open-source CNC motion controller firmware

Spindle collet	ER11 — accepts 1/8-inch (3.175 mm) and 6 mm shank bits
Connection	USB — communicates with your computer via Universal G-code Sender (UGS)
Materials (course)	Timber (pine, hardwood), MDF — all projects in this course
Materials (capable)	Soft metals (aluminium with appropriate settings), plastics, foam

The 400 × 400 mm work area defines the maximum size of any single piece you can cut. You will notice in Week 10 that the Tool Rack project approaches this maximum — that is intentional, and it teaches you to design with physical machine constraints in mind. Professional CNC work always begins with an understanding of what the machine can and cannot do.

> 📖 **Key Term: GRBL**
> Open-source firmware that controls the motion of CNC machines. It runs on the machine's controller board and interprets G-code commands into precise motor movements. Pronounced 'gerbil'.

The GRBL controller is what makes the CNC 4040 Pro Max respond to your computer's instructions. When you send a G-code file from UGS, GRBL reads each line and converts it into electrical signals that drive the stepper motors on each axis. Understanding that GRBL is the interpreter between your G-code and the machine's physical movement helps you make sense of the machine's behaviour — why it homes to a specific corner on start-up, why it stops if it receives an unexpected command, and why the settings matter as much as the toolpaths.

How This Course is Structured

The course runs for ten instructional weeks — 50 teaching days in total — followed by a certification week. Each week follows an identical structure: four days of teaching, followed by one project day. This rhythm is not arbitrary. It reflects a deliberate learning design principle: introduce new skills over four days, then immediately consolidate them by applying everything to a real, physical project on Day 5.

The Daily Structure

Each teaching day (Days 1–4 of each week) follows a consistent four-part structure. This consistency is intentional — when the structure of a lesson is predictable, you can focus your cognitive effort on the content rather than on figuring out what is expected of you.

Theory (20–30 min)	Your instructor introduces the day's concepts with explanations, diagrams, and demonstration. This is the 'what' and 'why' before the 'how'.
Demonstration (20 min)	A worked example is completed live, either by the instructor on a projected screen or in a video resource. Follow along but do not copy — watch first, then do.
Guided practice (30–40 min)	You work through the day's exercises with the textbook as your guide. Your instructor circulates and answers questions. Mistakes are expected and valuable here.
Independent exercise (20–30 min)	You complete a set exercise independently, without step-by-step guidance. This is where real learning happens — the slight discomfort of working without a safety net.

Project Day — Day 5 of Each Week

Every fifth day is a project day. You receive a project brief — a description of what you are making and the requirements it must meet — and you complete the entire design-to-machine cycle independently. There is no step-by-step walkthrough for project day. Your instructor is available for consultation, not instruction.

This distinction matters. On a project day, you are the designer. You make the decisions. If you are unsure of something, the first resource to consult is this textbook. The second is the FreeCAD documentation. The third is your instructor. Getting comfortable with that problem-solving sequence is itself an important skill that the course is deliberately building.

> 📋 **Note**
> Project day submissions include three components: your FreeCAD .FCStd file, your G-code .nc file, and a completed project reflection sheet. All three are required for a complete submission. Cutting the piece without submitting the files is an incomplete submission.

The Ten Projects and the Certification

The ten weekly projects follow a carefully designed escalating difficulty curve. Early projects build core skills in a low-stakes environment — a mistake costs a small piece of MDF, not a difficult multi-hour model. Later projects require you to synthesise multiple skill areas into a single, complex design. The certification project requires you to do this entirely independently, with no step-by-step guidance at all.

The table below shows all ten projects at a glance, along with the Bloom's Taxonomy level each week operates at — a concept explained fully in the next section.

Wk	Days	Module	Project	Bloom's Level
1	1–5	**Introduction to FreeCAD & Sketching**	*Personalised Name Badge*	Remember
2	6–10	**Text, Engraving & First CAM Operations**	*Engraved Timber Coaster*	Understand
3	11–15	**Parametric Design & Spreadsheets**	*Parametric Luggage Tag*	Understand → Apply
4	16–20	**Arrays, Patterns & Repetitive Geometry**	*Decorative Patterned Tile*	Apply
5	21–25	**Importing Artwork & Organic Curves**	*Artistic Knotwork Panel*	Apply → Analyse
6	26–30	**Multi-Body Design & Joinery**	*Two-Panel Finger Joint Box*	Analyse
7	31–35	**Addons, Gears & Mechanical Parts**	*Spur Gear Pair Display*	Analyse → Evaluate
8	36–40	**3D Modelling — Loft, Chamfer & Fillet**	*Organic Relief Carving Panel*	Evaluate
9	41–45	**Advanced CAM & Technical Drawing**	*Decorative Wall Clock Face*	Evaluate → Create
10	46–50	**Synthesis — Full Workflow**	*Workshop Tool Rack*	Create

After Week 10, the certification project — the Keepsake Memory Box — is completed in Week 11. It is the most complex project in the course, and it incorporates skills from all ten previous modules. It is completed entirely independently. Full details are provided in Chapter 24.

Bloom's Taxonomy — The Learning Framework Behind This Course

In 1956, an educational psychologist named Benjamin Bloom led a committee that produced one of the most influential frameworks in the history of education. Their work, later revised by Anderson and Krathwohl in 2001, described learning as a hierarchy of cognitive skills — from the most basic (remembering facts) to the most sophisticated (creating original work). This framework became known as Bloom's Taxonomy.

You will see Bloom's Taxonomy referenced throughout this textbook, in your lesson plans, and in your assessment criteria. Understanding it is not just academic trivia — it is genuinely useful for knowing where you are in the learning process and what is expected of you at any given point in the course.

The Six Levels

Bloom's Taxonomy describes six levels of cognitive skill, each one requiring a deeper and more sophisticated engagement with knowledge than the level below it. The levels build on each other: you cannot reliably Apply something you do not first Understand, and you cannot Evaluate something you have not yet Applied.

Level	What It Means	In This Course, You Will...
Remember	Recall and recognise information.	*Name interface elements, recall tool names, identify constraint types from a list.*
Understand	Explain and interpret information in your own words.	*Describe what a constraint does, explain why parametric design is useful, summarise a CAM operation.*
Apply	Use knowledge to carry out a task or solve a familiar problem.	*Use Sketcher tools to build a model to given dimensions, execute a CAM job, set up a parametric spreadsheet.*
Analyse	Break information down, compare, find patterns, differentiate.	*Compare ZigZag vs Adaptive pocket strategies, analyse why a joint fit is too tight, deconstruct a complex design into its constituent steps.*
Evaluate	Make judgements, critique quality, justify decisions.	*Assess whether a chosen feed rate is appropriate for the material, critique your own project output against design intent, justify CAM operation choices.*
Create	Produce original work by synthesising knowledge and skills.	*Design an original parametric model from a brief, plan a complete multi-operation CNC job from scratch, produce the certification project independently.*

How Bloom's Maps to This Course

The course is intentionally structured so that you ascend through Bloom's levels progressively. You do not begin at the Create level — not because you are not capable of creative thought, but because meaningful creativity in a technical domain requires a foundation of knowledge, understanding, and practice. A musician cannot improvise jazz without first knowing scales. A CNC designer cannot create original complex work without first knowing how to constrain a sketch.

In Weeks 1 and 2, you are primarily working at the Remember and Understand levels. You are learning the names of tools, understanding what they do, and building your first mental map of FreeCAD. The exercises in these weeks ask you to recall, identify, and explain.

By Weeks 3 through 5, you are firmly in the Apply level. You are not being told exactly what to do step by step — you are using the skills you have built to produce specific, required outcomes. The exercises ask you to use, execute, and demonstrate.

Weeks 6 through 8 introduce Analyse and Evaluate. You begin making design decisions that require you to compare options, identify why something worked or did not work, and justify your choices. You are no longer just following instructions — you are thinking critically about the process.

Weeks 9 and 10 operate at the Evaluate and Create levels. You are designing original work, planning complex multi-operation jobs, and producing complete documented CNC projects with increasing independence. The certification project in Week 11 operates entirely at the Create level — there is no step-by-step guidance, and the design is entirely your own.

> **Tip**
> Whenever you feel stuck or frustrated in this course, it can help to ask: which Bloom's level am I working at right now? If the task feels difficult, it may be because you are being stretched to a new level — which is exactly the point. Difficulty is not a sign that something is wrong. It is a sign that learning is happening.

How to Use This Textbook

This textbook is designed to be used actively, not read passively. It is not a book you read cover-to-cover before you start. It is a working companion that you consult during lessons, on project days, and when you need to look up a concept or procedure.

Each chapter corresponds to a set of lessons in the course. Part One (Chapters 1–3, which you are in now) covers everything you need before Day 1 of Week 1. Part Two covers Weeks 1 and 2. Each subsequent Part covers two more weeks of the course. The chapters are written in the order you will need them — follow the course sequence and the textbook sequence will naturally match.

Callout Boxes — Your Visual Signposts

Throughout the textbook, you will encounter a set of distinct callout boxes that signal different types of content. Learning to recognise these quickly will help you navigate the book efficiently during lesson time, when speed of reference matters.

	Tip — A practical shortcut, a useful habit, or a technique that makes the process easier or faster.		Note — Additional context, a clarification, or information you should be aware of but that is not critical to the immediate task.
	Important — A step, rule, or warning that must not be skipped or ignored. These relate to quality, accuracy, or safety.		Key Term — A defined vocabulary term that you should know and be able to use correctly.
	Stop and Think — A reflection prompt that asks you to connect what you are learning to broader ideas or your own experience.		Learning Objectives — What you should be able to do after completing this section or chapter.
	Project Brief — Marks the beginning of a project day brief. Read this carefully before beginning any design work.		Assessment — Describes the criteria against which your work will be assessed.

Working Through the Exercises

Each teaching day includes both guided exercises (where the textbook walks you through each step) and independent exercises (where you are given a goal and must reach it using your skills). Do not skip the independent exercises. They are the most important part of each lesson, because they are where you find out what you actually know versus what you thought you knew.

When you encounter an error during an exercise, resist the temptation to immediately ask for help. Spend at least five minutes trying to diagnose and fix the problem yourself. Consult the chapter's content, check the Key Vocabulary, look at the exercise instructions again. If you are still stuck after five genuine minutes of effort, then ask. This '5-minute rule' is not about making things harder — it is about building the diagnostic habit that every competent FreeCAD user needs.

The Vocabulary Lists

At the end of each chapter, you will find a vocabulary table listing every Key Term introduced in that chapter. These terms are the technical language of FreeCAD and CNC machining, and using them correctly is a mark of competence. When your instructor asks you to explain something, using the correct terminology demonstrates understanding far more clearly than vague descriptions.

A complete glossary of all terms used across the entire textbook is provided in Appendix F at the back of the book.

Chapter Summaries

Each chapter ends with a concise summary of the key points covered. Use these summaries as quick reference cards when you are on a project day and need to recall a concept without re-reading an entire section.

Companion Digital Resources

This textbook is accompanied by a set of digital files that support the course projects. These files are available from your course coordinator or the course's digital learning platform. Understanding what each type of file is and how to use it will save you significant time across the course.

Project .FCStd files	A completed FreeCAD project file for every weekly project. Use these as reference if you get stuck, or to compare your approach with the intended solution. Do not simply copy these files — working through the design process yourself is how you learn.
Project .nc files	The completed G-code files for every project, ready to load directly into UGS. Useful for understanding what a correct G-code output looks like, and for comparing against your own exports.

SVG artwork files	Five prepared SVG files for use in Week 5 (Importing Artwork). These have been verified to import correctly into FreeCAD's Draft workbench without modification.
Job Setup Sheet templates	Blank and completed example Job Setup Sheets (introduced in Week 9). Print a blank one for every cutting session you run on the machine.
Assessment rubrics	Detailed rubrics for all 11 projects. Reading these before you begin a project will help you understand exactly what is being assessed and to what standard.

> ⚠ **Important**
>
> The companion files are learning aids, not shortcuts. Using a provided .FCStd file instead of building your own is like copying the answers to a maths exercise — you produce a correct result but learn nothing.
>
> Always build your own design files for each project. The companion files are for reference after the fact, not as a starting template.

Chapter Summary

Chapter Summary

- FreeCAD is a free, open-source parametric 3D CAD application. It costs nothing to download and use, and is actively developed by a global community.
- Parametric design means that model dimensions are controlled by variables (parameters) rather than fixed numbers. Changing one value can update an entire model automatically.
- The CNC 4040 Pro Max is a desktop CNC router with a 400 × 400 mm work area, controlled by the GRBL firmware and operated from your computer via UGS (Universal G-code Sender).
- The course runs for 10 weeks (50 days) with four teaching days and one project day per week, followed by a certification week (Week 11).
- Bloom's Taxonomy is a six-level hierarchy of cognitive skills: Remember, Understand, Apply, Analyse, Evaluate, and Create. The course ascends through all six levels, beginning at Remember in Week 1 and reaching Create in the final weeks.
- The textbook is a working companion, not a passive read. Use the callout boxes, vocabulary tables, and chapter summaries as active reference tools during lessons and on project days.
- Companion digital files (FreeCAD project files, G-code files, SVGs, rubrics) are available from your course coordinator. They are reference tools, not shortcuts.

Key Vocabulary — Chapter 1

Term	Definition
FreeCAD	A free, open-source, parametric 3D CAD application used throughout this course for designing CNC-ready models.
Open-source	Software whose source code is publicly available and free to use, modify, and distribute.
Parametric design	A design approach in which model geometry is controlled by named variables (parameters) that can be changed to update the entire model automatically.
Parameter	A named variable in a FreeCAD model or spreadsheet that controls a dimension or property. For example, Width = 150 mm.
CAD	Computer-Aided Design. The use of software to create precise 2D drawings or 3D models of objects.
CNC	Computer Numerical Control. A machining process where a cutting tool is moved by a computer following a set of numerical instructions (G-code).
GRBL	Open-source CNC controller firmware that runs on the CNC 4040 Pro Max's control board and interprets G-code into motor movements.
G-code	The programming language used to control CNC machines. A G-code file contains a series of movement and spindle instructions for the machine to follow.
UGS	Universal G-code Sender. The free software application used in this course to send G-code files from your computer to the CNC 4040 Pro Max via USB.
Bloom's Taxonomy	A hierarchical framework for classifying learning objectives and cognitive skills, ranging from Remember (lowest) to Create (highest).
.FCStd	The native file format for FreeCAD project files. All your design work is saved in this format.
.nc	A common file extension for G-code files. These files contain the machine instructions generated by FreeCAD's CAM workbench and sent to the CNC machine via UGS.
CAM	Computer-Aided Manufacturing. The process of using software to generate machine toolpaths and G-code from a 3D model.
Work area	The maximum size of material the machine can cut in a single setup. The CNC 4040 Pro Max has a 400 × 400 mm work area.
ER11 collet	The tool-holding system on the CNC 4040 Pro Max spindle. It accepts cutting bits with 3.175 mm (1/8-inch) or 6 mm diameter shanks.

Looking Ahead — Chapter 2

Chapter 2 covers everything you need before Day 1: a complete guide to setting up your workspace, installing FreeCAD and UGS, understanding how the CNC 4040 Pro Max is physically set up, and the safety rules and procedures that apply every time you use the machine. Do not skip Chapter 2 — the safety content in particular is non-negotiable, and the software setup steps will save you from common first-run frustrations.

CHAPTER 2 of 24

PART ONE — BEFORE YOU BEGIN

Equipment, Software & Safety

Setting up your hardware and software, and the safety rules that apply every time you use the machine

Chapter Learning Objectives

By the end of this chapter, you will be able to:

- Identify the key physical components of the CNC 4040 Pro Max by name and function
- Install FreeCAD v1.1.0 on Windows, macOS, or Linux
- Install Universal G-code Sender (UGS) and establish a connection to the CNC machine
- Describe what GRBL is and the role it plays in the CNC workflow
- Set up a safe and functional physical workspace for CNC operation
- Apply the course safety rules consistently every time you use the machine
- Use the pre-cut safety checklist before every cutting session

The CNC 4040 Pro Max in Detail

Before you design anything or open any software, spend time getting familiar with the physical machine. Understanding each component will help you make better design decisions and operate the machine safely.

The CNC 4040 Pro Max is a moving-gantry desktop CNC router. That means the cutting tool — mounted on the gantry — moves across a stationary work surface. The gantry travels in the X axis (left and right), the Z axis (up and down), and a bridge carries the entire assembly in the Y axis (forward and backward). The combination of these three axes of movement allows the tool to reach any point within the machine's work envelope.

Machine Components

Knowing the name of each part matters because your instructor, this textbook, and the wider CNC community all use these names consistently. When a problem is described as 'Z-axis chatter' or 'Y-axis binding', you need to know which axis is which.

Frame	The rigid aluminium extrusion structure that holds everything together. The quality of the frame determines the accuracy and rigidity of the machine.
Gantry	The bridge-like assembly that carries the Z-axis and spindle. It moves along the Y axis (front to back).
X-axis rail	The horizontal rail along which the Z-axis carriage moves left and right.
Y-axis rails	The two parallel rails (one on each side of the machine bed) along which the gantry moves.

Z-axis assembly	The vertical component that moves the spindle up and down. This controls cutting depth.
Spindle	The electric motor that spins the cutting bit at high speed. On the CNC 4040 Pro Max, the spindle accepts ER11 collets.
ER11 collet	The tool-holding chuck on the spindle. It clamps around the cutting bit's shank. ER11 collets accept shanks from 1 mm to 7 mm.
Stepper motors	Three electric motors (one per axis) that drive the machine's movement in precise incremental steps. Each step is a fraction of a millimetre.
Control board	The electronics that interpret G-code and drive the stepper motors. It runs the GRBL firmware.
Limit switches	Small sensors at the end of each axis travel. When triggered, they tell GRBL the machine has reached its home position.
Work bed	The flat surface on which your material is clamped. The CNC 4040 Pro Max has a 400 × 400 mm work bed.
T-slot table	The aluminium extrusion work bed with T-shaped slots that accept M5 or M6 bolts for clamping fixtures.
USB cable	Connects the control board to your computer. GRBL communicates with UGS through this connection.

📖 **Key Term: Work envelope**

The three-dimensional space within which the machine's cutting tool can reach. For the CNC 4040 Pro Max: 400 mm (X) × 400 mm (Y) × approximately 70 mm (Z). Any part of your design that falls outside the work envelope cannot be cut in a single setup.

The Axes Explained

CNC machines use a coordinate system based on three axes: X, Y, and Z. If you have ever used a graph with X and Y axes, this is the same idea extended into three dimensions.

X axis	Left and right across the machine bed. Positive X moves the tool to the right.
Y axis	Forward and backward along the machine bed. Positive Y moves the tool away from you (toward the back of the machine).
Z axis	Up and down. Positive Z moves the tool upward, away from the material. Negative Z moves the tool downward, into the material. This is the depth axis — setting Z zero correctly is critical to every cut.
Origin (0,0,0)	The reference point from which all movements are measured in a given job. You set this manually at the start of each cutting session. It does not have to be the corner of the machine — it is wherever you decide your design starts.

> **Tip**
> A useful memory aid: X goes across the page (left–right), Y goes up the page (forward–back), and Z goes through the page (in–out of the material). Think of X and Y as your design's flat footprint, and Z as the depth of the cut.

Understanding GRBL

GRBL (pronounced 'gerbil') is the firmware that runs on the CNC 4040 Pro Max's control board. Firmware is software that is permanently installed on a piece of hardware, rather than on a general-purpose computer. GRBL's sole job is to receive G-code instructions from your computer and convert them into precise electrical pulses that drive the stepper motors.

You will never write GRBL commands directly. FreeCAD generates the G-code, UGS sends it, and GRBL executes it. Your interaction with GRBL is indirect — through the settings in UGS and the G-code your CAM operations produce. However, knowing what GRBL is and how it behaves helps you understand several things that will otherwise seem mysterious:

- Why the machine moves to a specific corner when you press Home — GRBL uses limit switches to find a known reference position at start-up.
- What 'Alarm state' means in UGS — GRBL has locked the machine's movement because something unexpected happened. Type $X in the UGS console to clear the alarm.
- Why soft limits matter — GRBL can be configured with soft limits that prevent the machine from moving beyond its physical boundaries, even if a G-code command asks it to.
- Why the machine stops if the USB connection drops — GRBL treats a communication loss as a safety event and halts all movement.

> **Key Term: Firmware**
> Software that is permanently embedded in a hardware device. Unlike applications on your computer, firmware runs directly on the device's microcontroller. GRBL is the firmware on the CNC 4040 Pro Max's control board.

The key GRBL concepts you will encounter in this course are covered as they arise. A full reference table of GRBL settings (the $ parameter system) is provided in Appendix D for those who want to explore the machine's configuration more deeply. For now, the important things to know are: GRBL is what makes the machine respond to your instructions, and if the machine behaves unexpectedly, GRBL's status in the UGS console is the first place to look.

Installing FreeCAD v1.1.0

FreeCAD is available for Windows, macOS, and Linux. The installation process is straightforward on all three platforms. Follow the steps for your operating system below. If FreeCAD is already installed on your course workstation, skip this section and go directly to the First Launch section.

> ⚠ **Important**
>
> Always install FreeCAD v1.1.0 specifically for this course. Other versions may have different menu locations, tool names, and available features.
>
> The download is approximately 600 MB. Ensure you have a stable internet connection before beginning.

Windows Installation

1 **Go to the FreeCAD website**
Open your web browser and go to freecad.org. Click the large Download button on the homepage. You will be taken to the releases page.

2 **Download the Windows installer**
Look for the release labelled v1.1.0. Under the Windows section, download the file ending in .exe (the installer). The filename will be something like FreeCAD_1.1.0-WIN-x64.exe.

3 **Run the installer**
Once the download is complete, double-click the .exe file. If Windows asks 'Do you want to allow this app to make changes to your device?', click Yes. Follow the installer prompts: accept the licence agreement, choose an installation folder (the default is fine), and click Install.

4 **Launch FreeCAD**
Once installation is complete, click Finish. FreeCAD should launch automatically. If it does not, find FreeCAD in your Start menu and open it from there.

5 **Verify the version**
In FreeCAD, go to Help → About FreeCAD. The version number should show 1.1.0. If it shows a different version, you may have installed the wrong release — return to freecad.org and download v1.1.0 specifically.

macOS Installation

1 **Download the macOS package**
Go to freecad.org → Download → Releases. Under the macOS section, download the file ending in .dmg for v1.1.0. The filename will be something like FreeCAD_1.1.0-macOS-x86_64.dmg (Intel Mac) or FreeCAD_1.1.0-macOS-arm64.dmg (Apple Silicon Mac).

2 **Open the disk image**
Once downloaded, double-click the .dmg file. A virtual disk will mount and a window will open showing the FreeCAD application icon alongside a shortcut to your Applications folder.

3 **Drag to Applications**
Drag the FreeCAD icon into the Applications folder. This copies FreeCAD to your system.

4 **Handle the security prompt**
The first time you open FreeCAD on macOS, you may see a warning that the application is from an unidentified developer. To bypass this: right-click (or Control-click) the FreeCAD icon in your Applications folder and select Open. Then click Open in the dialog that appears. You only need to do this once.

5 **Verify the version**
Go to FreeCAD → About FreeCAD in the menu bar. Confirm the version shows 1.1.0.

💡 **Tip**

On Apple Silicon Macs (M1, M2, M3 chips), download the arm64 build for best performance. On older Intel Macs, download the x86_64 build. If you are unsure which chip your Mac has, go to Apple menu → About This Mac and look at the Processor field.

Linux Installation

FreeCAD is available for Linux through several methods. The AppImage method is recommended for this course as it works across all distributions without requiring administrative privileges.

1 **Download the AppImage**
Go to freecad.org → Download → Releases. Under the Linux section, download the AppImage file for v1.1.0. The filename will end in .AppImage.

2 **Make it executable**
Open a terminal and navigate to the folder where you downloaded the file. Run: chmod +x FreeCAD_1.1.0*.AppImage

3 **Launch FreeCAD**
In the terminal, run: ./FreeCAD_1.1.0*.AppImage — or double-click the AppImage file in your file manager. If prompted to install a desktop integration, click Yes for convenience.

4 **Verify the version**
Go to Help → About FreeCAD and confirm v1.1.0.

📋 **Note**

Ubuntu and Debian users can also install FreeCAD via the package manager (sudo apt install freecad), but the packaged version may not be v1.1.0. The AppImage method guarantees the correct version regardless of distribution.

FreeCAD First Launch — Setting Up for This Course

When FreeCAD opens for the first time, you will see the Start workbench — a welcome screen with recent files and news. Before you do anything else, apply the following preference settings. These are not required for FreeCAD to work, but they will make the course experience significantly smoother.

1. Go to Edit → Preferences (Windows/Linux) or FreeCAD → Preferences (macOS).
2. Under General → General: set the Language to English (Australian) if available, or English. Enable Autosave every 5 minutes.
3. Under Display → 3D View: ensure Navigation Style is set to CAD. This gives you the most intuitive mouse navigation for this course (middle mouse button to orbit, Shift+middle to pan, scroll to zoom).
4. Under Part Design: check that the 'Allow multiple solids in Part Design Body' option is available in your version and set it as appropriate for your workflow.
5. Close Preferences. Your settings are saved automatically.

> **Tip**
>
> FreeCAD's interface can be customised extensively, but resist the urge to change things in the first few weeks. A consistent, default interface makes it much easier to follow the step-by-step instructions in this textbook. Customise once you are confident.

Installing Universal G-code Sender (UGS)

Universal G-code Sender (UGS) is the software you will use to send your finished G-code files from your computer to the CNC 4040 Pro Max. It is free, open-source, and runs on Windows, macOS, and Linux. UGS connects to the machine via USB and communicates with the GRBL firmware.

> **Key Term: Universal G-code Sender (UGS)**
>
> Free, open-source G-code sending software that connects to GRBL-based CNC machines via USB. It provides a 3D toolpath visualiser, machine position display, manual jog controls, and a console for sending commands directly to GRBL.

Downloading and Installing UGS

1 Go to the UGS website

Open your browser and go to: github.com/winder/Universal-G-Code-Sender. Click the Releases link on the right side of the page, or go directly to the Releases section.

2 Download UGS Platform

Under the latest release, download the file labelled ugs-platform-app-[version]-[your-OS].zip. The Platform version (not Classic) is recommended for this course — it has a more complete interface including the 3D visualiser.

3	**Extract the zip file** Extract the downloaded zip file to a location on your computer. UGS does not require a traditional installer — it runs directly from the extracted folder.

4	**Launch UGS** Open the extracted folder. On Windows, double-click ugsplatform.exe. On macOS, open the .app file. On Linux, run the ugsplatform shell script. UGS will open to its main interface.

5	**Install Java if prompted** UGS requires Java to run. If UGS displays an error about Java not being found, download and install the latest Java Runtime Environment (JRE) from java.com, then relaunch UGS.

Connecting UGS to the CNC 4040 Pro Max

Once UGS is installed, you need to connect it to the machine. Before attempting to connect, ensure the machine is powered on and the USB cable is connected to your computer.

1. In UGS, look at the connection bar at the top of the window. You will see a Port drop-down, a Baud Rate field, and a Connect button.
2. Click the small refresh icon next to the Port drop-down to detect available ports. On Windows, you are looking for a port labelled COM3, COM4, or similar. On macOS/Linux, look for /dev/ttyUSB0 or /dev/tty.wchusbserial.
3. Set the Baud Rate to 115200. This is the standard communication speed for GRBL.
4. Select the correct port from the drop-down, then click Connect.
5. The Machine Status area (top right) should change from Disconnected to Idle. If it shows Alarm, type $X in the console at the bottom of the UGS window and press Enter to clear the alarm.
6. If the status shows Connected but then immediately disconnects, check the USB cable is firmly seated at both ends and try a different USB port on your computer.

> ⚠ **Important**
>
> If Windows does not show a new COM port when you connect the machine, you may need to install a driver for the CH340 USB-to-serial chip on the machine's control board.
>
> Download the CH340 driver from: wch-ic.com/downloads/CH341SER_EXE.html — install it, restart your computer, and reconnect the USB cable.

The UGS Interface — A Quick Tour

UGS has several panels that you will use throughout the course. Familiarise yourself with these now so you are not hunting for them when you are at the machine.

Connection bar	Top of the window. Port selection, baud rate, Connect/Disconnect button, and machine status (Idle/Run/Alarm/Hold).
Machine position	Shows the current X, Y, Z coordinates of the tool in both machine coordinates and work coordinates (your set zero).

Jog controls	Arrow buttons for manually moving the machine. Set the step size and feed rate before jogging. Use small steps when close to the material surface.
3D visualiser	Displays a 3D preview of the loaded G-code toolpath. Use this to verify every job before sending. Rotate the view with mouse drag, zoom with scroll wheel.
Console	A text input where you can type GRBL commands directly (e.g., $H to home, $X to clear alarm, ? to query status).
File actions	Open button to load a .nc G-code file. Send button to begin execution. Pause and Stop buttons for mid-job control.
Override controls	Sliders that let you adjust feed rate and spindle speed in real time during a running job. Useful for fine-tuning on the fly.

> 💡 **Tip**
> Run a dry-run air cut before your first real cutting session. Load a G-code file, raise the Z axis so the bit is well clear of any material, and press Send. Watch the visualiser and the machine move through the entire job with no material present. This is the safest possible way to verify a new job.

Optional Software

The two tools below are not required to complete the core course, but they are useful for specific projects and worth installing if you have the time.

Inkscape — For SVG Artwork Preparation

Inkscape is a free, open-source vector graphics editor. You will use it in Week 5 (Importing Artwork & Organic Curves) to prepare SVG artwork files for import into FreeCAD. Specifically, Inkscape allows you to convert text to paths, simplify overly complex node structures, and remove fills and colours that FreeCAD does not need.

Download Inkscape from inkscape.org. The installation process is identical to a standard application install on all platforms. Inkscape is a large, capable program — you do not need to learn it deeply. For this course, you only need three of its features: Path → Object to Path, Path → Simplify, and File → Save.

Candle — An Alternative G-code Sender

Candle is a simpler, lighter-weight G-code sender than UGS. It has a cleaner interface and is slightly easier to use for beginners. The course uses UGS as its primary sender because of UGS's superior 3D visualiser and broader feature set, but Candle is an excellent backup option if UGS has connection issues on your specific machine.

Download Candle from github.com/Denvi/Candle. The connection procedure is identical to UGS: select port, set baud rate to 115200, click Connect.

> **Note**
> Either UGS or Candle will work for every project in this course. If your institution has standardised on one or the other, use what is installed. The G-code files FreeCAD produces are compatible with both.

Setting Up Your Workspace

The quality and safety of your CNC work depends significantly on how your physical workspace is organised. A cluttered, poorly lit workspace is not just inefficient — it is a safety risk. Take the time to set up properly before your first cut.

Physical Layout

The ideal CNC workspace has the machine positioned so that you can access all four sides without obstruction. You need to be able to reach the e-stop button, change tools, and place and remove workpieces without awkward reaching or leaning over the machine while it is running.

Machine clearance	Minimum 300 mm clear space on all four sides of the machine. This gives you room to clamp workpieces and change tools without reaching across the cutting area.
Computer position	Your computer should be within arm's reach but positioned so that the screen is visible while you are standing next to the machine. You should be able to pause a job in UGS without stepping away from the machine.
E-stop accessibility	The emergency stop button (if fitted) must be immediately accessible — never block it with material, tools, or cabling.
Lighting	The cutting area should be well lit. Poor lighting makes it difficult to see the tool position, check clamping, and inspect the work surface for debris.
Power access	The machine and computer should be on a dedicated power board with an accessible master switch. In an emergency, being able to cut all power quickly is important.
Cable management	USB and power cables must be routed away from the machine's movement axes. A cable caught in the gantry can damage the machine and create an electrical hazard.
Floor surface	Avoid smooth, slippery floors near the machine if possible. MDF dust and chips can make a smooth floor treacherous.

Dust Management

CNC routing — especially in MDF — produces significant quantities of fine dust. MDF dust in particular contains formaldehyde resin from the manufacturing process, and prolonged exposure is a health risk. Do not underestimate this.

- Use a dust extraction system connected to the machine if available. A workshop vacuum with a fine-particle filter is a minimum.
- Even with extraction, wear a P2-rated dust mask whenever the machine is cutting. The fine particles generated by MDF pass through surgical masks and basic paper masks.
- After each session, vacuum the machine bed, gantry rails, and the surrounding floor. Do not blow dust with compressed air — this disperses it into the air you are breathing.
- Sweep up or vacuum chips before they accumulate underfoot.

> ⚠ **Important**
>
> MDF dust contains respirable wood fibres and formaldehyde from the binding resin. Regular exposure without adequate respiratory protection is associated with respiratory irritation and, with chronic exposure, serious respiratory conditions. Wear your dust mask every time — not just occasionally.

Safety Fundamentals

Safety is not a bureaucratic formality. It is a discipline — a set of consistent habits that prevent incidents before they happen. The rules in this section are not negotiable.

The CNC 4040 Pro Max is a tool that moves at high speed with a spinning cutting bit. When operated correctly, it is safe. When operated without proper procedures — even once — the consequences can include broken bits, damaged material, damaged machines, and, in the worst case, physical injury. The good news is that every serious CNC incident is preventable, and the prevention comes from consistent habits rather than technical expertise.

You do not need to be afraid of the machine. You need to respect it. These are different things. Fear leads to hesitation and poor decisions. Respect leads to consistent, careful procedure — which is exactly what the machine requires.

Personal Protective Equipment (PPE)

PPE is not optional on project days. Before the machine starts for any reason, the following must be worn:

Safety Rule

SAFETY GLASSES — must be worn whenever the machine's spindle is running, or whenever you are within 1 metre of the machine during a cut. Flying chips from a broken or loose bit can cause serious eye injury. Standard glasses are not sufficient — use rated safety glasses or a face shield.

HEARING PROTECTION — the spindle operates at 12,000–24,000 RPM and produces significant noise. Prolonged exposure to CNC routing noise without hearing protection contributes to permanent hearing damage. Use earplugs or earmuffs rated for the noise level.

DUST MASK — wear a P2-rated dust mask whenever the machine is cutting. This applies even if dust extraction is in use.

TIED-BACK HAIR AND CLOTHING — long hair must be tied back. Loose sleeves, scarves, and dangling jewellery must be secured or removed. The machine's moving parts can catch loose material.

The Never-Do List

The following actions are prohibited under any circumstances in this course. These rules exist because each one represents a scenario where a single lapse in procedure has previously resulted in a machine incident or injury in a CNC environment.

1. Never reach into the cutting area while the spindle is running — not to adjust the workpiece, not to clear chips, not for any reason.
2. Never leave the machine unattended during a cutting job. If you need to leave the room, pause the job in UGS first.
3. Never send a G-code file to the machine that you have not first verified in the UGS 3D visualiser.
4. Never override or bypass the pre-cut safety checklist, even for a 'quick' job or a simple cut.
5. Never use the machine if the workpiece is not securely clamped. A workpiece that moves during cutting can shatter the bit and become a projectile.
6. Never operate the machine if the e-stop button is inaccessible or not functioning.
7. Never change a cutting bit while the spindle is running or powered. Ensure the spindle has fully stopped before touching any tooling.
8. Never cut materials the machine is not rated for without explicit instructor approval (e.g., carbon fibre, which produces toxic dust; stone; metals other than soft aluminium).
9. Never work alone in the workshop without another person present or within calling distance.

Emergency Procedures

Every student must know these procedures before their first cutting session. Do not wait until there is an emergency to learn them.

Spindle does not stop	Type M5 in the UGS console and press Enter. If the spindle still does not stop, cut power to the machine using the master switch or power board. Do not touch the machine until the spindle has fully stopped.
Machine moves unexpectedly	Press the e-stop button immediately, or click the Stop button in UGS. Do not attempt to grab the gantry or any moving part.
Tool breaks during cut	Press Stop in UGS. Wait for the spindle to stop completely. Inspect the machine bed and surrounding area for fragments before removing the workpiece or changing the tool.
Fire or smoke	Press Stop, cut all power, use the nearest fire extinguisher if safe to do so. Do not use water on an electrical fire. Evacuate the room if the fire is not immediately controllable.
UGS connection drops mid-job	The machine will stop (GRBL halts on communication loss). Do not panic. Wait for all movement to stop, then note the G-code line number shown in UGS before the disconnect. Re-connect, re-zero if necessary, and restart from that line number.
Workpiece comes loose	Press Stop immediately. A free workpiece being driven by the cutting bit is unpredictable and potentially dangerous. Do not attempt to restrain it by hand.

The Pre-Cut Safety Checklist

The pre-cut safety checklist is the most important routine in this course. It takes approximately two minutes to complete. In those two minutes, it catches the majority of errors that would otherwise result in a damaged workpiece, a broken bit, or — in the worst case — an unsafe machine operation.

The checklist below is reproduced on a laminated card that lives at the machine. Before sending any G-code job to the CNC 4040 Pro Max — for any reason, regardless of how simple the job seems — you complete this checklist. Every item. Every time.

☑ SECTION 1 — Before You Start (PPE & Environment)

- ☐ Safety glasses are on.
- ☐ Hearing protection is on.
- ☐ Dust mask is on.
- ☐ Hair is tied back. Loose clothing is secured.
- ☐ No bystanders are within 1 metre of the machine.
- ☐ The e-stop button is accessible and functional.
- ☐ Dust extraction is running (if available).

☑ **SECTION 2 — Material & Clamping**

- ☐ Workpiece dimensions have been measured and match the CAM stock settings.
- ☐ Workpiece is securely clamped at a minimum of two points.
- ☐ No clamps are within the cutting path of any operation.
- ☐ The workpiece surface is clean and free of previous cut debris.
- ☐ The correct cutting bit is installed and the collet nut is tightened firmly.

☑ **SECTION 3 — Machine & Software**

- ☐ Machine is powered on and UGS shows 'Idle' status.
- ☐ Machine has been homed ($H or Home button) since power-on.
- ☐ X/Y/Z zero has been set to the correct position on the workpiece.
- ☐ The correct G-code file is loaded in UGS.
- ☐ The G-code file has been inspected in the UGS 3D visualiser — all operations are visible and correct.
- ☐ The estimated run time is known and you are prepared to supervise the full duration.

☑ **SECTION 4 — Final Check Before Sending**

- ☐ Feed rate is set correctly in UGS (100% unless a specific override is required).
- ☐ Spindle speed is correct in the G-code (check M3 S_____ in the file if unsure).
- ☐ You know how to pause the job (Pause button in UGS) and stop it (Stop button).
- ☐ You are ready to supervise the entire job without leaving the room.
- ☐ All checks above are complete. You are clear to send the job.

📋 **Note**

Print and laminate this checklist. Attach it to the machine or the wall directly beside the machine. It is the last thing you look at before pressing Send — every single time.

Troubleshooting Installation

If FreeCAD or UGS is not behaving as expected after installation, the following table covers the most common first-run issues. If your problem is not listed here, the FreeCAD forum (forum.freecad.org) and the UGS GitHub issues page are excellent resources staffed by active, helpful communities.

Problem	Likely Cause	Solution
FreeCAD won't launch — Windows	*Missing Visual C++ runtime libraries*	Download and install the latest Microsoft Visual C++ Redistributable (x64) from microsoft.com, then re-launch FreeCAD.

FreeCAD shows wrong language	*Language not set during installation*	Go to Edit → Preferences → General → Language. Set to English and restart FreeCAD.
FreeCAD is very slow on first launch	*Loading workbench modules and indexing*	Give it 60–90 seconds on first launch. Subsequent launches will be much faster.
macOS says 'app is damaged'	*macOS Gatekeeper quarantine*	In Terminal: xattr -cr /Applications/FreeCAD.app — then relaunch.
UGS won't launch — 'Java not found'	*Java Runtime not installed*	Download and install Java from java.com. Restart your computer, then relaunch UGS.
UGS shows no ports in the list	*Driver not installed / USB not connected*	Ensure USB cable is connected. Install CH340 driver (Windows) or check USB permissions (Linux: sudo usermod -aG dialout $USER, then log out and back in).
UGS connects but machine shows Alarm	*GRBL alarm state (limit switch triggered or power cycle)*	In UGS console, type $X and press Enter. If the alarm persists, type $H to re-home the machine.
Machine moves in wrong direction	*Axis direction inverted in GRBL settings*	Check GRBL $3 setting (direction port invert mask). Consult Appendix D for details.
FreeCAD CAM shows 'no post processor'	*Post processor files not in expected location*	Go to CAM → Job → Output File tab and manually navigate to the GRBL post processor. It is typically in /path/to/FreeCAD/Mod/CAM/Path/Post/Scripts/.
G-code file opens in text editor, not UGS	*File association on Windows*	In UGS, use File → Open to navigate to your .nc file rather than double-clicking it from Explorer.

Chapter Summary

- The CNC 4040 Pro Max is a moving-gantry CNC router with a 400 × 400 mm work area, GRBL controller, and ER11 collet spindle. Understanding its physical components by name is essential for effective design and safe operation.
- GRBL is the firmware that translates G-code into motor movements. You interact with GRBL indirectly through UGS. The UGS console allows direct GRBL commands when needed (e.g., $H to home, $X to clear alarms).
- FreeCAD v1.1.0 is installed from freecad.org. Verify the version via Help → About FreeCAD after installation.
- UGS connects to the machine at 115200 baud via USB. The correct COM port must be selected before connecting. Windows users may need to install the CH340 driver.
- Inkscape (for SVG preparation) and Candle (alternative G-code sender) are optional but useful tools available for free.
- A safe workspace has clear access to all four sides of the machine, accessible e-stop, good lighting, dust extraction, and managed cabling.
- PPE — safety glasses, hearing protection, and dust mask — must be worn whenever the spindle is running. Long hair and loose clothing must be secured.
- The Never-Do List contains nine absolute prohibitions. Memorise them. The pre-cut safety checklist contains 20 items across four sections and must be completed before every single cutting session.

Key Vocabulary — Chapter 2

Term	Definition
Gantry	The bridge-like assembly that carries the spindle across the machine bed. It moves in the Y axis (forward and backward).
Stepper motor	An electric motor that moves in precise incremental steps. CNC machines use stepper motors on each axis to achieve accurate, repeatable positioning.
Limit switch	A sensor at the end of each axis's travel. When triggered during homing, it tells GRBL the axis has reached its reference position.
Work envelope	The 3D volume within which the machine's tool can reach. For the CNC 4040 Pro Max: 400 × 400 × ~70 mm.
ER11 collet	The tool-holding chuck on the CNC 4040 Pro Max spindle. Accepts shank diameters from 1 mm to 7 mm.
GRBL	Open-source firmware running on the machine's control board. Interprets G-code into precise stepper motor pulses.
Firmware	Software permanently embedded in a hardware device's microcontroller, as opposed to software installed on a general-purpose computer.
UGS	Universal G-code Sender. Free software that sends G-code files to GRBL-based CNC machines via USB connection.
Baud rate	The speed at which data is transmitted over a serial connection. GRBL communicates at 115200 baud.
COM port	The Windows name for a serial communication port. On macOS/Linux, equivalent ports are named /dev/ttyUSB0 or similar.
CH340	A USB-to-serial chip commonly used on CNC controller boards. Windows requires a driver to communicate with it.
Idle	GRBL's status when the machine is connected, ready, and not executing any movement. The normal state before and after a job.
Alarm state	GRBL's locked status when an unexpected event has occurred. Cleared with $X in the UGS console.
Homing	The machine's start-up procedure of moving each axis to its limit switch to establish a known reference position. Triggered with $H.
E-stop	Emergency stop. A button (physical or software) that immediately halts all machine movement. Must be accessible at all times during operation.
P2 dust mask	A respirator with a P2 particle filter rating, capable of filtering fine particles including MDF dust. Standard surgical masks are not sufficient for CNC dust.
Air cut / dry run	Running a G-code job with the bit raised well above the material surface. Used to verify a toolpath safely before cutting.
Work coordinates	The coordinate system set relative to your defined zero point on the workpiece, as distinct from machine coordinates (measured from the homing position).

Looking Ahead — Chapter 3

Chapter 3 dives deeper into G-code and the UGS workflow — reading a G-code file line by line, understanding what each command does, setting up zeros correctly for different job types, handling tool changes, and building the pre-cut verification habits that will serve you throughout the course. By the end of Chapter 3, you will be fully prepared to begin Week 1 of the course proper.

CHAPTER
3
of 24

PART ONE — BEFORE YOU BEGIN

Understanding G-code & the UGS Workflow

Reading machine instructions, setting zeros, managing tool changes, and verifying every job before you cut

Chapter Learning Objectives

By the end of this chapter, you will be able to:

- Explain what G-code is and describe the role of the most common commands
- Read a .nc G-code file and identify what each section instructs the machine to do
- Carry out the complete UGS pre-cut workflow: connect, home, zero, load, visualise, send
- Set X, Y, and Z zero correctly using three different methods
- Perform a tool change in UGS, including re-zeroing the Z axis
- Use the feed rate and spindle speed override controls during a running job
- Recover a paused or interrupted job and resume from a known G-code line
- Measure and evaluate a finished piece against the original design

What is G-code?

G-code is the language of CNC machines. It is not programming in the sense that most people fear — you will never write it by hand. But understanding it will make you a far more confident and effective CNC operator.

G-code is a set of standardised instructions that tells a CNC machine exactly how to move. Each line of G-code contains one instruction — move here, spin at this speed, turn the spindle on. The machine reads these instructions in sequence, from the first line to the last, and executes each one precisely.

The name G-code comes from the 'G' commands (preparatory commands) that define movement types — straight lines, arcs, and rapid traverses. But a complete G-code file also contains 'M' commands (miscellaneous functions) that control non-movement actions like the spindle and coolant, as well as parameter words like F (feed rate), S (spindle speed), X, Y, and Z (coordinates).

In this course, you will never write G-code from scratch. FreeCAD's CAM workbench generates it automatically from your 3D model and toolpath operations. Your role is to understand what the generated G-code is doing so that you can verify it is correct before you send it to the machine. A G-code file that looks reasonable in the UGS visualiser but contains an error in its commands can damage your workpiece, your cutting bit, or your machine. Understanding G-code is your quality control.

📖 **Key Term: G-code**

A standardised programming language used to control CNC machines. Each line contains a command that instructs the machine to move, spin, or perform an action. FreeCAD's CAM workbench generates G-code automatically from your 3D model.

🤔 **Stop and Think**

You do not need to memorise G-code commands to use this course successfully. But as you read through the examples in this chapter, you will notice that G-code is far more readable than it looks at first glance. A line like 'G0 X50 Y30 Z5' means: 'move rapidly to X=50, Y=30, Z=5'. Even without training, that sentence describes an action. Keep that in mind as you read the examples.

The G-code Commands You Will Encounter

The following table covers every G-code and M-code command that appears in the G-code files this course generates. GRBL supports a subset of the full G-code standard — the commands below represent the complete working vocabulary for CNC routing work with the 4040 Pro Max.

Command	Type	What it does	Example
G0	**Motion**	Rapid move — moves at the machine's maximum speed. Used to position the tool quickly when NOT cutting. The tool must be clear of the material during any G0 move.	`G0 X0 Y0 Z5`
G1	**Motion**	Linear feed move — moves in a straight line at the specified feed rate (F). This is the command used for all cutting moves.	`G1 X50 Y30 Z-2 F800`
G2	**Motion**	Clockwise arc — moves the tool in a clockwise circular arc. Requires I and J values (arc centre offsets) or R (radius).	`G2 X10 Y0 I5 J0 F600`
G3	**Motion**	Counter-clockwise arc — same as G2 but in the opposite direction.	`G3 X0 Y10 R5 F600`
G17	**Plane**	Select XY plane for arc moves. This is the default for flat CNC routing and is set at the start of most FreeCAD-generated files.	`G17`
G20	**Units**	Set units to inches. Rarely used in Australian workshops — G21 (metric) is standard.	`G20`
G21	**Units**	Set units to millimetres. This line appears near the top of every G-code file generated by FreeCAD with metric settings.	`G21`
G28	**Home**	Return to machine home position. Moves all axes to the homing switches. Use with caution — this moves at rapid speed.	`G28`
G54	**Coords**	Use work coordinate system 1 — the coordinate system offset from the machine home to your set work zero. This is the default coordinate system for all jobs.	`G54`

G80	**Cancel**	Cancel any fixed cycle (e.g., drilling cycles). Included as a safety reset at the start of many job files.	`G80`
G90	**Mode**	Absolute positioning mode — all coordinates are measured from the work zero (origin). This is the standard mode for this course. X50 always means 50 mm from zero.	`G90`
G91	**Mode**	Incremental positioning mode — coordinates are measured relative to the current position. X10 means 'move 10 mm in X from wherever I am now'. Rarely used in standard operations.	`G91`
G94	**Feed**	Feed rate in units per minute mode — the standard feed rate mode for this course. F800 means 800 mm per minute.	`G94`
M3	**Spindle**	Spindle on — clockwise rotation. The S word sets the speed in RPM. This command starts the spindle. Always appears before any cutting move.	`M3 S15000`
M5	**Spindle**	Spindle off. Appears at the end of the job and after each tool change. Always wait for the spindle to fully stop before touching the machine.	`M5`
M6	**Tool**	Tool change. Signals that a different tool is required. UGS will pause the job and wait for you to confirm the tool has been changed before continuing.	`T2 M6`
M30	**Program**	Program end — signals the end of the G-code file. The machine stops, the spindle turns off, and the program rewinds to the start.	`M30`
F	**Param**	Feed rate — sets the cutting speed in mm/min. Used with G1, G2, G3. Remains in effect until changed by another F word.	`F800`
S	**Param**	Spindle speed in RPM. Used with M3. Remains in effect until changed.	`S15000`
T	**Param**	Tool number — selects the tool to be used. FreeCAD generates T1, T2 etc. based on the tools in your CAM job. Used immediately before M6.	`T1 M6`
X Y Z	**Param**	Coordinate values for each axis. Used with all motion commands. Values are in mm (when G21 is active) and measured from work zero (when G90 is active).	`X50 Y30 Z-2`

> **Tip**
>
> You do not need to memorise this table. Use it as a reference when reading through a G-code file before a cut. The most important commands to recognise on sight are: G0 (rapid — not cutting), G1 (cutting move), M3 (spindle on), M5 (spindle off), M6 (tool change), and M30 (end of file).

Reading a G-code File — An Annotated Walkthrough

One of the most useful habits you can develop as a CNC operator is the ability to quickly read through a G-code file and understand what it is going to do before you send it. This takes practice, but the fundamentals are straightforward. Let's walk through a real example — a simple name badge engraving job, exactly the type you will produce in Project 1.

The file below is a simplified extract from a FreeCAD-generated GRBL G-code file. It engraves the letter 'A' using a 60° V-bit, then cuts around the outside of a 100 × 60 mm rectangular badge with a 6 mm end mill. Read through it once without worrying about every detail, then read the annotations on the right.

G-code	What it means
`; FreeCAD GRBL Post Processor`	Start of file — this is a comment (ignored by the machine)
`; Project: NameBadge.FCStd`	Comment — the source FreeCAD file name
`G21`	Use millimetres for all coordinates
`G90`	Use absolute positioning (all coords from work zero)
`G17`	Select the XY plane for arc moves
`G94`	Feed rate is in mm per minute
`; --- Operation 1: V-Bit Engraving ---`	Comment marking the start of Operation 1
`T1 M6`	Tool change: load Tool 1 (V-bit). UGS will pause here.
`G0 Z5`	Rapid move: raise Z to 5 mm (clear of material)
`M3 S15000`	Spindle ON at 15,000 RPM — clockwise rotation
`G0 X12.5 Y22.0`	Rapid move: position over the start of the letter 'A'
`G1 Z-0.8 F200`	Feed down to Z = -0.8 mm at 200 mm/min (plunge into material)
`G1 X18.2 Y38.5 F800`	Cutting move: trace the left diagonal stroke of 'A' at 800 mm/min
`G1 X23.8 Y22.0 F800`	Cutting move: trace the right diagonal stroke of 'A'
`G0 Z5`	Rapid move: lift Z to 5 mm (clear material for repositioning)
`G0 X14.8 Y30.2`	Rapid move: position over the crossbar of 'A'
`G1 Z-0.8 F200`	Plunge again for the crossbar
`G1 X21.2 Y30.2 F800`	Cutting move: trace the horizontal crossbar of 'A'
`G0 Z5`	Lift clear again
`M5`	Spindle OFF — end of engraving operation
`; --- Operation 2: Profile Cutout ---`	Comment marking the start of Operation 2
`T2 M6`	Tool change: load Tool 2 (6 mm end mill). UGS pauses again.
`G0 Z5`	Raise Z to safe height
`M3 S15000`	Spindle ON at 15,000 RPM
`G0 X-3.0 Y-3.0`	Rapid to start position (outside badge corner — lead-in clearance)
`G1 Z-1.5 F300`	Plunge to first pass depth: 1.5 mm at 300 mm/min
`G1 X103.0 Y-3.0 F1000`	Cut along bottom edge of badge
`G1 X103.0 Y63.0 F1000`	Cut along right edge
`G1 X-3.0 Y63.0 F1000`	Cut along top edge
`G1 X-3.0 Y-3.0 F1000`	Cut along left edge — first pass complete
`G1 Z-3.0 F300`	Plunge to second pass depth: 3.0 mm
`G1 X103.0 Y-3.0 F1000`	Second pass — bottom edge
`; ... (remaining passes omitted for brevity) ...`	Remaining depth passes follow the same pattern
`G0 Z10`	Raise Z well clear of material — job finishing
`G0 X0 Y0`	Rapid back to work zero (optional but tidy)
`M5`	Spindle OFF
`M30`	Program end — file complete

Reading through this file, several things are worth noting. First, the file always begins with unit and mode setup (G21, G90, G17, G94). These four lines ensure the machine is in a known state before any movement happens. Second, every operation begins with a tool change (T_ M6), raising to safe height (G0 Z5), and then spindle on (M3 S_). Third, every plunge into the material uses a slow feed rate (F200 or F300) — this is the plunge rate. Horizontal cutting moves use the faster feed rate (F800 or F1000). Fourth, the file ends with the spindle off (M5) and program end (M30). These are the structural patterns of every well-formed G-code file.

> ⚠ **Important**
> Never send a G-code file to the machine before checking these four things in the file:
>
> 1. G21 is present (metric units). If you see G20, the file is in inches.
> 2. M3 appears before any cutting move (spindle on before cutting).
> 3. M5 appears at the end (spindle off).
> 4. All Z depths are negative numbers (into the material). A positive Z cutting move means the tool is rising — something is wrong.

The Complete UGS Workflow — Step by Step

The UGS workflow is the sequence of actions you perform every time you use the CNC machine, from connecting the machine to handing off the finished piece. Perform it in order, every time, without shortcuts.

There is a temptation, especially as you gain experience, to skip steps in the workflow when a job seems simple. Resist this. The vast majority of CNC incidents occur on jobs that seemed simple — a quick test cut, a part the operator had cut dozens of times before. The workflow exists precisely because familiar tasks are the most dangerous ones. Familiarity breeds complacency, and complacency is where mistakes happen.

Phase 1 — Connect and Verify

1	**Power on the machine** Switch on the CNC 4040 Pro Max using its power switch. Wait five seconds for the control board to initialise.
2	**Open UGS and connect** Launch UGS on your computer. In the connection bar at the top, select the correct COM port from the drop-down (refresh the list if needed). Verify the baud rate is 115200. Click Connect. The Machine Status should change to Idle. If it shows Alarm, type $X in the console and press Enter.

3 **Home the machine**

Click the Home button in UGS, or type $H in the console and press Enter. The machine will move all three axes to their limit switches to establish the machine coordinate reference. Keep hands clear during homing. The status will return to Idle once complete.

☑ **Check This**

After homing, the machine coordinates in UGS should show values near zero for all three axes. If they show large unexpected values, homing may have not completed correctly — try homing again.

Phase 2 — Prepare the Material

4 **Install the cutting bit**

With the spindle off and stationary, insert the correct cutting bit into the ER11 collet. Tighten the collet nut firmly using the collet spanner — finger-tight is not sufficient. The bit should not move at all when gripped.

5 **Clamp the workpiece**

Place your material on the machine bed. Clamp it securely at a minimum of two points. Check that no clamp is within the path of any cutting operation — consult the UGS visualiser if unsure.

💡 **Tip**

Double-sided tape is a useful supplement to mechanical clamps for thin or small workpieces. Apply it to the full underside of the material, press firmly onto the bed, and use clamps for added security. Tape alone is not sufficient for profile cuts where the part may break free.

Phase 3 — Set X/Y/Z Zero (Work Origin)

Setting the work zero is the most critical step in the entire CNC workflow. It tells the machine where your design starts on the physical material. If your zero is wrong by even one millimetre, every feature of your design will be shifted by that amount. If your Z zero is wrong, your engraving depth will be incorrect or the tool will fail to touch the material at all.

FreeCAD generates toolpaths relative to the origin you defined in your CAD model. The most common origin positions are: the front-left corner of the design, or the centre of the design. Before zeroing at the machine, check your FreeCAD model to confirm where the origin is — the red dot at the intersection of the X, Y, and Z axes in FreeCAD's 3D view is your origin point.

6	**Jog to the X/Y origin** Using the jog controls in UGS, move the spindle so that the cutting bit is directly above the point on your material that corresponds to your CAD origin. Use small step sizes (0.1–1 mm) when close to the target position. Take your time — precision here pays dividends for the entire job.

7	**Set X and Y zero** In UGS, click the Zero X button and the Zero Y button (or right-click the work coordinates display and select Zero X/Y). The work coordinate display for X and Y should now show 0.000.

8	**Set Z zero** Lower the Z axis until the cutting bit just touches the top surface of the material. Use the paper method (see below) or the touch-off method. Once the bit is touching the surface, click Zero Z. The Z work coordinate should now show 0.000. Raise Z by 5–10 mm after zeroing.

The Paper Method for Z Zero

The paper method is the most reliable way to set Z zero for beginners. Slide a standard piece of printer paper between the cutting bit and the material surface. Lower the Z axis slowly in small increments (0.1 mm at a time). When you feel the paper gripping slightly — some resistance when you try to slide it — the bit tip is at approximately 0.1 mm above the surface. Set this as Z zero. The paper is exactly 0.08–0.10 mm thick, so this method consistently gives you an accurate surface reference.

> 💡 **Tip**
> Always use the paper method with the spindle completely off and stationary. Lower the Z slowly using the 0.1 mm step size in UGS jog controls. Never lower Z quickly when close to the material surface — one accidental overshoot can damage your bit or your workpiece.

Corner Zero vs Centre Zero

Corner zero (front-left)	Set zero at the front-left corner of your material. Most common for rectangular parts. In FreeCAD, ensure your model's origin is at the front-left corner of the design outline.
Centre zero	Set zero at the centre of the material. Used for circular parts (e.g., the clock face in Week 9) or when the CAD model is designed with the origin at the model's centre. In FreeCAD, centred sketches use this approach.
Custom zero	For some projects, the origin is at a specific feature rather than a corner or centre. Always confirm zero position from your FreeCAD model before cutting.

Phase 4 — Load, Visualise, and Send

9	**Load the G-code file** In UGS, click the Open (folder) button and navigate to your .nc G-code file. Select it and click Open. The file will load and UGS will display the toolpath in the 3D visualiser.

10	**Verify in the 3D visualiser** Spend at least 60 seconds reviewing the 3D visualiser. Rotate and zoom to inspect from multiple angles. Check: (a) all expected toolpath lines are visible; (b) rapid moves (typically shown in a different colour) do not pass through your workpiece or clamps; (c) the overall shape matches your design; (d) the zero point of the visualiser aligns with where you zeroed on the machine.

11	**Put on your PPE** Safety glasses, hearing protection, dust mask. Check that bystanders are clear.

12	**Send the job** Click the Send (play) button in UGS. The machine will begin executing the G-code. Stand by the machine, within reach of the Pause button, for the entire duration of the job.

> ⚠ **Important**
>
> Never walk away from the machine while a job is running.
>
> Keep your hand near (but not touching) the Pause button throughout the job.
>
> If anything sounds or looks wrong — unusual chatter, burning smell, unexpected movement — press Pause immediately.

Performing a Tool Change

Many jobs in this course require more than one cutting bit — for example, a V-bit for engraving followed by an end mill for profiling. When FreeCAD generates G-code for a multi-tool job, it inserts a tool change command (T_ M6) between operations. When UGS encounters this command, it pauses the job and waits for you to confirm that the tool has been changed before continuing.

The most important rule of tool changes is this: only Z zero needs to be re-set after a tool change. X and Y zero do not change when you swap bits, because the spindle is still in the same horizontal position. Re-zeroing X or Y mid-job is one of the most common — and most damaging — mistakes in CNC operation. If you zero X or Y again after a tool change, every subsequent operation will be offset from all previous ones.

● **Safety Rule**

After a tool change, you MUST re-zero Z only.

NEVER re-zero X or Y after a tool change in a mid-job tool swap.

Re-zeroing X or Y mid-job will cause all subsequent operations to be misaligned.

Step-by-Step Tool Change Procedure

1. When UGS pauses for a tool change, click Pause in UGS if the machine hasn't already stopped. The spindle will turn off (M5 was sent before the tool change command).
2. Wait for the spindle to fully stop rotating. Do not touch the tool until it has completely stopped.
3. Jog the Z axis upward to give yourself comfortable working clearance — at least 30 mm above the material.
4. Using the collet spanner, loosen the collet nut. Remove the current cutting bit and set it safely aside.
5. Insert the new cutting bit into the collet. Ensure the bit is inserted to the correct depth — not too shallow (risk of slipping) and not so deep that the cutting flutes are inside the collet. As a rule, insert until the bit's shank shoulder is 5–10 mm above the collet.
6. Tighten the collet nut firmly with the collet spanner. Do not overtighten — ER11 collets can crack if overtightened.
7. Set Z zero for the new tool using the paper method: lower the Z axis until the new bit just touches the material surface, then click Zero Z in UGS.
8. Raise Z by 5 mm after zeroing.
9. In UGS, click Resume (or the Continue button that appeared when the job paused). The machine will read the next M3 command in the G-code and restart the spindle for the next operation.

Tip

Label your cutting bits with a permanent marker on the shank — T1, T2 etc. matching your FreeCAD CAM tool library. This eliminates the risk of installing the wrong tool after a change, especially when you have several bits of similar diameter in your tool roll.

Feed Rate and Spindle Speed Override

UGS provides real-time override sliders that let you adjust the machine's feed rate and spindle speed during a running job without stopping it. This is a professional feature that gives you control over the cutting process when something is not going quite as expected.

The override sliders work as a percentage of the programmed value. At 100%, the machine runs at the feed rate and spindle speed set in the G-code. Sliding to 80% reduces the feed rate to 80% of the programmed value. The machine slows down in real time.

Feed rate override — reduce	If the machine is chattering (vibrating and making a harsh rattling sound), the feed rate is likely too high for this material, bit condition, or depth of cut. Reduce feed rate to 70–80% and observe whether the chatter resolves.
Feed rate override — increase	If the cut looks clean and the machine sounds healthy, you can carefully increase feed rate to improve efficiency. Increase in 5–10% increments and listen for any change in machine sound.
Spindle speed override	Reducing spindle speed generates less heat, which can help with materials prone to melting or scorching (e.g., some plastics). Increasing spindle speed on hardwoods can improve surface finish.
Returning to 100%	Always return overrides to 100% before the end of the job so they are reset for the next run. Leaving an override in place is a common cause of unexpected behaviour on the next job.

> **Tip**
> Listen to your machine. A healthy cut sounds smooth and consistent — a low, steady tone from the spindle and material. Chatter sounds like vibration and rattling. Burning sounds like reduced spindle noise with a slight sizzling quality, often accompanied by a smoke smell. Learn to distinguish these sounds — they are more informative than any display.

Job Recovery — What to Do When a Job Is Interrupted

Occasionally, a CNC job will be interrupted before it is complete. This can happen due to a power flicker, a dropped USB connection, an accidental press of the Stop button, or an e-stop event. Knowing how to recover and resume from a known point — rather than starting the entire job again — saves significant time and material.

When to Recover vs When to Start Over

The decision depends on what point in the job the interruption occurred and what operations are left to complete. If the job stopped during an engraving pass and the profile cut has not yet started, recovery is straightforward. If the job stopped mid-profile cut, recovery is more complex — the part may be partially free, and the risk of the bit hitting a tab or edge on resume must be carefully assessed. When in doubt, consult your instructor before resuming.

The Recovery Procedure

1. When the interruption occurs, note the G-code line number shown in the UGS display before the connection drops or the job stops. This line number is your recovery point. If UGS closed completely, the file can be re-opened — UGS remembers the last-executed line.
2. Do not move the workpiece. Leave everything exactly as it was when the job stopped. If clamps or the workpiece have shifted, recovery is much more difficult.

3. Assess the machine state. Is the spindle off? Is the machine in Alarm state? Clear any alarm with $X. Re-home the machine if the position is uncertain.
4. Check that the work coordinates (X, Y, Z zero) are still correct. If the machine was homed but not re-zeroed, you will need to re-zero from the original zero position on the workpiece. If the workpiece has not moved, your original zero marks on the material will still be valid.
5. In UGS, open the G-code file. Find the line number where the job stopped (use Edit → Find in most text editors to locate the line). Scroll to a point a few lines before the interruption — enough to ensure the tool is in the correct position and the spindle is running before it reaches the interrupted point.
6. In UGS, use the Start From Line feature (available in the File menu or right-click context) to begin execution from the selected line.
7. Watch the first few moves carefully to confirm the tool is moving as expected before standing back.

> ⚠ **Important**
>
> If the workpiece has moved, or if you cannot confidently determine the original zero position, do not attempt recovery. Starting the job over with the material re-clamped and re-zeroed is always safer than resuming from an uncertain position. An incorrectly resumed job can drive the tool into a tab or partially cut edge at full feed rate.

Post-Cut Measurement and Evaluation

Every time you cut a piece, your job is not finished when the machine stops. The final step is evaluating what was produced against what was designed. This practice — measuring, comparing, and recording — is what separates a developing CNC operator from a competent one. Without it, you repeat the same errors across multiple projects. With it, you improve steadily.

What to Measure

You do not need to measure every dimension of every piece. Identify the critical dimensions — the ones that matter most for the piece to work correctly — and measure those. For a nameplate, this might be the overall width and height, and the engraving depth. For a joinery project, it is the joint clearance. For a gear, it is the pitch diameter and bore.

Vernier caliper	The primary measuring tool for CNC work. Reads to 0.02 mm or better. Use for all critical dimensions — length, width, pocket depth, bore diameter.
Steel rule	For quick overall dimension checks. Not precise enough for joint tolerances, but fine for overall size verification.
Depth gauge	For measuring pocket and engraving depth. A vernier caliper with a depth rod works for most purposes.
Feeler gauges	For measuring joint clearance — how much gap exists between two mating surfaces. Useful for joinery projects from Week 6 onwards.

Completing the Project Reflection

Every project in this course requires a completed project reflection sheet alongside the FreeCAD and G-code files. The reflection is not a formality — it is a structured thinking exercise that cements learning and builds the self-evaluation skills that sit at the Evaluate level of Bloom's Taxonomy.

A strong reflection answers four questions honestly: What did I design and why did I make the choices I made? What worked exactly as intended? What did not work as intended, and why? What would I do differently if I were to make this piece again? The last question is the most valuable — it requires you to diagnose a gap between intention and outcome, which is the core skill of the Analyse and Evaluate levels.

> **Tip**
>
> Photograph your finished piece before removing it from the machine bed. A photo of the piece in context — still clamped, with the machine visible — provides useful reference for future jobs, especially when you are troubleshooting an issue that only becomes apparent after the piece is released and cleaned up.

The Full Post-Cut Procedure

1. Turn off the spindle (M5 in console if needed) and raise Z well clear of the material.
2. Jog the machine away from the workpiece so you have clear access.
3. Remove clamps carefully. For profile cuts with tabs, use a sharp chisel or craft knife to release the piece from the tabs — do not lever or force it.
4. Brush or vacuum all dust and chips from the piece and the machine bed.
5. Measure critical dimensions with a vernier caliper. Record measurements on the reflection sheet.
6. Compare dimensions against the FreeCAD model. Note any discrepancies greater than 0.5 mm.
7. Assess surface quality: are the engraving lines clean? Are profile edges smooth? Is the material surface free of burning or excessive tearout?
8. Complete the project reflection sheet and submit it with your FreeCAD and G-code files.

Chapter Summary

- G-code is the machine instruction language for CNC routers. FreeCAD generates it automatically — your role is to verify it before cutting. The key commands to recognise are G0 (rapid), G1 (cutting move), M3 (spindle on), M5 (spindle off), M6 (tool change), and M30 (end of file).
- A well-formed G-code file begins with unit and mode setup (G21, G90, G17, G94), starts every operation with a tool change and spindle on, and ends with spindle off (M5) and program end (M30).
- The UGS workflow has four phases: Connect & Verify, Prepare Material, Set Zero, and Load/Visualise/Send. Perform all phases in order, every time.
- Setting X/Y/Z zero is the most critical step in CNC operation. X and Y define where the design starts horizontally; Z defines the material surface reference. Use the paper method for Z zero. Corner zero and centre zero are the two most common strategies.
- After a tool change, re-zero Z only. Never re-zero X or Y mid-job — doing so misaligns all subsequent operations.
- Feed rate and spindle speed overrides in UGS let you adjust cutting parameters in real time. Use them to respond to chatter or surface quality issues. Return them to 100% at the end of every job.
- When a job is interrupted, record the G-code line number, assess the machine state, verify zero is intact, then use UGS's Start From Line feature to resume. If zero is uncertain or the workpiece has moved, start the job over.
- Post-cut evaluation — measuring, comparing to design, and completing the reflection sheet — is a required part of every project submission and is as important as the cutting itself.

Key Vocabulary — Chapter 3

Term	Definition
G-code	The standardised machine instruction language used by CNC routers. FreeCAD's CAM workbench generates G-code automatically from toolpath operations.
G0	Rapid positioning command. Moves the tool at maximum machine speed. The tool must be clear of all material during a G0 move.
G1	Linear feed command. Moves the tool in a straight line at the programmed feed rate (F). Used for all cutting moves.
G2 / G3	Clockwise and counter-clockwise arc commands. Used for curved cutting moves. Require arc centre (I, J) or radius (R) parameters.
G21	Sets the unit system to millimetres. Must appear at the start of every metric G-code file.
G90	Absolute positioning mode. All coordinates are measured from the work zero origin. The standard mode for this course.
M3	Spindle on command — clockwise rotation. Always accompanied by an S value (spindle speed in RPM). Must appear before any cutting move.
M5	Spindle off command. Appears after the last cutting move of each operation and at the end of the file.
M6	Tool change command. Pauses the job in UGS and waits for confirmation that the new tool is installed and Z has been re-zeroed.

M30	Program end command. Signals the end of the G-code file. The machine stops and the program rewinds.
F word	Feed rate parameter in G-code. Specifies the cutting speed in mm/min. Remains in effect until changed by another F value.
S word	Spindle speed parameter. Specifies RPM when used with M3.
Work zero	The reference point (0,0,0) defined by the operator at the start of each job. All G-code coordinates are measured from this point.
Paper method	A technique for setting Z zero precisely. A piece of paper (0.08–0.10 mm thick) is slid between the bit and material; Z is lowered until the paper grips, then set as Z=0.
Corner zero	Setting the work zero at the front-left corner of the workpiece. Common for rectangular designs.
Centre zero	Setting the work zero at the centre of the workpiece. Common for circular designs and FreeCAD models centred on the origin.
Feed rate override	A real-time control in UGS that adjusts the feed rate as a percentage of the programmed value during a running job.
Start From Line	A UGS feature that allows a G-code job to begin execution from a specified line number rather than from the beginning. Used for job recovery.
Plunge rate	The feed rate (F value) used for Z-axis moves when the tool is entering the material. Always slower than the horizontal cutting feed rate.
Chatter	A vibrating, rattling sound during cutting caused by the tool deflecting under excessive cutting forces. Usually resolved by reducing feed rate or depth per pass.

Looking Ahead — Part Two and Chapter 4

With Chapters 1, 2, and 3 complete, you have everything you need before Day 1. Part Two begins the course proper: Chapter 4 introduces the FreeCAD v1.1.0 interface in detail, Chapter 5 dives deep into the Sketcher, and Chapter 6 covers Part Design — Pad, Pocket, Chamfer, Fillet, and ShapeString. By the end of Week 1 (Day 5), you will have designed and cut your first physical project: the Personalised Name Badge.

PART TWO

Foundations of FreeCAD

Weeks 1 & 2 • Chapters 4 through 7 • Bloom's Levels: Remember → Understand

📖 About Part Two

Part Two covers Weeks 1 and 2 of the course — Days 1 through 10. These are the foundational weeks. You are building the mental model of FreeCAD from scratch: how the interface is organised, how 2D geometry is drawn and constrained, how a 3D body is built from 2D sketches, and how your first CAM operations are configured and sent to the machine.

These weeks operate at the Remember and Understand levels of Bloom's Taxonomy. You are not yet expected to make independent design decisions — you are learning the tools and concepts that will make independent design possible. Be patient with yourself during these weeks. The pace of learning will feel uneven: some things will click immediately, others will take several repetitions. Both are normal.

CHAPTER
4
of 24

PART TWO — FOUNDATIONS

The FreeCAD Environment

Interface layout, workbenches, navigation, the Model tree, Properties panel, and file management

Chapter Learning Objectives

By the end of this chapter, you will be able to:

- Name and describe every major area of the FreeCAD v1.1.0 interface
- Switch between workbenches and explain the purpose of each workbench used in this course
- Navigate the 3D view confidently using orbit, pan, and zoom
- Read the Model tree and explain what the feature hierarchy means
- Use the Properties panel to inspect and modify feature data
- Save, name, and manage FreeCAD project files using consistent conventions
- Apply the recommended preference settings for this course

Opening FreeCAD for the First Time

Open FreeCAD now. This chapter is designed to be read with FreeCAD open in front of you. Every concept is more meaningful when you can see it on your screen at the same time.

When FreeCAD v1.1.0 first opens, you see the Start workbench — a welcome screen with recent files, FreeCAD news, and a few shortcut buttons. The Start workbench is not a working environment; it is a launch pad. You will move away from it almost immediately every session.

Before exploring the interface, create a new document: go to File → New, or press Ctrl+N. A new, empty document opens. The screen changes from the Start workbench to a grey 3D view. Now the real interface is visible, and we can start exploring it.

Try It Now

Go to File → New now. Your screen should show a grey 3D view with a small coordinate indicator (three coloured arrows) in the bottom-left corner and a model tree panel on the left. If your screen looks different, your workbench may not be set to Part Design — we will fix this in a moment.

The FreeCAD Interface — A Guided Tour

The FreeCAD interface is divided into six functional areas. Understanding what each area does — and how they relate to each other — is the foundation of working efficiently in FreeCAD. The diagram below shows the layout. Each area is described in detail in the sections that follow.

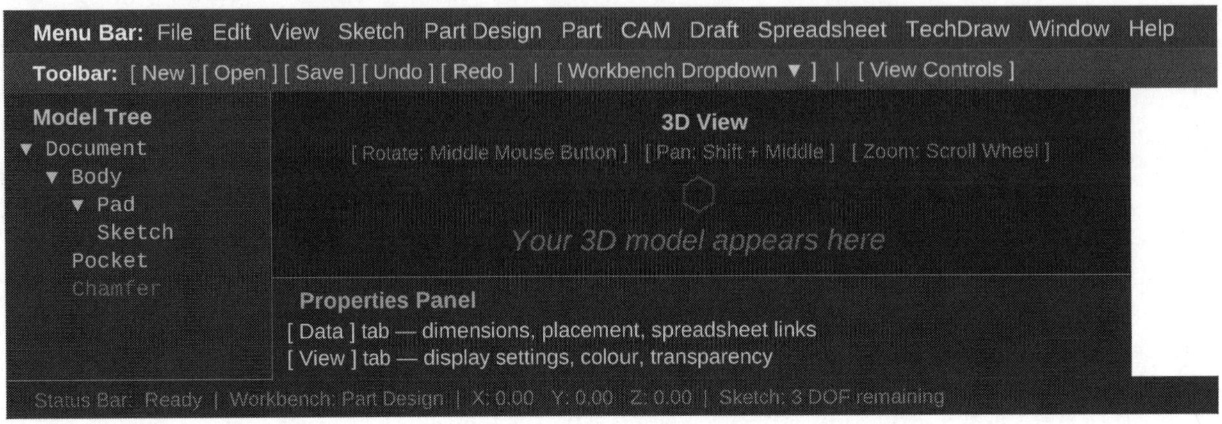

> 📋 **Note**
>
> FreeCAD's interface is highly customisable — panels can be moved, docked, and hidden. If your screen layout looks different from the diagram, go to View → Panels and ensure that Model, Properties, and 3D View are all visible (ticked). If the 3D view has disappeared entirely, go to View → Standard Views → Home.

The Menu Bar

The menu bar runs along the very top of the FreeCAD window. It contains every command available in FreeCAD, organised into drop-down menus. Many of these commands change depending on which workbench is active — the Sketch menu only appears when you are in the Sketcher, for example. The File, Edit, View, and Help menus are always present regardless of workbench.

In the early weeks of this course, you will use the menu bar frequently because it gives you access to all commands by name. As you become more fluent, you will find yourself using keyboard shortcuts and toolbar buttons more often — but the menu bar is always the reliable fallback when you cannot remember where a command is.

Keyboard Shortcuts

`Ctrl+N`	New document
`Ctrl+O`	Open document
`Ctrl+S`	Save document
`Ctrl+Z`	Undo
`Ctrl+Y`	Redo (Ctrl+Shift+Z on some systems)
`Ctrl+Q`	Quit FreeCAD

The Toolbar

The toolbar sits below the menu bar and provides one-click access to the most frequently used commands. Like the menu bar, the toolbar changes depending on the active workbench — different tools appear for Sketcher, Part Design, CAM, and so on.

The most important element in the toolbar is the Workbench drop-down. This is typically displayed as the name of the current workbench with a small arrow. Clicking it opens a list of all available workbenches. Switching workbenches is something you will do many times in every session — finding this control quickly is an early priority.

> **Tip**
> If you cannot find a toolbar button, use the menu bar first. Every toolbar button has an equivalent menu command. Once you find the command in the menu, right-click the toolbar to see if it can be added as a button.

The Model Tree

The Model tree (also called the model panel or feature tree) occupies the left side of the interface. It shows every object, body, and feature in your current document as a hierarchical list. Understanding the Model tree is essential because it represents the complete history of every decision you have made in your model.

The hierarchy works from top to bottom: the Document contains Bodies; each Body contains Features (Pad, Pocket, Chamfer, etc.); each Feature has a Sketch attached to it. This is called a feature-based, history-based modelling approach — your model is not just a shape, it is a record of how that shape was built.

Document	The top-level container. Corresponds to your .FCStd file. Everything in your project lives inside the Document.
Body	A container for a set of features that together form one solid part. A single FreeCAD file can contain multiple Bodies (you will use this from Week 6 onwards for joinery projects).
Sketch	A 2D drawing attached to a plane or face. Sketches are the foundation of all Part Design features — every Pad, Pocket, and other feature starts with a Sketch.

Pad	A feature that extrudes a Sketch in 3D to create or add solid geometry. Shown as 'Pad' or 'Pad001', 'Pad002' etc. if there are multiple.
Pocket	A feature that removes material from an existing solid. Shown as 'Pocket' in the tree.
Chamfer/Fillet	Edge treatment features. Always appear below the feature that created the edges they modify.

The order of features in the Model tree matters. FreeCAD builds the model from top to bottom — every feature is applied in the sequence it appears in the tree. If you delete a feature that a later feature depends on, the later feature will fail (it will turn red or yellow in the tree). This cause-and-effect relationship is important to understand — we will explore it in detail in Chapter 6.

Try It Now

In your empty document, look at the Model tree. You should see only 'Document' (the file name). Go to Part Design → Body in the menu bar to add a Body. Now go to Part Design → Create a new Sketch. A dialog will ask you to select a plane — click on 'XY Plane' and click OK. Look at the Model tree again. You should now see: Document → Body → Origin → Sketch. You have just built the first two levels of a FreeCAD model hierarchy.

The Properties Panel

The Properties panel lives below the Model tree on the left side of the interface. It has two tabs: Data and View. This panel is context-sensitive — it shows the properties of whatever is selected in the Model tree or 3D view.

Data tab	Shows the functional properties of the selected feature — dimensions, positions, spreadsheet links, and calculation values. This is where you edit the length of a Pad, the depth of a Pocket, or see what formula is driving a dimension. You will use the Data tab constantly once you start working with parametric design in Week 3.
View tab	Shows the display properties of the selected feature — colour, transparency, line width, point size. These properties affect how the feature looks in the 3D view but have no effect on the geometry itself. You will use the View tab occasionally to change the colour of a body for visual clarity when working with multiple parts.

Tip

The Properties panel is one of the most underused features in FreeCAD by beginners. Any time something about a feature does not look right — wrong size, wrong position, wrong depth — click on that feature in the Model tree and check the Data tab first. The answer is almost always there.

The 3D View

The 3D view occupies the large central area of the FreeCAD interface. This is where your model appears in three dimensions. You can rotate, pan, and zoom to inspect your design from any angle. The 3D view is not just a display — it is also an interactive workspace. You click directly on faces, edges, and points in the 3D view to select geometry for operations like Pocket, Chamfer, and CAM toolpath selection.

In the bottom-left corner of the 3D view, there is a small coordinate indicator showing three coloured arrows: red for X, green for Y, and blue for Z. This indicator always shows the orientation of the current view relative to the model's coordinate system. It rotates as you orbit the view, so you always know which direction is which.

The Status Bar

The status bar runs along the very bottom of the FreeCAD window. It displays context-sensitive information about what FreeCAD is doing or what it expects you to do next. When you are in the Sketcher, the status bar shows the number of degrees of freedom remaining in your sketch. When you hover over a toolbar button, it shows a brief description. When an operation fails, it often shows an error message here. Develop the habit of glancing at the status bar when something unexpected happens — it usually tells you what went wrong.

Workbenches — FreeCAD's Modular Design

FreeCAD is organised into specialised workbenches rather than one giant toolbox. This is one of its most distinctive features, and understanding why it works this way will save you a lot of confusion in the early weeks.

A workbench is a set of tools, menus, and toolbar buttons designed for a specific type of task. When you switch to the Sketcher workbench, FreeCAD replaces the toolbar with Sketcher-specific tools and adds a Sketch menu to the menu bar. When you switch to the CAM workbench, a completely different set of tools appears. The 3D view stays the same — only the available tools change.

This modular approach means that FreeCAD does not overwhelm you with every possible tool at once. In the early weeks, you will mostly use three workbenches: Part Design (for building 3D models), Sketcher (for drawing 2D geometry, accessed from within Part Design), and CAM (for creating toolpaths). The others are introduced progressively as the course advances.

📖 Key Term: Workbench

A collection of tools, menus, and toolbar buttons in FreeCAD that is activated together for a specific type of task. Switching workbenches changes the available tools but does not affect your model or document.

Workbench	Primary Purpose	Key Tools	Used In
Sketcher	Draw and constrain 2D geometry — lines, circles, rectangles, splines	*Sketch tools, Constraint tools, Construction geometry*	All projects — accessed from Part Design
Part Design	Build 3D solid models from 2D sketches using features	*Pad, Pocket, Chamfer, Fillet, Linear Pattern, Polar Pattern, Datum Plane*	All projects — primary modelling workbench
Part	Direct solid operations — Booleans, ShapeString, Loft, Primitives	*Part Cut, Part Fuse, Part Loft, Part ShapeString*	Projects 3, 8, 9, 10, 11, 12
CAM	Generate machine toolpaths and G-code from 3D models	*Job, Engraving, Pocket Shape, Profile, Drilling, 3D Surface, Post Process*	All projects — for machine output
Spreadsheet	Create parameter spreadsheets that drive model dimensions	*Spreadsheet, Cell aliases, Refresh*	Projects 1, 3, 5, 6, 7, 10
Draft	2D/3D drawing, SVG import, arrays and object manipulation	*Draft Array, Draft Move, Draft Scale, SVG Import*	Projects 3, 4, 5, 7
TechDraw	Generate 2D engineering drawings from 3D models	*Insert Page, Insert View, Insert Dimension, Export PDF*	Projects 9, 10, 11, Certification
Mesh	Work with STL mesh files — import, repair, convert	*Import Mesh, Evaluate & Repair, Smooth, Shape from Mesh*	Projects 8, 12 (STL workflow)

💡 Tip

Do not try to memorise all eight workbenches now. Focus on Part Design and Sketcher this week. You will meet each additional workbench at the moment in the course where it becomes relevant — and at that point, the context will make it immediately meaningful.

Switching Workbenches

To switch workbenches, click the Workbench drop-down in the toolbar (it shows the name of the current workbench with a small arrow). This opens a list of all available workbenches. Click the one you want. The toolbar will change immediately to reflect the new workbench's tools.

There is one important exception: the Sketcher workbench is almost always accessed from within Part Design rather than directly from the drop-down. When you create a new Sketch in Part Design (Part Design → Create a new Sketch), FreeCAD automatically switches to the

Sketcher workbench, and when you close the sketch (Sketch → Close Sketch), it returns to Part Design automatically. You rarely need to manually switch to or from the Sketcher.

Try It Now

Practice switching workbenches: click the Workbench drop-down and switch to CAM. Note how the toolbar changes completely. Now switch to Draft. Now switch back to Part Design. You have not changed your model at all — you have only changed which tools are visible. This is the workbench concept in action.

Navigating the 3D View

Fluent 3D navigation is one of the first skills to develop in any CAD software. In FreeCAD, navigation is controlled primarily with the mouse — specifically the middle mouse button. The default navigation style for this course is CAD mode, which you set in Preferences during Chapter 2's first-launch setup.

Mouse Action	Navigation Mode: CAD	What It Does
Orbit (rotate)	`Hold Middle Mouse Button + Drag`	Rotates the view around the model. Hold the middle button and drag the mouse to spin the view freely in three dimensions.
Pan	`Shift + Hold Middle Mouse Button + Drag`	Moves the view horizontally and vertically without rotating. Essential for repositioning the model when it has scrolled off-screen.
Zoom	`Scroll Wheel (forward/back)`	Zooms in and out. Scroll toward yourself to zoom out, away from yourself to zoom in. The zoom centres on the cursor position.
Zoom to fit	`Numpad 0 or V then F`	Resets the view to show the entire model. Invaluable when you have zoomed in too far and lost the model.
Front view	`Numpad 1`	Snaps to a direct front view (looking along the Y axis).
Top view	`Numpad 2`	Snaps to a direct top view (looking down the Z axis). Most useful for reviewing a sketch layout.
Right view	`Numpad 3`	Snaps to a direct right-side view.
Isometric view	`Numpad 0`	A 3D view at a standard angle showing depth on all three axes. The best general-purpose view for inspecting a model.
Select a face/edge	`Left click`	Selects the face or edge under the cursor. Selected items highlight in green.
Multi-select	`Ctrl + Left click`	Adds to the selection. Ctrl+click multiple faces or edges to select several at once — used in Chamfer, Fillet, and CAM operations.

> **Tip**
>
> If you orbit too aggressively and lose track of your model, press the V key followed by F (View → Standard Views → Fit All) or press Numpad 0. This always resets the view to show everything. This keyboard sequence will save you many minutes of frustration in the first few weeks.

> **Try It Now**
>
> Open your FreeCAD document with the Body and Sketch you created earlier. Click on the grey floor plane of the 3D view to give it focus. Now: hold the middle mouse button and drag — the view rotates. Scroll the wheel — the view zooms. Hold Shift + middle mouse button and drag — the view pans. Practice until these three actions feel natural. They are the foundation of all 3D CAD work.

Working With the Model Tree

The Model tree does more than display your model's history — it is an interactive control panel. Every item in the Model tree responds to clicks and right-clicks in ways that are essential to efficient FreeCAD work.

Single-click	Selects the feature. The corresponding geometry highlights in the 3D view, and the Properties panel updates to show this feature's data.
Double-click	Opens the feature for editing. This is one of the most important interactions in FreeCAD. Double-click a Sketch to re-enter the Sketcher and modify the geometry. Double-click a Pad to change its depth. Double-click a Pocket to adjust its settings. Almost every feature can be edited this way.
Right-click	Opens a context menu with actions for the selected item: Toggle visibility (show/hide), Rename, Delete, Mark to recompute, and others depending on the feature type.
Eye icon (Space bar)	Toggles the visibility of a feature in the 3D view. Hiding a body or feature can be useful when working on complex multi-body files — hide bodies you are not currently working on to reduce visual clutter.
Drag to reorder	In some cases, features can be reordered by dragging them in the tree. However, reordering features that depend on each other can cause failures. Only reorder features if you are confident the dependencies will still be satisfied.

The colour of an item in the Model tree carries meaning. A normal grey icon means the feature is healthy. A yellow warning icon means the feature has a problem but is still producing geometry. A red error icon means the feature has completely failed and is producing no geometry. A red feature in the tree is always something to investigate and fix before proceeding.

> ⚠ **Important**
>
> If a feature turns red in the Model tree, do not continue building on top of it.
>
> FreeCAD will attempt to build subsequent features from the last valid state, which may produce unexpected geometry.
>
> Fix the red feature first — double-click it to investigate, read the status bar for error messages, and check the Properties panel for incorrect values.

File Management for CNC Projects

Good file management is a professional habit, not just a nice-to-have. A CNC workshop generates many files — FreeCAD project files, G-code files, TechDraw drawings, SVG artwork, Job Setup Sheets — and a disorganised file system will cost you real time as the course progresses. Set up a clean, consistent structure from Day 1.

The .FCStd Format

FreeCAD saves its native project files in the .FCStd format. Technically, an .FCStd file is a compressed ZIP archive containing XML files that describe every object, feature, and setting in your document. You do not need to understand the internal structure, but it is useful to know that .FCStd files can sometimes be manually repaired by unzipping them if they become corrupted.

Every FreeCAD project in this course should be saved as a .FCStd file. Do not save projects as STEP, IGES, or other exchange formats for your working files — these formats do not preserve the feature history and parametric links that are essential to this course's design approach.

> 📖 **Key Term: .FCStd**
> FreeCAD's native project file format. A compressed archive containing the full model history, feature tree, spreadsheet data, and all settings. Always use this format for working FreeCAD files in this course.

Recommended Folder Structure

Create the following folder structure on your computer or network drive before starting Week 1. You will add to it every week of the course.

CNC_Projects/	Root folder for all CNC course work
CNC_Projects/Week_01/	Week 1 daily exercises and Project 1 files
CNC_Projects/Week_01/ NameBadge/	Sub-folder for Project 1: NameBadge.FCStd, NameBadge_v1.nc, NameBadge_TechDraw.pdf
CNC_Projects/Week_02/	Week 2 files, same sub-folder structure
CNC_Projects/Reference/	Companion files from this course: example .FCStd files, SVG artwork, G-code references
CNC_Projects/Certification/	Certification project files — kept separate from weekly work

File Naming Conventions

Consistent file naming prevents the inevitable confusion of 'which version was the good one?' Apply these rules to every file you create in this course:

- Always use the project name as the base filename: NameBadge, FingerJointBox, SpurGear.
- Add a version number for major updates: NameBadge_v1.FCStd, NameBadge_v2.FCStd. Never overwrite v1 with v2 — keep both.
- Add the material thickness to G-code filenames: NameBadge_12mm_v1.nc. If you cut the same design in different thicknesses, the filename tells you which file to use.
- Add the date to certification project files for easy chronological sorting: CertBox_2024-11-15_v1.FCStd.
- Never use spaces in filenames. Use underscores instead. Spaces cause problems in some G-code senders.

> **Tip**
>
> FreeCAD's autosave creates a backup file every few minutes (if enabled in Preferences). Autosave files are stored in your system's temporary folder and are named with a tilde (~). They will save you from a catastrophic file loss if FreeCAD crashes. Set autosave to 5 minutes in Edit → Preferences → General → Document.

Recommended Preferences for This Course

If you have not already applied these preferences from Chapter 2, do so now. These settings are not required for FreeCAD to function, but they make the course experience significantly more reliable and less frustrating.

Applying Preferences — Step by Step

1. Go to Edit → Preferences (Windows/Linux) or FreeCAD → Preferences (macOS). The Preferences dialog opens.
2. General → General: Set Language to English. Enable Autosave every 5 minutes. Set the number of recent files to 10.
3. Display → 3D View: Set Navigation Style to CAD (this is the mouse control scheme described in this chapter). Enable Anti-Aliasing for smoother model edges if your computer supports it.
4. Display → Colours: The default dark background with coloured axes is recommended. Do not change this early in the course — instructors and course materials refer to the default colour scheme.
5. Sketcher → General: Enable Show grid. Set Grid spacing to 10 mm. Enable Constrain to grid for early practice (you can turn this off later).
6. Part Design → General: Leave defaults. Do not enable experimental features.
7. Close Preferences. Your settings are saved automatically and will persist across sessions.

> **Note**
>
> FreeCAD's preferences are saved per-user on your computer. If you use a shared workstation, check the preferences at the start of each session — another user may have changed them.

Week 1, Day 1 — A Worked Example

This section shows exactly what your first lesson session looks like in practice. Use it as a reference during Day 1 if you are unsure of the expected flow.

Day 1 of Week 1 is deliberately straightforward. The goal is not to produce anything — it is to become comfortable in the FreeCAD environment. By the end of Day 1, you should be able to: open FreeCAD, create a new document, switch workbenches, navigate the 3D view, and save a file. Nothing more, and nothing less.

The day follows the standard four-part lesson structure introduced in Chapter 1: theory, demonstration, guided practice, and independent exercise. Here is how Day 1 uses that structure:

Time	Activity	What You Are Doing
0–25 min	**Theory: Interface Tour**	Your instructor walks through the FreeCAD interface using the diagram in this chapter. No FreeCAD is open yet — this is conceptual. You label the six areas of the interface on a printed diagram (handout provided).
25–45 min	**Demonstration: First Steps**	Your instructor opens FreeCAD, creates a new document, switches through several workbenches, and navigates the 3D view. You watch without doing — this is the overview pass.
45–75 min	**Guided Practice: Navigate**	Open FreeCAD on your own workstation. Follow along with this chapter: create a new document, add a Body, create a Sketch on the XY plane. Do not draw anything yet — just practice opening, creating, and closing the sketch. Then practice 3D navigation: orbit, pan, zoom until they feel natural.
75–90 min	**Independent Exercise**	Without following any step-by-step instructions: create a new FreeCAD document, set it up with a Body, save it as Day01_Practice.FCStd in your Week_01 folder, and close FreeCAD. Reopen the file and verify it loaded correctly. This verifies that your file save and load workflow is working.

🤔 **Stop and Think**

At the end of Day 1, pause and notice how you feel about the interface. What feels confusing? What feels natural? Identifying the things that confuse you now makes it much easier to focus your learning in the days ahead. Write two things down: one thing that clicked today, and one thing that is still unclear.

Chapter Summary

- The FreeCAD interface has six areas: Menu bar, Toolbar (with Workbench drop-down), Model tree, Properties panel, 3D view, and Status bar. Each serves a distinct purpose in the design workflow.
- Workbenches are modular toolsets in FreeCAD. The eight workbenches used in this course are: Sketcher, Part Design, Part, CAM, Spreadsheet, Draft, TechDraw, and Mesh. Each is introduced at the point in the course where it becomes relevant.
- 3D navigation in CAD mode uses three mouse actions: middle mouse button drag (orbit), Shift + middle drag (pan), and scroll wheel (zoom). Press V then F to fit the model to screen at any time.
- The Model tree shows every feature in the order it was created. Double-clicking a feature opens it for editing. Red features indicate errors that must be fixed before continuing.
- The Properties panel has two tabs: Data (functional values — dimensions, links, positions) and View (display settings — colour, transparency).
- FreeCAD files are saved as .FCStd. Use a consistent folder structure and file naming convention from Day 1. Enable autosave every 5 minutes in Preferences.
- Day 1 of Week 1 is about comfort, not production. The goal is to navigate the interface, create a document, and save a file. Everything else builds from here.

Key Vocabulary — Chapter 4

Term	Definition
Workbench	A modular set of tools, menus, and toolbar buttons in FreeCAD activated for a specific type of task. Switching workbenches changes the available tools but not the model.
Model tree	The hierarchical list on the left side of the FreeCAD interface showing every object, body, and feature in the document. Also called the feature tree or model panel.
Properties panel	A panel below the Model tree showing the properties of the selected feature. The Data tab contains functional values; the View tab contains display settings.
3D view	The large central area of the FreeCAD interface where the 3D model is displayed and interacted with.
Status bar	The bar at the bottom of the FreeCAD window displaying context-sensitive messages, degree-of-freedom counts, and error information.
Body	A Part Design container that holds a set of features forming one solid part. A FreeCAD document can contain multiple Bodies.
Feature	A single operation in Part Design that adds or removes geometry — Pad, Pocket, Chamfer, Fillet, etc. Features are listed in the Model tree in creation order.
Feature tree	Another name for the Model tree, emphasising that it shows the ordered history of features applied to build the model.
Orbit	Rotating the 3D view around the model. Activated by holding the middle mouse button and dragging.
Pan	Translating (sliding) the 3D view without rotation. Activated by holding Shift + middle mouse button and dragging.

Context menu	A menu that appears when you right-click an item. In the Model tree, the context menu offers actions specific to the selected feature.
.FCStd	FreeCAD's native file format. A compressed archive containing the full model history, feature tree, and all settings for the project.
CAD navigation	FreeCAD's navigation preset for this course. Uses middle mouse button to orbit, Shift + middle to pan, and scroll wheel to zoom.
Autosave	A FreeCAD preference that automatically saves a backup of your document at a set interval. Recommended at 5 minutes to prevent data loss.
Start workbench	FreeCAD's default opening screen. Not a modelling environment — it is a launch pad showing recent files and news. A new document switches away from it automatically.

Looking Ahead — Chapter 5: The Sketcher

Chapter 5 dives deep into the Sketcher workbench — the 2D drawing environment where every FreeCAD model begins. You will learn every drawing tool, every constraint type, and the golden rule of FreeCAD modelling: the sketch must be fully constrained (all green) before you proceed. By the end of Chapter 5, drawing precise, constrained 2D geometry will be as natural as freehand sketching on paper.

CHAPTER
5
of 24

PART TWO — FOUNDATIONS

The Sketcher

Drawing and constraining 2D geometry — the foundation of every FreeCAD model

Chapter Learning Objectives

By the end of this chapter, you will be able to:

- Open a Sketch on the correct plane within Part Design
- Use all major Sketcher drawing tools: Line, Rectangle, Circle, Arc, Polyline, and Regular Polygon
- Identify what each sketch colour (white, yellow, green, red, blue) means
- Explain the concept of degrees of freedom and describe what it means for a sketch element to be fully constrained
- Apply all geometric constraints: Coincident, Horizontal, Vertical, Equal, Symmetric, Tangent, Parallel, Perpendicular
- Apply all dimensional constraints: Horizontal Distance, Vertical Distance, Radius, Angle, Distance
- Toggle lines between normal and construction mode
- Identify and fix an over-constrained sketch
- Apply the green light rule before closing every sketch
- Draw the most common CNC sketch profiles using efficient constraint strategies

The Sketcher — Where Every Model Begins

Every solid shape in FreeCAD starts as a 2D sketch. The quality of your sketch — how precisely and efficiently it defines the geometry — determines the quality of everything that follows.

The Sketcher is FreeCAD's 2D drawing environment. It is accessed from within the Part Design workbench rather than as a standalone workbench — when you create a new Sketch in Part Design, FreeCAD automatically enters the Sketcher and changes the toolbar and menus to Sketcher-specific tools. When you close the sketch, you return to Part Design automatically.

Sketching in FreeCAD is a deliberately two-phase process. In Phase 1, you draw the approximate geometry — lines, circles, rectangles — without worrying about exact positions or sizes. The shapes will look roughly right but nothing is precise. In Phase 2, you apply constraints — rules that define exact positions, sizes, and relationships — until the sketch is fully defined. Only when the sketch is fully defined (all green) do you proceed to 3D operations.

This two-phase approach is the opposite of what most beginners expect. Most people who have used drawing software before try to draw precisely from the start — clicking exactly at the right coordinates, trying to get shapes to snap to the right size. In FreeCAD's Sketcher, this is the wrong approach. Draw approximately, then constrain precisely. Once you internalise this, your sketching speed will increase dramatically.

> 📐 **The Two-Phase Sketching Rule**
> Phase 1: Draw the approximate shape — do not try to be precise. Rough positions and sizes are fine. Phase 2: Apply constraints until every element is fully constrained (green). Then, and only then, close the sketch. Never try to achieve precision by careful clicking. Apply it through constraints.

Entering the Sketcher

You always enter the Sketcher through Part Design, not directly. The process has three steps: ensure a Body exists in your document, create a new Sketch, and select the plane the sketch will be attached to.

Step 1 — The Body

Every Part Design sketch must be attached to a Body. If your document does not yet have a Body, create one first: Part Design → Body. A Body object will appear in the Model tree. Without a Body, many Part Design tools — including Create a new Sketch — will be greyed out and unavailable.

Step 2 — Create a New Sketch

With a Body in place, go to Part Design → Create a new Sketch (or click the Sketch icon in the Part Design toolbar). A task panel will appear on the left side of the interface asking you to select the attachment plane — the flat surface the sketch will be drawn on.

Step 3 — Choose a Plane

For most projects in this course, you will sketch on the XY plane — this represents the flat top surface of your material, looking straight down. Select XY Plane in the task panel and click OK. FreeCAD will switch to the Sketcher workbench and orient the view to look directly down at the XY plane.

XY Plane	The horizontal plane — flat, looking from above. Used for designing the top face of any flat part. This is the standard starting plane for CNC projects in this course.
XZ Plane	A vertical plane running front to back. Used for designing side profiles — for example, the cross-section of a moulding or an extruded shape.
YZ Plane	A vertical plane running left to right. Similar use cases to XZ — designing vertical profiles.

A model face	Any flat face of an existing solid body can be used as a sketch attachment plane. This is how you add pockets, countersinks, and text to the top face of an already-extruded model.

Tip

After selecting XY Plane and clicking OK, the 3D view will snap to look straight down at the sketch plane. The grid becomes visible. The toolbar changes to Sketcher tools. You are now in the Sketcher. The red X axis runs horizontally and the green Y axis runs vertically. The intersection of the axes — the small orange dot — is the sketch origin (0, 0).

Try It Now

Open FreeCAD with a new document. Create a Body (Part Design → Body). Then go to Part Design → Create a new Sketch. When asked for a plane, select XY Plane and click OK. Notice how the toolbar completely changes. Look at the bottom of the screen — the status bar shows 'Sketcher in edit mode'. You are in the Sketcher. Do not draw anything yet — just look at the environment and locate the X axis (red), Y axis (green), and origin point (orange dot at the centre).

Sketcher Drawing Tools

The Sketcher's drawing tools create geometric elements — lines, arcs, circles, rectangles, polygons. These are the raw shapes that your constraints will then define precisely. All drawing tools are found under Sketch → Sketcher geometries in the menu bar, or as toolbar buttons in the Sketcher toolbar.

To use any drawing tool: click it once to activate it. Then click in the 3D view to place geometry. Right-click or press Escape to deactivate the tool and return to selection mode. Most tools remain active until you deactivate them — one activation lets you draw multiple elements of the same type.

Tool	Menu Path (Sketch →)	What It Draws	Best Used For
Line	*Geometries → Create Line (L)*	A straight line segment from one point to another	All straight-edged profiles — the most used Sketcher tool
Rectangle	*Geometries → Create Rectangle (R)*	A rectangle defined by two opposite corners	Rectangular panels, sign outlines, stock profiles — use this before Line for any rectangular shape
Centred Rect.	*Geometries → Create Centered Rectangle*	A rectangle centred on the first click point	When your design is centred on the origin — reduces the number of symmetry constraints needed
Circle	*Geometries → Create Circle (C)*	A circle defined by a centre point and one edge point	Holes, bores, circular outlines, round pockets

Arc by Centre	*Geometries → Create Arc by Centre*	An arc defined by centre, start angle, and end angle	Rounded corners, decorative curves — more controllable than the three-point arc
Arc by 3 Points	*Geometries → Create Arc by 3 Points*	An arc passing through exactly three specified points	When you know three points the arc must pass through
Polyline	*Geometries → Create Polyline (P)*	A connected series of lines drawn continuously	Complex outlines where many connected line segments are needed — efficient for multi-sided shapes
Regular Polygon	*Geometries → Create Regular Polygon*	A polygon with N sides, all equal length	Hexagonal pockets, octagonal shapes, gear-like outlines — the N is set before clicking
B-Spline	*Geometries → Create B-Spline*	A smooth freeform curve controlled by draggable points	Organic shapes, artistic curves — covered in detail in Week 5
Slot	*Geometries → Create Slot*	An oblong slot: two parallel lines with semicircular ends	Mounting slots, elongated holes for bolted connections
Point	*Geometries → Create Point*	A single reference point with no geometry	Reference points for symmetric constraints, auxiliary construction locations

Tip

For rectangular CNC parts, always use Create Rectangle or Create Centered Rectangle rather than drawing four separate lines. The rectangle tool creates all four lines in one operation AND automatically applies Horizontal and Vertical constraints to them — saving you four constraint applications immediately.

Try It Now

In the Sketcher, try each of these tools once:

1. Press R and click twice to draw a rectangle. Note that it appears white/yellow.

2. Press C and click twice to draw a circle. Note the unconstrained white colour.

3. Press L and click to draw a few connected line segments. Right-click to end.

4. Draw a Regular Polygon (Sketch → Sketcher geometries → Create Regular Polygon). FreeCAD will ask for the number of sides — type 6 for a hexagon. Click to set the centre and drag to set the size.

Notice that none of these shapes are precise yet. That is correct — precision comes from constraints in the next section.

Understanding Sketch Colours — Your Real-Time Feedback System

The colour of sketch elements is not decorative. It is FreeCAD's primary feedback mechanism, telling you precisely how constrained each element is at every moment.

Unlike most software where colour choices are aesthetic, every colour in the FreeCAD Sketcher is meaningful and functional. Learning to read sketch colours instantly is one of the most valuable early skills you can develop. Once you can glance at a sketch and immediately know which elements still need constraining, your workflow speed increases enormously.

Colour	State	Meaning
	Unconstrained	The element has degrees of freedom remaining. It can still be moved or resized. This is normal immediately after drawing — you have not applied constraints yet.
	Partially constrained	The element is partially constrained. Some of its degrees of freedom have been removed, but not all. You are in progress.
	Fully constrained	The element has no remaining degrees of freedom — all its position, orientation, and size are completely defined. This is what you are aiming for. Every element in the sketch should be green before you close it.
	Over-constrained	A conflict exists — one or more constraints are contradicting each other. FreeCAD cannot satisfy all constraints simultaneously. The conflicting constraint must be deleted before proceeding.
	Construction geometry	A line or shape set to construction mode. It is used for reference and positioning but will not be machined. See Section 7 of this chapter.
	External geometry	Geometry projected from outside the current sketch (from a model face or another sketch) for reference. Read-only — it cannot be directly edited within this sketch.

The single most important colour is green. Green means the element is completely defined — its position, size, and orientation are all fully specified. A sketch where every element is green is a fully constrained sketch, and a fully constrained sketch is the only kind that produces predictable, correct 3D geometry.

The most dangerous colour to ignore is red. Red means there is a conflict in the sketch's constraints — two constraints are asking for incompatible things. FreeCAD will not let you close a red sketch and proceed to a Pad or Pocket. More importantly, even if FreeCAD did allow it, a red sketch would produce unpredictable geometry because the software cannot determine which constraint to satisfy and which to violate.

> ⚠ **Important**
>
> Never proceed with a red sketch. Find the conflicting constraint, delete it, and re-apply a correct version.
>
> Deleting the most recently applied constraint is almost always the correct fix — that is the constraint most likely to have introduced the conflict.

Degrees of Freedom — The Fundamental Concept

Degrees of freedom is the concept that explains why constraints are necessary and how many you need. Once you understand it, the entire constraint system will make logical sense.

A degree of freedom (DOF) is one dimension of possible movement or change in a geometric element. A single point on a 2D plane has 2 DOF — it can move left-right (X) and up-down (Y). A line segment has 4 DOF — it has two endpoints, each with 2 DOF (though in practice, the line's DOF is counted as the combination of position, length, and angle). A circle has 3 DOF — X position, Y position, and radius.

When you apply a constraint, you remove one or more DOF. A Horizontal constraint applied to a line removes the rotation DOF — the line can no longer rotate. A dimensional constraint fixing the line's length removes one more DOF. A Coincident constraint fixing one endpoint to the origin removes 2 more DOF (X and Y position of that endpoint). Each applied constraint eliminates a degree of freedom until none remain.

The Sketcher shows you the remaining DOF count in the status bar at the bottom of the screen. A message like 'Sketch has 5 degrees of freedom' means you need five more constraint applications before the sketch is fully constrained. When the count reaches 0 — and the sketch is all green — you are done.

> 📖 **Key Term: Degree of Freedom (DOF)**
>
> One dimension of possible movement or change in a geometric element. A fully constrained sketch element has zero DOF remaining — its position, orientation, and size are completely and unambiguously defined.

The table below shows the DOF reduction process for a single horizontal line, starting from a freshly drawn unconstrained state and ending at fully constrained. Follow this example in FreeCAD as you read it:

Stage	DOF left	Sketch colour
Just drawn — no constraints applied	5	Unconstrained. The line can move, rotate, and stretch freely.
Horizontal constraint applied	4	One DOF removed (rotation locked). Still movable and resizable.
One endpoint fixed to origin (Coincident)	2	Two more DOF removed (position partially defined). Getting closer.
Length constraint applied (100 mm)	1	One more DOF removed. Only the along-axis position is free.
Second endpoint fixed (Coincident)	0	Fully constrained. This line cannot move. Ready to proceed.

Stop and Think

A common beginner experience: you have applied what feels like many constraints, but the sketch is still showing some yellow elements. The status bar says '1 degree of freedom'. This is frustrating — one DOF left should be easy to find. In practice, that last DOF is often the most revealing: it forces you to think about what is still undefined. Is the whole sketch floating — is it constrained in shape but not anchored to the origin? Is one circle's centre still free to slide along a line? Finding the last DOF is a genuine problem-solving exercise, and solving it builds your constraint thinking skills.

Geometric Constraints — Defining Relationships

Geometric constraints define the relationships between sketch elements — not their sizes, but their relative positions, orientations, and connections. They are applied by selecting one or two elements and then choosing the constraint. Most geometric constraints remove one DOF.

All geometric constraints are found under Sketch → Sketcher constraints in the menu bar. Each constraint also has a keyboard shortcut — learn the most common ones (listed below) and your sketching speed will increase significantly.

Constraint	What It Does	How to Apply	CNC Use Case
Coincident	Forces two points to occupy the same location — they become a single shared point. Removes 2 DOF.	*Select two points → Sketch → Sketcher constraints → Constrain coincident (shortcut: C then N, or the toolbar)*	Connecting line endpoints to form a closed profile. Attaching a shape corner to the sketch origin.
Horizontal	Forces a line to be perfectly horizontal (parallel to the X axis). Removes 1 DOF (rotation).	*Select a line → Sketch → Sketcher constraints → Constrain horizontal (shortcut: H)*	Ensuring the top and bottom edges of a nameplate are perfectly horizontal.

Vertical	Forces a line to be perfectly vertical (parallel to the Y axis). Removes 1 DOF.	*Select a line → Sketch → Sketcher constraints → Constrain vertical (shortcut: V)*	Ensuring the left and right edges of a panel are perfectly vertical.
Equal	Forces two lines to be the same length, or two circles/arcs to have the same radius. Removes 1 DOF.	*Select two lines (or two circles/arcs) → Sketch → Sketcher constraints → Constrain equal (shortcut: Q)*	Making the four sides of a square equal. Making two mounting holes the same diameter.
Symmetric	Forces two points to be mirror images of each other about a third element (a line or point). Removes 2 DOF.	*Select two points and a symmetry axis line or point → Sketch → Sketcher constraints → Constrain symmetric*	Centring a slot or text on a panel. Making left and right features mirror each other.
Tangent	Forces a line and an arc, or two arcs, to meet at a point without a kink — a smooth continuous transition. Removes 1 DOF.	*Select a line and an arc (or two arcs) at their meeting point → Sketch → Sketcher constraints → Constrain tangent*	Creating a slot (oblong) profile. Rounding the transition between a straight edge and an arc.
Parallel	Forces two lines to be parallel to each other. Removes 1 DOF.	*Select two lines → Sketch → Sketcher constraints → Constrain parallel*	Ensuring two edges of a trapezoid are parallel. Making guide lines parallel for layout purposes.
Perpendicular	Forces two lines to be at exactly 90° to each other. Removes 1 DOF.	*Select two lines → Sketch → Sketcher constraints → Constrain perpendicular*	Ensuring a vertical wall is exactly perpendicular to a horizontal base. T-joint geometry.
Fix (Block)	Completely pins an element to its current position — removes all remaining DOF at once. Use with caution.	*Select an element → Sketch → Sketcher constraints → Block*	Pinning a reference point to a specific location when other constraints would be complex. Rarely needed if the sketch is well-designed.
Constrain Internal Angle	Sets the interior angle at a vertex of a polygon to a specific value.	*Select two lines meeting at a vertex → Sketch → Sketcher constraints → Constrain internal angle*	Setting the angled side walls of a dovetail joint (Week 9) or a chamfered corner.

Tip

When connecting line segments end-to-end, always use the Coincident constraint — never rely on visual 'snapping' to connect points. Two points that look connected but are not coincident will leave the profile open, and FreeCAD's Pad and Pocket operations require a fully closed profile. A missing Coincident is one of the most common beginner errors.

Try It Now

Draw two separate lines in the Sketcher so they are close to each other but not connected. Now select the endpoint of one line and the endpoint of the other. Go to Sketch → Sketcher constraints → Constrain coincident. The two endpoints jump to the same location and merge. Now both lines share that endpoint. This is how a closed profile is built — connecting every endpoint to its neighbour with Coincident constraints.

Dimensional Constraints — Defining Exact Sizes

While geometric constraints define relationships (horizontal, equal, coincident), dimensional constraints define exact numerical values — the width is exactly 150 mm, the radius is 25 mm, the angle is 45°. Dimensional constraints are the precision tools that turn a roughly-shaped sketch into an exactly-dimensioned design.

When you apply a dimensional constraint, a dialog box appears asking for the value. Type the number and press Enter. The geometry immediately snaps to that exact dimension. The constraint value appears as a number on the sketch — it is displayed in the 3D view and can be double-clicked to change.

From Week 3 onwards, you will replace the typed numbers in dimensional constraints with spreadsheet references (e.g., =Spreadsheet.SignWidth) instead of fixed values. For now, type the numbers directly.

Constraint	Menu Path	What It Constrains	Example
Constrain Horizontal Distance	*Sketch → Sketcher constraints → Constrain horizontal distance*	The horizontal distance between two points, or the width of a line	Width = 150 mm — sets the width of a nameplate outline
Constrain Vertical Distance	*Sketch → Sketcher constraints → Constrain vertical distance*	The vertical distance between two points, or the height of a line	Height = 80 mm — sets the height of a panel
Constrain Radius	*Sketch → Sketcher constraints → Constrain radius*	The radius of a circle or arc	Radius = 40 mm — a circle representing an 80 mm diameter hole
Constrain Diameter	*Sketch → Sketcher constraints → Constrain diameter*	The diameter of a circle (alternative to Radius — same effect, different input)	Diameter = 8 mm — a mounting hole for an M8 bolt

Constrain Distance	*Sketch → Sketcher constraints → Constrain distance*	The distance between two selected points (can be at any angle — Euclidean distance)	Distance = 60 mm — the distance between two hole centres at any angle
Constrain Angle	*Sketch → Sketcher constraints → Constrain angle*	The angle of a line relative to the X axis, or the angle between two lines	Angle = 45° — a chamfered corner cut at 45 degrees
Constrain Internal Angle	*Sketch → Sketcher constraints → Constrain internal angle*	The interior angle at the junction of two lines	Internal angle = 80° — a dovetail joint wall at 10° from vertical
Fix X Coordinate	*Sketch → Sketcher constraints → Fix X coordinate (Lock)*	Fixes the X coordinate of a point to an exact value	X = 0 — locks a corner to the horizontal axis
Fix Y Coordinate	*Sketch → Sketcher constraints → Fix Y coordinate (Lock)*	Fixes the Y coordinate of a point to an exact value	Y = 0 — locks a corner to the vertical axis

> **Tip**
>
> To constrain the overall width of a rectangle: click on one of the vertical side lines (not a corner point). Then go to Sketch → Sketcher constraints → Constrain horizontal distance. The dialog will offer to constrain the length of the line's horizontal projection — which, for a vertical line, means the width of the bounding box. For a cleaner approach, select the leftmost and rightmost corner points of the rectangle and apply Constrain horizontal distance between them.

Construction Geometry — Reference Lines and Circles

Construction geometry is sketch geometry that exists only as a reference. It helps you position and constrain other elements but is not included when the sketch is used for a Pad, Pocket, or CAM operation. Construction geometry is displayed in blue, distinguishing it clearly from normal (to-be-machined) geometry.

The most common use of construction geometry in CNC work is creating a reference circle to position features at a specific radius from the centre — the technique used to place hour markers on the clock face in Week 9, and to position teeth on a gear. You draw the reference circle, make it a construction line, constrain its radius to the exact position you need, and then constrain your actual geometry to lie on it.

Toggling an Element to Construction Mode

To convert any sketch element to construction geometry: select it, then go to Sketch → Sketcher geometries → Toggle construction geometry (or press the blue dotted-square toolbar button). The selected element will turn blue. To convert it back to normal geometry, select it and toggle again.

You can also create construction geometry directly by activating the toggle before drawing — click the Toggle construction geometry button in the toolbar, draw your reference lines or circles, then click the toggle again to return to normal geometry mode.

Reference circle	A circle set to construction mode, used to position holes, features, or text at an exact radius from the centre. Constrain a point to lie on the construction circle using Constrain Point on Object (Sketch → Sketcher constraints → Constrain point on object).
Centreline	A horizontal or vertical construction line through the origin, used as an axis for the Symmetric constraint. Very commonly used to centre designs on the sketch origin.
Offset line	A construction line offset from an edge at a fixed distance, used to position features (like a border pocket inset from the outer edge) relative to another element.
Diagonal	A construction line from corner to corner of a rectangle, used to find the geometric centre when you need to position something at the centre of an existing shape.

📖 **Key Term: Construction geometry**

Sketch geometry displayed in blue that is used for reference and positioning only. It is not included in Pad, Pocket, or CAM operations. Toggle any sketch element to/from construction mode using Sketch → Sketcher geometries → Toggle construction geometry.

Try It Now

Draw a circle in the Sketcher. Select it. Go to Sketch → Sketcher geometries → Toggle construction geometry. The circle turns blue — it is now construction geometry. Draw a small rectangle near the edge of the circle. Select one corner point of the rectangle. Go to Sketch → Sketcher constraints → Constrain point on object. Now click on the blue construction circle. The corner snaps to lie on the circle's circumference. This is exactly how you would position 12 hour markers equally spaced on a clock face — using one reference circle to position all of them at the same radius.

The Green Light Rule and Over-Constraining

📐 **The Green Light Rule**

Before closing any sketch, every element must be green. Do not close a sketch that still has white, yellow, or red elements. The time it takes to fully constrain a sketch before closing is always less than the time it takes to diagnose problems in a 3D model built from an underconstrained sketch.

The green light rule is the single most important habit to develop in FreeCAD. An underconstrained sketch — one where some elements are still white or yellow — will appear to work when you Pad or Pocket it. But the resulting 3D geometry will be fragile: if anything changes upstream, the underdefined elements can shift to unexpected positions, breaking the model. In a CNC workflow, a silently broken model can produce incorrect G-code that cuts the wrong shapes.

Over-Constraining — When You Go Too Far

If you apply a constraint that conflicts with an already-constrained element, some of the sketch's elements will turn red. This is called over-constraining, and it means the sketch now has contradictory instructions — more constraints than the geometry's degrees of freedom require, and at least one constraint is incompatible with the others.

Over-constraining always has a simple cause: the most recently applied constraint contradicts something that was already constrained. The fix is equally simple: undo the last constraint (Ctrl+Z) and think about what you are trying to achieve. Usually you will realise that the constraint you just applied was already implied by an earlier one — for example, constraining a line's length after already constraining both its endpoints to specific coordinates fully defines the line, so an additional length constraint is redundant and causes a conflict.

Symptom	Some sketch elements turn red after you apply a constraint.
Cause	The most recently applied constraint conflicts with one already in place. Common examples: applying a length constraint to a line whose both endpoints are already fixed to specific coordinates; applying Equal to two elements where other constraints already force them to be the same size; applying Horizontal to a line that already has a fixed angle.
Fix	Press Ctrl+Z to undo the most recent constraint. The sketch will return to its previous state. Consider whether the constraint was actually needed. If you still need it, check which earlier constraint it conflicts with and decide which one is the correct one to keep.
Prevention	Develop a consistent constraint sequence: geometric constraints first (connections, orientations), then dimensional constraints (sizes). Applying constraints in this sequence reduces the chance of conflicts.

> **Tip**
>
> FreeCAD also highlights the specific conflicting constraints in the Sketcher task panel on the left when over-constraining occurs — a list called 'Redundant constraints' or 'Conflicting constraints' will appear. This list identifies which constraints are in conflict, making it easier to decide which to remove.

Common CNC Sketcher Workflows

Most CNC profiles are variations of a small number of fundamental sketch patterns. Master these four patterns and you will be able to sketch the majority of CNC-ready profiles you will ever need.

Workflow 1 — The Centred Rectangle

The centred rectangle is the starting sketch for the majority of flat CNC parts in this course: nameplates, coasters, boxes, panels. The rectangle is centred symmetrically on the sketch origin (0, 0), which makes it easy to set work zero at the centre of the physical material.

1 **Draw the rectangle**
Go to Sketch → Sketcher geometries → Create Centered Rectangle (not the standard rectangle — use the Centred version). Click once on the origin (the orange dot) to set the centre. Drag out and click again to set the corner. A rectangle appears, centred on the origin.

2 **Apply Horizontal Distance for width**
Click one vertical side line of the rectangle. Go to Sketch → Sketcher constraints → Constrain horizontal distance. Type your width (e.g., 150) and click OK. The line turns partially constrained.

3 **Apply Vertical Distance for height**
Click one horizontal side line. Go to Sketch → Sketcher constraints → Constrain vertical distance. Type your height (e.g., 80) and click OK.

4 **Verify — all green**
The rectangle should now be fully green. The status bar should show 0 DOF. If it shows 1 DOF, check that the Centered Rectangle tool correctly placed the centre at the origin — if not, add a Symmetric constraint between opposite corners about the origin.

📋 **Note**

The Centred Rectangle tool in FreeCAD v1.1.0 should automatically add Symmetric constraints about the origin. If your rectangle is not centred after applying the two dimensional constraints, manually add a Symmetric constraint: select both left and right corner points and the Y axis as the symmetry line.

Workflow 2 — A Circle with a Fixed Centre

Used for circular parts (clock faces, coasters, gear blanks) and for holes (mounting holes, dowel holes, shaft bores). The key step is fixing the circle's centre at an exact position before constraining the radius.

1 **Draw the circle**
Press C to activate the Circle tool. Click at the approximate centre position and drag out to set a rough radius. The circle appears in white.

2 **Fix the centre**
If the circle centre should be at the origin (0, 0): select the centre point of the circle and the origin point together (Shift+click both). Apply Sketch → Sketcher constraints → Constrain coincident. The centre snaps to the origin. Two DOF removed.

3 **Constrain the radius**
With only the circle selected, go to Sketch → Sketcher constraints → Constrain radius. Type the radius value (e.g., 50 for a 100 mm diameter circle) and click OK. The circle turns green. 1 DOF removed — now 0 remain.

> **Tip**
> For holes whose centres should be at specific coordinates relative to the origin: apply Fix X coordinate and Fix Y coordinate to the circle's centre point rather than Coincident with the origin. For example, a mounting hole at X=60, Y=30 from the centre — fix its centre to X=60, Y=30.

Workflow 3 — A Slot (Oblong Profile)

A slot is a rectangle with semicircular ends. It is used for mounting slots that allow adjustment, elongated holes, and decorative oblong shapes. FreeCAD's dedicated Slot tool creates this profile in one operation.

1 — Use the Slot tool
Go to Sketch → Sketcher geometries → Create Slot. Click to set one end centre point, then drag and click to set the other end centre point. A slot profile (two parallel lines capped with semicircles) appears.

2 — Fix the position
Constrain the centre of the slot to the origin using Coincident (or to specific X, Y coordinates using Fix X/Y). This removes 2 DOF.

3 — Constrain the length
Select the two arc centres. Apply Constrain horizontal distance (for a horizontal slot) to set the distance between centres. This determines how long the slot's straight section is.

4 — Constrain the width
Select one of the semicircular arcs. Apply Constrain radius to set the end radius. Since both arcs are automatically constrained Equal by the Slot tool, constraining one constrains both. Slot is now fully green.

Workflow 4 — An Annular Profile (Two Concentric Circles)

An annular sketch — a large circle with a smaller circle inside it — defines a ring-shaped region. In Part Design, pocketing an annular sketch creates a recessed ring (like the border pocket in Project 2). This same pattern is used for rings, frames, and border insets throughout the course.

1 — Draw the outer circle
Press C. Click the origin to set the centre. Drag and click to set a rough size. Apply Coincident to fix the centre at the origin. Apply Constrain radius for the outer radius (e.g., 50 mm).

2	**Draw the inner circle** Press C again. Click the origin again for the centre. Apply Coincident — both circles now share the origin as their centre. Apply Constrain radius for the inner radius (e.g., 38 mm, giving a 12 mm wide ring).

3	**Check — both circles green** Both circles should be green. The sketch contains two fully defined concentric circles. When you close this sketch and apply a Pocket, FreeCAD understands the ring-shaped region between the two circles as the area to remove.

Week 1, Days 2 and 3 — Lesson Content

Days 2 and 3 of Week 1 cover the Sketcher material in this chapter. Day 2 focuses on drawing tools and sketch colours (Remember level), while Day 3 introduces constraints in depth (Understand level). The following overview describes the key activities for each day.

Day 2 — Drawing Tools and Sketch Colours (Remember)

The objective for Day 2 is simple: become comfortable drawing any shape in the Sketcher. Precision does not matter today — the focus is on using the tools, observing the colour states, and developing physical fluency with the drawing tools before constraint theory is introduced.

Guided activity 1	Draw 10 different sketch elements using 10 different tools: a line, a rectangle, a centred rectangle, a circle, an arc, a polyline, a regular polygon (6 sides), a slot, and a B-spline. Do not apply any constraints. Observe the colours — everything is white/yellow.
Guided activity 2	Draw a rough house outline using lines: a rectangle for the walls and a triangle for the roof. Do not constrain it. Now slowly apply Coincident constraints to connect every endpoint pair. Observe how the colour changes as you connect elements.
Independent exercise	Without looking at the textbook: draw a profile that consists of a rectangle with a circle centred on it (not concentric — the circle is positioned within the rectangle, not at the rectangle's centre). Make everything Coincident where elements should connect. Do not add dimensional constraints yet.
Exit ticket	Label the four sketch colours and what each means. This is a Remember-level task — the answer is in the colour table in this chapter.

Day 3 — Constraints and the Green Light Rule (Understand)

Day 3 builds on Day 2 by introducing the constraint system. The objective is to fully constrain at least two different sketches — achieving all-green — and to experience and fix at least one instance of over-constraining. Understanding constraints is not just about applying them; it is about understanding why each one removes the degree of freedom it does.

Guided activity 1	Take the house outline from Day 2 (or draw a fresh one). Apply constraints systematically: Horizontal to the base line, Vertical to each wall line, Coincident to all connecting points, Horizontal Distance for the width, Vertical Distance for the wall height and roof height. Work until the sketch is fully green.
Guided activity 2	Draw a rectangle and apply dimensional constraints. Then deliberately over-constrain it: after constraining all four sides, try to add another dimensional constraint for the full width. Observe the red elements. Read the Conflicting Constraints list. Press Ctrl+Z to undo. Discuss with your neighbour what caused the conflict.
Guided activity 3	Draw a centred rectangle (200 × 120 mm) with a centred circle (radius 20 mm) for a door sign. Fully constrain both elements. This is the sketch for an upcoming exercise in Chapter 6 — save the file as DoorSign_Sketch.FCStd.
Independent exercise	Without step-by-step instructions: draw a sketch for a rectangular panel (150 × 80 mm) with four mounting holes, each 6 mm diameter, positioned 15 mm from each corner. Fully constrain the entire sketch — panel, all four holes — until the status bar reads 0 DOF. Save as MountingPanel.FCStd.

Tip

The four-hole mounting panel is a classic CNC exercise. There are several ways to constrain it: Fix X/Y for each hole centre, use Symmetric to mirror pairs of holes, or combine Equal diameter constraints for all four circles with X/Y position constraints for each centre. Try at least two different approaches and notice which requires fewer total constraints.

Chapter Summary

- The Sketcher is entered through Part Design, not directly. Always create a Body first, then a Sketch attached to a plane (usually XY). The Sketcher automatically opens and closes as needed.
- Sketching is a two-phase process: draw the approximate shape first, then apply constraints to define it precisely. Never try to achieve precision by careful clicking.
- Sketch colours are a real-time feedback system: white/yellow = unconstrained, green = fully constrained, red = over-constrained, blue = construction geometry. Only a fully green sketch should be closed.
- Degrees of freedom (DOF) quantifies how much freedom an element has to move. Every constraint removes one or more DOF. The goal is zero DOF remaining — the status bar reports the count.
- Geometric constraints define relationships: Coincident (shared point), Horizontal, Vertical, Equal, Symmetric, Tangent, Parallel, Perpendicular. Apply geometric constraints before dimensional ones.
- Dimensional constraints define exact values: Horizontal Distance, Vertical Distance, Radius, Diameter, Angle, Distance. From Week 3 these will be driven by spreadsheet values rather than typed numbers.
- Construction geometry (blue) is reference-only. Toggle any element to construction mode using Sketch → Sketcher geometries → Toggle construction geometry. It is never machined.
- Over-constraining occurs when a new constraint conflicts with existing ones. Fix: press Ctrl+Z to undo the last constraint. Prevention: apply constraints in order — geometric first, dimensional second.
- The four fundamental CNC sketch workflows are: centred rectangle, circle with fixed centre, slot, and annular profile (two concentric circles). Master these and you can sketch most CNC profiles you will ever need.

Key Vocabulary — Chapter 5

Term	Definition
Sketcher	FreeCAD's 2D drawing environment. Accessed through Part Design → Create a new Sketch. The Sketcher automatically opens when a sketch is created and closes when the sketch is closed.
Degree of freedom (DOF)	One dimension of possible movement or change in a geometric element. Each constraint removes one or more DOF. A fully constrained sketch element has zero DOF remaining.
Fully constrained	A sketch state in which every element has zero degrees of freedom remaining — all positions, sizes, and orientations are completely defined. Displayed as all-green.
Over-constrained	A sketch state where conflicting constraints exist — more constraints than the geometry's DOF require, and at least one constraint is incompatible with others. Displayed as red.
Geometric constraint	A constraint that defines a relationship between sketch elements — coincident, horizontal, vertical, equal, symmetric, tangent, parallel, or perpendicular.
Dimensional constraint	A constraint that defines an exact numerical value — horizontal distance, vertical distance, radius, diameter, angle, or distance.

Coincident	A geometric constraint that forces two points to occupy the same location, creating a shared endpoint. Essential for building closed profiles.
Construction geometry	Sketch geometry set to reference-only mode. Displayed in blue. Used to position and constrain other elements but is not included in Pad, Pocket, or CAM operations.
Green light rule	The course rule that every sketch element must be fully constrained (green) before closing the sketch. Never close a sketch with white, yellow, or red elements.
Closed profile	A sketch outline where every endpoint is connected by Coincident constraints to form a continuous, unbroken loop. Required for Pad and Pocket operations.
Attachment plane	The plane on which a sketch is drawn. Chosen when creating a new Sketch. The most common choice for CNC projects is the XY plane.
Annular profile	A sketch consisting of two concentric circles (or rectangles) — a large outer shape and a smaller inner shape. Produces a ring-shaped pocket when used with the Pocket operation.
Toggle construction	The action of switching a sketch element between normal geometry mode (white/yellow/green) and construction mode (blue). Found under Sketch → Sketcher geometries → Toggle construction geometry.
Constrain point on object	A constraint that forces a point to lie on a line, circle, or arc (but not necessarily at a specific position along it). Used with construction circles to place features at a specific radius.
Two-phase sketching	The recommended FreeCAD sketching approach: first draw the approximate shape freely, then apply constraints to define it precisely. The opposite of trying to draw accurately from the start.

Looking Ahead — Chapter 6: Part Design

Chapter 6 builds on your Sketcher skills by adding the 3D dimension. You will learn Pad (extruding a sketch into a solid), Pocket (removing material), Chamfer, Fillet, and ShapeString. By the end of Chapter 6, you will have the complete toolkit to design a 3D CNC-ready model — and Chapter 7 will show you how to set up the CAM operations that turn that model into machine instructions.

CHAPTER
6
of 24

PART TWO — FOUNDATIONS

Part Design

Pad, Pocket, Chamfer, Fillet, ShapeString — turning 2D sketches into 3D CNC-ready models

Chapter Learning Objectives

By the end of this chapter, you will be able to:

- Apply Part Design Pad to extrude a sketch into a solid body with correct depth and options
- Apply Part Design Pocket to remove material from an existing solid
- Create a new Sketch attached to a model face rather than a coordinate plane
- Edit any feature by double-clicking it in the Model tree
- Apply Chamfer and Fillet to selected edges
- Explain why feature ordering in the Model tree matters
- Use Part ShapeString to place 3D text on a model face
- Diagnose and fix the most common Part Design errors

From Flat Sketch to 3D Solid

Chapter 5 gave you the ability to define precise 2D geometry. Chapter 6 gives it a third dimension. By the end of this chapter, you will have everything you need to design any flat CNC part.

Part Design is FreeCAD's primary 3D modelling workbench. It uses a feature-based, history-based approach — you build a model by stacking operations on top of each other, each recorded in the Model tree. A Pad extrudes a sketch into a solid. A Pocket carves material from that solid. Chamfers and Fillets refine the edges. ShapeString adds text. Together, these five tools produce every flat CNC part in the first six projects of this course.

The key mental model to carry into this chapter: Part Design does not give you a fixed shape. It gives you a process — a recorded sequence of decisions that can be edited, reversed, and rebuilt at any point. This is why double-clicking a feature in the Model tree and changing a value is so powerful. You are not editing a shape; you are editing the decision that produced the shape.

Key Term: Feature-based modelling

A 3D modelling approach where the model is built from a sequence of operations (features) applied in order. Each feature is recorded in the Model tree and can be edited independently. Changing a feature updates all subsequent features automatically.

The Part Design Workflow

Every Part Design model follows the same sequence of steps regardless of complexity. Understanding this sequence as a mental workflow — not just a set of button clicks — is what allows you to approach any design problem in FreeCAD systematically.

Sketch
2D outline drawn on XY plane

→

Pad
Sketch extruded to 12 mm thick solid

→

Sketch001
Circle drawn on top face of Pad

→

Pocket
Circle extruded downward to remove material

→

Chamfer
Angled edge applied to top rim

The diagram above shows a simple five-feature model: a Sketch extruded by a Pad, followed by a second Sketch on the Pad's top face, which is then removed by a Pocket, followed by a Chamfer applied to the top rim edge. Reading the Model tree from top to bottom tells you the complete history of the model — exactly the decisions made and in what order.

The workflow for any new CNC part always starts the same way:

1 **Create a Body**
Part Design → Body. Every Part Design model starts with a Body container. Without a Body, Part Design features are unavailable.

2 **Create a Sketch on the XY Plane**
Part Design → Create a new Sketch → XY Plane → OK. Draw your base outline. Fully constrain it (all green). Close the sketch.

3 **Pad the Sketch**
Part Design → Pad. Set the length to your material thickness. Click OK. You now have a 3D solid.

4 **Add more features**
Click a face of the solid. Create a new sketch on that face. Draw the next feature (a pocket, a hole, text). Apply Part Design → Pocket, Chamfer, Fillet as needed.

5 **Review the Model tree**
Read the tree from top to bottom. Does it tell the story of your design? Are all features green (healthy)? Are the names meaningful?

Pad — Extruding a Sketch into a Solid

The Pad operation extrudes a 2D sketch profile in a direction perpendicular to the sketch plane, creating a 3D solid. A sketch drawn on the XY plane is extruded upward (in the Z direction) by default. The result is a solid body with a footprint matching the sketch profile and a height equal to the Pad length.

To apply a Pad: draw a fully constrained sketch, then close it. Go to Part Design → Pad. The Pad dialog opens with a preview showing the extrusion. Set the Length value and click OK.

Pad Dialog — Key Options		
Option	**Value**	**Effect**
Length	*Numeric value (mm)*	The height of the extrusion. For CNC work, this is typically your material thickness — 6, 9, 12, 15, or 18 mm depending on your MDF or timber stock.
Symmetric to plane	*Checkbox*	When ticked, the Pad extrudes equally in both directions from the sketch plane — half the length above and half below. Useful for symmetric parts like spacers.
Reversed	*Checkbox*	When ticked, the Pad extrudes in the opposite direction to the default. Toggle this if your Pad appears to be going the wrong way.
Type: Blind	*Drop-down selection*	Extrudes to an exact specified length. This is the standard mode for most CNC work.
Type: Through All	*Drop-down selection*	Extrudes until it passes completely through any existing geometry. Not commonly used for the base Pad but useful for holes that must go through regardless of material thickness changes.
Type: Two Dimensions	*Drop-down selection*	Extrudes in both directions by independently specified distances. Used for parts that extend both above and below the sketch plane by different amounts.
Taper angle	*Numeric value (degrees)*	Applies a draft angle to the extruded walls — they are not vertical but angled inward or outward. A value of 0 produces vertical walls. Rarely needed for standard CNC routing but useful for mould-making.

Try It Now

Open your FreeCAD document. If you have the DoorSign_Sketch.FCStd from Chapter 5's Day 3 exercise, open it. Otherwise, create a new document with a Body, sketch a centred rectangle (200 × 120 mm) on the XY plane, fully constrain it, and close the sketch.

Now go to Part Design → Pad. A dialog opens with a green arrow showing the extrusion direction.

Set Length to 12. Click OK.

A 3D solid appears — 200 mm wide, 120 mm deep, 12 mm tall. This is your first 3D object.

Now try editing it: double-click 'Pad' in the Model tree. The dialog reopens. Change the length to 18. Click OK. The solid immediately updates to 18 mm tall.

Change it back to 12. This edit-by-double-click habit is one of the most important techniques in FreeCAD.

What Pad Requires — Common Issues

The Pad operation has two absolute requirements. First, the sketch must be fully closed — every endpoint connected by Coincident constraints, forming a single unbroken loop with no gaps and no open ends. Second, the sketch must be fully constrained — all green. If either condition is not met, Part Design → Pad will be greyed out or will produce an error.

Common Error

ERROR: Pad is greyed out (not available) — Most likely cause: the sketch is not closed (at least one endpoint is not connected), or no sketch is selected. Fix: double-click the sketch in the Model tree to re-enter it. Check for any white endpoints that are not coincident. Apply Coincident to connect them.

ERROR: Pad produces a twisted or unexpected shape — Possible cause: the sketch has an open loop that FreeCAD filled in an unexpected way, or the sketch has self-intersecting lines. Fix: re-enter the sketch and inspect carefully — look for any line that crosses another.

Pocket — Removing Material from a Solid

Pocket is the inverse of Pad. Where Pad adds material by extruding a sketch outward from a face, Pocket removes material by extruding a sketch inward. Pockets create holes, recesses, channels, and cavities in an existing solid body.

To apply a Pocket: click the face of the existing solid where you want to remove material. Create a new Sketch on that face and draw the shape to be removed. Fully constrain the sketch and close it. Go to Part Design → Pocket. Set the depth and click OK.

This workflow — click a face, create a sketch on that face, apply Pocket — is the sequence you will use dozens of times in this course. It becomes muscle memory within a few sessions.

Pocket Dialog — Key Options		
Option	**Value**	**Effect**
Depth	*Numeric value (mm)*	How deep the pocket cuts into the solid. Must be less than or equal to the Pad height (you cannot pocket deeper than the solid exists). For recessed borders, text backgrounds, etc.
Type: Blind	*Drop-down selection*	Cuts to the exact specified depth. Most common mode for pockets in CNC work.
Type: Through All	*Drop-down selection*	Cuts all the way through the solid regardless of thickness. Used for through-holes, cutouts, and any pocket that must go completely through. For CNC, the CAM Profile operation also cuts through — but Through All Pocket is useful for modelling holes that appear through the body.
Type: Symmetric	*Drop-down selection*	Cuts equally in both directions from the sketch face. Rarely used in standard CNC but useful for symmetric cavities.
Reversed	*Checkbox*	Reverses the pocket direction. Useful if the preview shows the pocket going the wrong way (into air rather than into the solid).
Taper angle	*Numeric value (degrees)*	Applies an angle to the pocket walls — angled outward for draft. 0 = vertical walls (standard for CNC routing).

Try It Now

Continuing with your DoorSign or rectangle body from the Pad exercise:

Click the top face of your solid. It highlights green.

Go to Part Design → Create a new Sketch. The sketch attaches to the top face.

Draw a smaller centred rectangle inside the existing one — try 170 × 100 mm (15 mm inset from each edge). Fully constrain it.

Close the sketch. Go to Part Design → Pocket.

Set Depth to 3. Click OK.

A 3 mm deep recessed panel appears on the top face. This is the border pocket technique used in Project 2.

Now add a through-hole: click the top face, create a sketch, draw a centred circle radius 10 mm. Close sketch. Pocket → Through All. Click OK. A hole passes all the way through the solid.

Inspect the Model tree. You should see: Body → Sketch → Pad → Sketch001 → Pocket → Sketch002 → Pocket001. The tree tells the complete story.

Sketching on a Model Face

Every sketch you have created so far has been attached to the XY coordinate plane — a flat abstract plane in space. From this point in the course, you will also attach sketches directly to faces of existing solid models. This is what allows you to add pockets, holes, text, and details to any face of a 3D object.

How to Select a Face

To create a sketch on a model face, move your cursor over the face in the 3D view. The face highlights in green when your cursor is over it. Left-click to select it — the face remains highlighted. Now go to Part Design → Create a new Sketch. FreeCAD automatically attaches the new sketch to the selected face and orients the view to look directly at it from above.

The sketch coordinate system on the face uses the face's own local X and Y axes — not the global coordinate axes. The origin of the sketch is at the face's own reference point (often the face's midpoint or corner, depending on the face). This local coordinate system is important to understand when you are constraining geometry relative to the face's edges.

> 💡 **Tip**
>
> When you create a sketch on a model face, use Sketch → Sketcher geometries → Create external geometry to project the face's edges into the sketch as reference (purple) lines. These external geometry references make it much easier to constrain your sketch elements relative to the existing body — for example, to inset a pocket 10 mm from each edge, project the face edges and constrain your inner rectangle to be 10 mm from them.

External Geometry — Referencing Existing Edges

External geometry allows you to bring the edges or vertices of the existing solid into your current sketch as reference elements (shown in purple). These references are read-only within the current sketch — you cannot move them or directly edit them — but you can constrain your new sketch elements relative to them.

To add external geometry: in the Sketcher (while editing a sketch on a face), go to Sketch → Sketcher geometries → Create external geometry. Then click on any edge of the existing model. That edge appears in the sketch as a purple reference line. Now you can apply Coincident, Distance, or other constraints between your new sketch elements and these purple reference lines.

Try It Now

In your model, click the top face of the solid. Create a new sketch on it.

Go to Sketch → Sketcher geometries → Create external geometry.

Click each of the four edges of the top face. Four purple lines appear, tracing the face boundary.

Press Escape to exit the external geometry tool.

Now draw a rectangle. Constrain its corner points to be exactly 15 mm away from each purple edge (use Constrain distance). This creates an inset rectangle perfectly positioned relative to the face — without needing to know the exact coordinates.

This is the technique used to create border pockets throughout the course.

Editing Features — The Power of Double-Click

One of Part Design's most powerful capabilities is that every feature in the Model tree is completely editable after it has been created. You are never locked in to an earlier decision. This is the practical benefit of feature-based modelling — your model is not a fixed shape, it is an editable history.

The primary way to edit a feature is double-clicking it in the Model tree:

Double-click a Sketch	Re-enters the Sketcher for that sketch. You can change the geometry, modify constraints, or add new elements. When you close the sketch, all features that depend on it automatically update.
Double-click a Pad	Re-opens the Pad dialog. You can change the length, mode, or other options. The solid updates immediately when you click OK.
Double-click a Pocket	Re-opens the Pocket dialog. Change the depth, mode, or taper angle.
Double-click a Chamfer	Re-opens the Chamfer dialog. Change the distance or add/remove edges.
Double-click a Fillet	Re-opens the Fillet dialog. Change the radius or add/remove edges.

The key insight about editing upstream features: if you double-click Sketch (the first sketch in the tree) and change the rectangle from 200 × 120 mm to 250 × 150 mm, everything that was built on top of it updates automatically. The Pad becomes 250 × 150 mm. The Pocket that was inset 15 mm from the edges repositions itself to maintain the same relative inset from the new (larger) face. This cascading update is the feature-based model working as designed.

> ⚠ **Important**
>
> Editing an upstream feature can cause downstream features to fail if the upstream change removes geometry that a downstream feature depends on.
>
> Example: if your Pocket is constrained to a specific face edge, and you change the Pad sketch so that face no longer exists in the same place, the Pocket will turn red in the Model tree.
>
> Fix: double-click the failed feature, reattach it to the correct geometry, and click OK.

Reordering Features in the Model Tree

Features in the Model tree are applied in top-to-bottom order. In some situations — particularly when adding a feature to a model you have been building for a while — you may find that a newly created feature needs to be applied before an existing one. Part Design allows limited reordering of features, but it must be done carefully.

To move a feature up or down in the tree: right-click the feature and look for Move Up / Move Down in the context menu (availability depends on the FreeCAD version and feature type). Alternatively, you can use Part Design → Move object to other body or use the Recompute mechanism.

> **Tip**
>
> The safest approach to feature ordering is to plan your build sequence before you start. Ask: what is the base shape? What material comes out next? What edge treatments come last? Edge treatments (Chamfer, Fillet) should always be the final features in the tree — applied after all Pads and Pockets are complete. Applying a Chamfer and then trying to add a Pocket below it in the tree frequently causes failures.

Chamfer — Angled Edge Treatment

A Chamfer replaces a sharp edge with an angled flat surface. In the physical world, chamfered edges catch less light, are less likely to chip or splinter, and feel more finished to the touch. For CNC joinery, a small chamfer (0.3–0.5 mm) on mating edges acts as a lead-in that makes assembly significantly easier.

Applying a Chamfer

1. Ensure you are in the Part Design workbench and your body is selected in the Model tree.
2. In the 3D view, click the edge you want to chamfer. It highlights green. To select multiple edges, hold Ctrl and click each additional edge.
3. Go to Part Design → Chamfer. The Chamfer dialog opens with a preview showing the angled cut.
4. Set the Size (distance) value. For a lead-in chamfer on joinery: 0.3–0.5 mm. For a decorative chamfer on a display piece: 1–3 mm.
5. Click OK. The Chamfer feature appears in the Model tree below the features you applied it to.

Chamfer Dialog — Options

Option	Value	Effect
Size	*Numeric value (mm)*	Equal distance from the edge on both faces. The most common mode — produces a 45° chamfer when the two faces are at 90°.
Size & angle	*Size (mm) + Angle (°)*	Allows an asymmetric chamfer — the cut is not at 45° but at the specified angle. Useful for specific visual effects or functional tapers.
Use all edges	*Checkbox*	Applies the chamfer to every edge of the selected face or body simultaneously. Efficient for finishing a complete panel but may apply to edges you do not want chamfered.

Joinery lead-in	0.3–0.5 mm chamfer on all mating edges. Allows joint to start entering the socket without perfect alignment. Makes assembly fast and reduces the chance of corner damage.
Visual finish	1–2 mm chamfer on all visible top edges of a display piece. Softens the appearance and catches light pleasingly.
Safety chamfer	2–3 mm chamfer on any edge that will be handled frequently — reduces the risk of splinter or chip injury.

Functional taper	5–10° chamfer angle on a wedge-shaped feature. Used for parts that need to lock together under load.

> **Try It Now**
>
> On your model, select all four top edges of the solid (click the first edge, then Ctrl+click the remaining three).
>
> Go to Part Design → Chamfer. Set Size to 1.5. Click OK.
>
> A clean 45° chamfer appears around the top perimeter. Inspect it by orbiting the 3D view.
>
> Now double-click the Chamfer in the Model tree and change the size to 0.5. Click OK. The chamfer is much more subtle — closer to a standard finishing chamfer.
>
> Notice that the Chamfer appears at the bottom of the Model tree, below all the Pads and Pockets. This is correct feature ordering — edge treatments always come last.

Fillet — Curved Edge Treatment

A Fillet replaces a sharp edge with a smooth curved radius. Where a Chamfer creates a flat angled surface, a Fillet creates a tangentially smooth transition — a continuous curve that blends the two adjacent faces. Fillets are generally more difficult for a CNC router to cut than chamfers (they require multiple passes or a ball-nose bit for true 3D edges), so in CNC work they are used primarily on top-face profiles in 2D rather than on 3D extruded edges.

Applying a Fillet

1. Select one or more edges in the 3D view (same selection process as Chamfer).
2. Go to Part Design → Fillet.
3. Set the Radius value. For a visible rounded corner on a CNC panel: 3–8 mm. For a subtle visual softening: 0.5–1 mm.
4. Click OK. The Fillet feature appears at the bottom of the Model tree.

Chamfer vs Fillet — When to Use Each

Chamfer — use for	Joinery lead-ins (0.3–0.5 mm), clean angular visual finish, any edge the CNC machine will cut in 2D (straight pass), decorative angled bevels.
Fillet — use for	Rounded corners on the top profile of a panel (cut by following the 2D profile radius), decorative rounded corners, removing stress concentration at inside corners in structural parts.
Fillet in the Sketcher vs Fillet in Part Design	A Sketcher fillet (applied to a 2D sketch corner) produces a rounded 2D corner that is then extruded by Pad — this is the correct way to create a CNC-machinable rounded corner profile. A Part Design Fillet applied to a 3D extruded edge produces a 3D curved surface that requires 3D CAM to machine. Use Sketcher fillet for simple rounded rectangular parts; use Part Design Fillet for complex 3D display work.

Preferred for CNC beginners	Use Sketcher fillet (create the rounded corner in the sketch itself) for most Projects 1–6. Part Design Fillet becomes relevant in Projects 8–12 with 3D CAM.

> **Tip**
>
> The most common way to create a rounded-corner panel for CNC work is to apply fillets or arcs in the Sketcher before closing the sketch — not as a Part Design Fillet on the 3D solid. In the Sketcher, go to Sketch → Sketcher tools → Create fillet, click a corner point, and set the radius. This produces a rounded 2D profile that the CNC Profile operation can follow in a single pass.

ShapeString — Adding 3D Text to Your Model

ShapeString is FreeCAD's tool for creating 3D text geometry. It generates a solid text shape that can be used as a sketch profile for engraving (in the CAM workbench) or extruded/pocketed as part of a 3D model. ShapeString is in the Part workbench — you need to switch to Part briefly to access it, then return to Part Design.

Accessing ShapeString

ShapeString is found under Part → Part ShapeString in the menu bar. However, it is only available when the Part workbench is active — you must switch from Part Design to Part using the Workbench drop-down, use ShapeString, then switch back to Part Design.

> **Note**
>
> In FreeCAD v1.1.0, there is also a Draft → Draft ShapeString option that can be used within the Draft workbench. Both tools produce similar results. For this course, use Part → Part ShapeString — the Part version integrates more cleanly with Part Design bodies.

The ShapeString Dialog — Step by Step

1. Switch to the Part workbench (Workbench drop-down → Part).
2. Go to Part → Part ShapeString. The ShapeString dialog opens.
3. In the String field, type the text you want to appear on your model — your name, a word, or any short text. Keep it brief for CNC — long strings take longer to machine.
4. Set the Height field to the desired text height in millimetres. For a 100 × 60 mm nameplate, 25–30 mm is a good starting height. For a workshop sign, 40–60 mm.
5. Click the small folder icon next to the Font file field. A file browser opens. Navigate to a font file (.ttf or .otf) on your computer.
6. For Windows: C:/Windows/Fonts/ — choose Arial.ttf (good for CNC), Impact.ttf (bold and clear), or Verdana.ttf.
7. For macOS: /Library/Fonts/ — choose Arial.ttf or similar.
8. For Linux: /usr/share/fonts/ — find a suitable .ttf font.

9. Set the X and Y position fields to place the text approximately where you want it. (You can reposition it precisely after creation using the Data panel.)
10. Click OK. The ShapeString appears in the 3D view as a 3D text object.

Choosing a CNC-Appropriate Font

Not all fonts are equal for CNC engraving. The characteristics that make a font read well on screen are often the characteristics that make it difficult to machine cleanly.

Best for V-bit engraving	Bold, simple sans-serif fonts: Arial Bold, Impact, Verdana Bold, Franklin Gothic Heavy. Wide strokes, no thin serifs, no hairlines. The V-bit follows the outline of each letter — wider letters give more depth and visual impact.
Avoid for V-bit engraving	Thin serif fonts (Times New Roman, Garamond), script/cursive fonts, any font with very thin strokes. The V-bit cannot cut a thin line at significant depth without the walls collapsing or burning.
Best for pocket engraving	Simple, moderately-sized text. When using an end mill to pocket the text background (creating raised letters), cleaner, more open fonts work better — avoid fonts with very small enclosed spaces (like the dot of 'i' or the counter of 'e') that are too small for the bit to enter.
Letter height guideline	Minimum practical height for a 60° V-bit at 0.8 mm depth: approximately 8–10 mm per letter height. Below this, the letters are too small for the bit to trace accurately. 20–40 mm is the practical sweet spot for most CNC nameplates.

Positioning the ShapeString

After creating the ShapeString, it appears in the Model tree. Its position is controlled through the Placement property in the Data tab of the Properties panel. Select the ShapeString in the Model tree, then look at the Properties panel (Data tab). You will see Placement with X, Y, Z position values.

To centre the text on a nameplate: you need to know the text's overall width. The easiest approach is to adjust the X position by trial and look — move it left or right by changing the X value in the Data panel and watching the 3D view update. A more precise method uses a construction geometry line in the Sketcher to find the centre position.

> **Tip**
> A quick centring trick: create the ShapeString, note its approximate width by reading the bounding box in the Properties panel. Set X position to negative half that width. For a text string that is approximately 120 mm wide on a 200 mm wide panel, set X to -60. The text will be approximately centred. Adjust by ±5 mm to fine-tune.

Using ShapeString in CAM

For V-bit engraving, the ShapeString text edges are used directly as the base geometry for the CAM Engraving operation. In the CAM workbench, when you create an Engraving operation and click Add for base geometry, you click on the outline edges of the ShapeString letters in the 3D view. FreeCAD then generates a toolpath that follows each letter's outline at the specified depth.

This workflow is covered in full in Chapter 7 (CAM Operations) and in the Project 1 walkthrough. For now, the key understanding is that ShapeString produces edges that the CAM workbench can follow — it is not just a visual decoration, it is functional geometry.

Common Part Design Errors — Diagnosis and Fixes

Part Design is forgiving of many errors — it shows you what has gone wrong through the Model tree colours and error messages — but a few categories of error come up repeatedly. The table below covers the errors you are most likely to encounter in the first six weeks of the course.

Error / Symptom	Likely Cause	How to Fix
Pad greyed out	*Sketch is not closed or not selected*	Re-enter the sketch (double-click). Check for any white endpoints — apply Coincident to connect them. Ensure the sketch is selected in the Model tree before clicking Pad.
Pad produces no geometry	*Sketch profile is self-intersecting*	Re-enter the sketch. Look for any lines that cross over each other. Remove or fix the crossing lines. A self-intersecting profile is not a valid closed loop.
Pocket fails (red in tree)	*Pocket sketch is outside the face area, or the pocket depth exceeds the solid thickness*	Re-enter the Pocket sketch and ensure all geometry is within the bounds of the face. Reduce the Pocket depth if it exceeds the Pad height.
Chamfer fails	*Edge no longer exists (upstream change removed it) or edge has zero length*	Double-click the Chamfer in the tree. Deselect the missing edge and select the correct current edge. If the edge is gone, check whether the earlier Pad or Pocket needs to be corrected.
Feature turns yellow	*Upstream geometry changed — feature is recomputing in a degraded state*	Right-click the yellow feature and select Mark to recompute. Then go to Edit → Refresh model. If still yellow, double-click to open the feature and re-check its settings.
ShapeString not visible in 3D view	*ShapeString may be positioned at Z=0, below the model face*	In the Data panel, increase the Z value of the ShapeString's Placement by the Pad height (e.g., if your Pad is 12 mm, set ShapeString Z to 12.1). This places it just above the top face.

Sketch on face not available	*No face is selected before Part Design → Create new Sketch*	Click the face first (it must be highlighted green), then create the sketch. The face click and the new sketch command must be sequential.
Model tree has duplicate Body	*Two Bodies created accidentally*	Right-click the empty Body and delete it. All features must be inside one Body. If features are in the wrong Body, they will need to be recreated inside the correct one.
Fillet produces 'edges lost' error	*Fillet radius is too large for the geometry*	Reduce the Fillet radius. The radius cannot be larger than the smallest adjacent face dimension. Start with a small value (0.5 mm) and increase until the error appears.

> **Tip**
>
> When any feature turns red or yellow, do not continue building. Stop, diagnose, and fix the issue. Building on top of a failed feature almost always makes the situation worse — you end up with a cascade of failures that are very difficult to unwind.

Week 1, Day 4 — Lesson Content

Day 4 of Week 1 is the first day students produce a complete 3D object using all the tools in this chapter. It is the most productive day of Week 1 — students arrive having mastered the Sketcher and leave with a 3D model that is ready for CAM setup in Chapter 7.

Day 4 Activities

Guided activity 1: First Pad	Using the DoorSign sketch from Day 3 (or a fresh centred rectangle 200 × 120 mm), apply Part Design → Pad at 12 mm. Inspect the result. Double-click the Pad and change to 15 mm. Change back to 12 mm. Practice the edit workflow.
Guided activity 2: Pocket and face sketch	Click the top face of your DoorSign solid. Create a new Sketch on the face. Project the face edges using external geometry. Draw an inset rectangle 15 mm from each edge (170 × 90 mm). Apply Pocket at 3 mm depth. Inspect the recessed panel.
Guided activity 3: Chamfer	Select the four top rim edges of the solid (Ctrl+click each). Apply Chamfer at 1 mm. Inspect from different angles. Double-click and change to 0.5 mm.
Guided activity 4: ShapeString	Switch to Part workbench. Apply Part ShapeString. Type your name. Set height to 30 mm. Choose Arial Bold or Impact font. Position on the top face. Switch back to Part Design. Inspect the Model tree.
Independent exercise	Without instructions: design a completely new sign — your workshop name, your room number, or any short text — as a fully 3D model with a Pad body, a recessed border Pocket, a Chamfer around the top rim, and a ShapeString. Save as WorkshopSign.FCStd. This is the design you will take to CAM in Chapter 7.

Tip

Day 4's independent exercise is the first time students design something entirely their own. The result of today's exercise becomes the source file for Chapter 7's CAM walkthrough. Encourage students to design something they genuinely want to make — personal investment in a project dramatically increases engagement in the CAM and machine setup steps.

Chapter Summary

- Part Design uses a feature-based, history-based approach. Every operation — Pad, Pocket, Chamfer, Fillet — is recorded in the Model tree in creation order and can be edited at any time by double-clicking.
- Pad extrudes a fully closed, fully constrained sketch into a 3D solid. For CNC parts, the Pad length matches your material thickness. Blind mode (exact depth) is the standard choice.
- Pocket removes material from an existing solid by extruding a sketch inward. The sketch is drawn on a selected model face. Through All mode cuts completely through regardless of thickness.
- External geometry (purple) allows sketch elements to reference existing model edges — essential for accurately constraining inset pockets and face features relative to existing geometry.
- Every feature is editable by double-clicking it in the Model tree. Editing an upstream feature automatically updates all downstream features. Edge treatments (Chamfer, Fillet) must always be the last features in the tree.
- Chamfer produces an angled edge (45° by default). Use it for joinery lead-ins (0.3–0.5 mm) and visual finishing. Fillet produces a curved radius edge. For CNC work, prefer Sketcher fillets (rounded corners in the 2D profile) over Part Design fillets on 3D extruded edges.
- ShapeString creates 3D text geometry in the Part workbench. Use bold, simple sans-serif fonts (Arial Bold, Impact) for V-bit engraving. Position using the Placement X, Y, Z values in the Properties panel. ShapeString edges are used directly as base geometry in CAM Engraving operations.
- The most common Part Design errors involve open sketch profiles (Pad greyed out), sketch attachment failures after upstream changes (Pocket fails), and chamfer/fillet size exceeding adjacent face geometry. Always fix red or yellow tree items before continuing.

Key Vocabulary — Chapter 6

Term	Definition
Pad	A Part Design operation that extrudes a closed, fully constrained sketch profile into a 3D solid in a direction perpendicular to the sketch plane.
Pocket	A Part Design operation that removes material from an existing solid by extruding a sketch profile inward. Requires an existing solid body and a sketch on a model face.

Sketch on face	The process of attaching a new Sketch directly to a flat face of an existing solid, rather than to a coordinate plane. Enables features on non-origin surfaces.
External geometry	Read-only reference geometry (displayed in purple) projected from existing model edges into the current sketch. Used to constrain new sketch elements relative to existing geometry.
Feature ordering	The sequence in which Part Design features are applied, recorded top-to-bottom in the Model tree. Each feature is applied to the result of all features above it in the tree.
Chamfer	A Part Design operation that replaces a sharp edge with an angled flat surface. Used for joinery lead-ins, visual finishing, and functional tapers.
Fillet	A Part Design operation that replaces a sharp edge with a smooth curved radius. Used for visual softening, stress reduction, and decorative rounded profiles.
Sketcher fillet	A fillet applied within the Sketcher (Sketch → Sketcher tools → Create fillet) to produce a rounded 2D corner in the sketch profile. Preferred over Part Design Fillet for most CNC work.
ShapeString	A Part workbench tool that generates 3D text geometry from a typed string, font file, and height value. The resulting text edges are used as base geometry for CAM Engraving operations.
Through All	A Pocket (or Pad) type option that cuts completely through the solid regardless of its thickness. Used for holes and cutouts that must pass completely through any material thickness.
Feature fails	A Part Design state where a feature (shown as red in the Model tree) cannot be computed — either because its sketch is invalid, its dependent geometry has changed, or its settings conflict with the current model state.
Placement	A property in the Data panel that controls the X, Y, Z position and orientation of an object. Used to reposition ShapeString objects and other Part-level geometry.
Model recompute	The process by which FreeCAD recalculates all features in the Model tree after an upstream change. Triggered automatically on feature edits, or manually via Edit → Refresh model.
Lead-in chamfer	A very small chamfer (0.3–0.5 mm) applied to mating edges of joinery parts to allow easier assembly. One of the most practical uses of the Chamfer feature in CNC woodworking.
History-based model	A 3D model built as an ordered sequence of feature operations, each recorded in the Model tree. Allows any earlier decision to be edited and all subsequent features to update automatically.

Looking Ahead — Chapter 7: First CAM Operations & the UGS Workflow

Chapter 7 takes the 3D model you have built in Chapters 4–6 and generates the machine instructions to cut it. You will set up a CAM Job, configure Stock, add tools to the Tool Library, create V-bit Engraving and Profile operations, run the CAM Simulator, and post-process to GRBL G-code. Then you will load the file in UGS, verify it, and — on Project Day (Day 5) — cut your first physical piece on the CNC 4040 Pro Max.

CHAPTER 7 of 24

PART TWO — FOUNDATIONS

First CAM Operations & the UGS Workflow

Jobs, Stock, Tools, Engraving, Profile, Tabs, Simulator, Post-processing — and your first two physical projects

Chapter Learning Objectives

By the end of this chapter, you will be able to:

- Create a CAM Job linked to a Part Design body
- Configure Stock dimensions to match your physical material
- Add a V-bit and an end mill to the Tool Library with correct settings
- Select appropriate feeds and speeds for MDF and timber
- Create a CAM Engraving operation targeting ShapeString or sketch edges
- Create a CAM Profile operation with holding tabs
- Run and interpret the CAM Simulator
- Post-process a job to GRBL G-code and verify the output
- Complete the full UGS workflow from connection to job completion
- Cut Project 1 (Personalised Name Badge) and Project 2 (Engraved Coaster) on the CNC 4040 Pro Max

The Bridge Between Design and Machine

You have a 3D model. The machine needs G-code. The CAM workbench is what builds the bridge between them.

CAM stands for Computer-Aided Manufacturing. In FreeCAD's CAM workbench, you take the 3D model you built in Part Design and generate the toolpaths — the precise sequences of machine movements — that will produce that model as a physical object. The CAM workbench does not change your design. It reads it, understands which faces and edges need to be cut, and generates the G-code instructions that tell the CNC 4040 Pro Max exactly how to move the cutting bit to produce the result.

The CAM workbench introduces four new concepts — Job, Stock, Tools, and Operations — that together form the complete structure of a CNC manufacturing job. Understanding these four concepts before touching any buttons will make the workflow feel logical rather than arbitrary.

JOB	STOCK	TOOLS	OPERATIONS
The master container. Links the CAM operations to your 3D model. Every CAM project starts with a Job.	The virtual block of material the machine will cut from. Must match your physical material dimensions exactly.	Your cutting bits, defined in the Tool Library: diameter, type, material. Every operation uses a named tool.	The actual cutting instructions: Engraving, Profile, Pocket, Drilling. Each operation uses one tool and targets specific model geometry.

Think of it this way: the Job is the project folder. The Stock is the blank piece of material on the machine bed. The Tools are the cutting bits in your tool roll. The Operations are the instructions you write for each cut. Every CNC job you run in this course — from a simple name badge to the certification project — uses this exact four-part structure.

Creating a CAM Job

The CAM Job is the container that holds all your toolpaths, tool settings, and stock configuration. Creating the Job is always the first step in any CAM workflow — before tools, before operations, before anything else.

Step-by-Step: Creating Your First CAM Job

1. Ensure your Part Design model is open and the Body is visible in the Model tree. If FreeCAD is not already in the CAM workbench, switch to it using the Workbench drop-down.
2. Go to CAM → Job. The Job dialog opens.
3. In the Model field, FreeCAD should automatically detect your Body. If it does not, click the drop-down and select your Body manually.
4. Leave the other settings at their defaults for now. Click OK.
5. A Job object appears in the Model tree. Expand it — you will see three sub-items: Model (the link to your Part Design body), Stock, and Tools.

> 📋 **Note**
> The CAM Job is independent of your Part Design model — it reads the model but does not modify it. You can have multiple Jobs for the same model (for example, one for MDF at 12 mm and another for hardwood at 18 mm) without affecting the design.

Configuring Stock

The Stock represents the physical block of material that will be placed on the CNC machine bed. It must exactly match — or slightly exceed — your actual material dimensions. The Stock tells the CAM workbench where the material boundaries are, which affects how the toolpaths are generated and whether the Simulator can correctly show material removal.

Opening Stock Settings

In the Model tree, expand your Job and double-click on Stock. The Stock dialog opens. It shows the current stock dimensions and position relative to the model.

X / Width	The left-to-right dimension of your material. Set this to your actual material width, or 1–2 mm larger to include edge margin. For a 100 mm wide name badge: set X to 102 mm.
Y / Depth	The front-to-back dimension. Same logic — match your material or add a small margin.
Z / Height	The thickness of your material. Set this exactly — Z height is critical for depth calculations.
Origin	Where the stock sits relative to the model origin. The default (centred on the model) is correct when your FreeCAD sketch is centred on the origin. If your model's origin is at the front-left corner, the stock origin should also be at the front-left corner.

> 💡 **Tip**
> For your first few projects, set the Stock dimensions to exactly match your material — no extra margin. This keeps the toolpath visualisation clean and makes it easier to verify that the profile cut will reach the material edges. Once you are more experienced, adding 2–5 mm of margin is common practice to allow for material edge irregularities.

The Tool Library — Defining Your Cutting Bits

The Tool Library stores definitions for each cutting bit you will use. A tool definition tells the CAM workbench the bit's physical properties — its diameter, shape, and cutting geometry. FreeCAD uses these properties to calculate accurate toolpaths. An incorrect tool definition produces incorrect toolpaths — the offset from the edge, the width of the cut, and the depth calculations all depend on the tool geometry being correct.

Opening the Tool Library

Go to CAM → Tools → Tool Library. The Tool Library dialog opens. It may be empty on first use, or it may contain example tools from FreeCAD's defaults. For this course, you will add your own tools.

Adding a V-Bit (Tool 1)

1. In the Tool Library, click the + button (Add Tool) or go to CAM → Tools → Add Tool.
2. In the Tool dialog, set the following fields:
3. Name: V-Bit 60deg
4. Tool Type: V-Bit
5. Material: HSS (High Speed Steel) or Carbide if you know your bit's material
6. Diameter: 3.175 (the shank diameter — 1/8 inch is 3.175 mm)
7. Cutting Edge Angle: 60 (the included angle of the V-bit — check your actual bit, common values are 60° and 90°)
8. Cutting Edge Height: leave as default
9. Click OK. Tool 1 (T1) is now defined.

Adding a 6 mm End Mill (Tool 2)

1. Click + again.
2. Name: End Mill 6mm
3. Tool Type: End Mill
4. Diameter: 6
5. Flutes: 2 (two-flute end mills are standard for wood routing)
6. Cutting Edge Height: 18 (or however long the cutting flutes are on your actual bit — check the bit's specification)
7. Click OK. Tool 2 (T2) is defined.

T1 — V-Bit 60°	Diameter: 3.175 mm | Angle: 60° | Used for: engraving ShapeString text and 2D profile edges at shallow depths (0.5–1.5 mm)
T2 — End Mill 6 mm	Diameter: 6 mm | Flutes: 2 | Used for: pocket clearing, profile cutting, border recesses
T3 — End Mill 3 mm	Diameter: 3 mm | Flutes: 2 | Used for: fine detail pockets, small holes, thin profile cuts — introduced in Week 3 onwards

> **Tip**
>
> Build your Tool Library once and reuse it across all projects. The Tool Library persists between FreeCAD sessions — you do not need to re-enter tool definitions for every new job. Always name tools descriptively (V-Bit 60deg, not Tool 1) so you know exactly which bit to load when a tool change is required.

Feeds and Speeds — The Four Key Settings

Feeds and speeds are the most frequently asked-about topic in CNC machining. The numbers matter — wrong settings can burn the material, break bits, or produce rough, poor-quality surfaces.

Every cutting operation in the CAM workbench uses four settings that together define how fast and how aggressively the machine cuts. These four settings — feed rate, plunge rate, depth per pass, and spindle speed — are the fundamental parameters of all CNC machining. Understanding what each one does gives you the ability to adjust for different materials and bit conditions rather than just following a table blindly.

Feed rate (F)	The horizontal cutting speed in mm/min. How fast the tool moves sideways through the material. Too fast: chatter, rough surface, bit deflection. Too slow: burning, inefficiency. Units: mm/min.
Plunge rate	The vertical entry speed in mm/min. How fast the tool descends into the material at the start of each pass. Always slower than feed rate — typically 20–40% of feed rate. Too fast: bit breakage on entry. Units: mm/min.
Depth per pass	How deep the tool cuts in a single pass in mm. For multi-pass operations (like a profile through 12 mm MDF), the tool makes multiple passes at increasing depths. Reducing depth per pass reduces cutting forces and improves surface quality.
Spindle speed (S)	The RPM of the cutting bit. Higher RPM = more cuts per mm of travel = finer finish. Too high for the feed rate causes burning. Too low causes chip load issues. Units: RPM.

The table below provides recommended settings for the CNC 4040 Pro Max cutting timber and MDF — the materials used throughout this course. These values are safe starting points. Your specific bits, material batch, and machine condition may require small adjustments. Use the feed rate override in UGS during cutting to fine-tune.

Bit / Operation	Material	Feed (mm/min)	Plunge (mm/min)	Depth/Pass	Spindle (RPM)
V-Bit 60° — engrave	MDF	**800**	200	0.8 mm	15,000
V-Bit 60° — engrave	Pine	**700**	180	0.6 mm	15,000
V-Bit 60° — engrave	Hardwood	**600**	160	0.5 mm	18,000
End Mill 6mm — pocket	MDF	**1100**	320	1.5 mm	15,000
End Mill 6mm — pocket	Pine	**900**	260	1.2 mm	15,000
End Mill 6mm — pocket	Hardwood	**700**	200	1.0 mm	18,000
End Mill 6mm — profile	MDF	**1000**	300	1.5 mm	15,000
End Mill 6mm — profile	Pine	**850**	240	1.2 mm	15,000
End Mill 3mm — profile	MDF	**700**	200	1.0 mm	21,000
End Mill 3mm — profile	Hardwood	**500**	160	0.8 mm	24,000

⚠ **Important**

These values are starting points, not guarantees. Always run an air-cut (Z raised 20 mm) and inspect the toolpath before cutting real material.

If you hear chatter (rattling vibration) during a cut, reduce feed rate to 70–80% using UGS override. If you see burning or smoke, reduce spindle speed or increase feed rate.

MDF produces fine dust — always wear your P2 dust mask.

CAM Engraving Operation

The Engraving operation makes the V-bit follow a 2D edge or profile at a set depth, tracing its exact path. Unlike a Pocket operation (which fills an area), Engraving follows a line — the outline of a letter, a decorative curve, a border edge. This is how you engrave text, logos, and fine details onto a CNC piece.

Creating an Engraving Operation

1. In the CAM workbench, go to CAM → Engraving. The Engraving task panel opens on the left.
2. In the Base Geometry section, click Add. Then click on the edges of your ShapeString letters in the 3D view. Each letter's outline will be added as a path for the V-bit to follow. Shift+click to select multiple edges. For a name with 4 letters, you will typically click 4–8 edge groups.
3. In the Tool section, select your V-Bit (T1) from the drop-down.
4. Set the following Depth settings:
5. — Start Depth: 0 (the top of the material surface)
6. — Final Depth: -0.8 (the engraving depth — negative because it goes into the material)
7. In Speeds & Feeds, set Feed Rate to 800 and Plunge Rate to 200 (for MDF). Spindle speed is typically set in the G-code header, not in the operation — leave it if a separate field is not shown.
8. Click OK. Coloured toolpath lines will appear on your model showing the V-bit's path through each letter.

> **Tip**
>
> If no toolpath lines appear after creating the Engraving operation, check that the ShapeString edges are correctly selected as Base Geometry, and that the ShapeString's Z position is at the top face of your model (not buried below it). The most common cause of a missing toolpath is the engraving geometry being at Z=0 when the model face is at Z=12.

Checking the Engraving Result

After creating the Engraving operation, look at the 3D view. You should see coloured lines tracing the outline of each letter of your ShapeString. These lines represent the exact path the V-bit will follow. Zoom in to verify that:

- Every letter's outline is covered by toolpath lines.
- The lines stay within the face of your model — no toolpath line runs off the edge.
- The path follows the full letter outline — partial coverage suggests some edges were missed when selecting Base Geometry.

CAM Profile Operation

The Profile operation cuts along the outer (or inner) edge of a profile, producing a cut-out or contour. It is how the CNC machine cuts the actual shape of your part — the outer rectangle of a name badge, the circle of a coaster, the gear outline. The Profile operation is the last operation in any job that produces a cut-out piece, because once it completes its final pass, the part is free from the stock.

Creating a Profile Operation

1. Go to CAM → Profile. The Profile task panel opens.
2. In the Base Geometry section, click Add and select the outer edge of your model body (click the outer perimeter edge in the 3D view).
3. Set Side to Outside — the tool cuts around the outside of the profile, producing a cut-out piece.
4. In the Tool section, select your End Mill 6mm (T2).
5. Set Depth settings:
6. — Start Depth: 0
7. — Final Depth: -12.5 (0.5 mm below your 12 mm material — ensures the cut goes fully through)
8. — Step Down: 1.5 (each pass goes 1.5 mm deeper, so 12.5 ÷ 1.5 = 9 passes approximately)
9. Set Feed Rate to 1000 and Plunge Rate to 300.
10. Click OK. The profile toolpath appears — concentric lines offset from the model edge showing the multiple passes.

Understanding Profile Offset

The Profile operation automatically offsets the toolpath from the model edge by half the tool diameter. For a 6 mm end mill, the toolpath is offset 3 mm from the model edge. This is correct — the centre of the bit is 3 mm from the edge, meaning the bit's cutting flutes are exactly on the model edge. FreeCAD handles this offset calculation automatically. You do not need to manually offset your sketch.

> 📖 **Key Term: Profile offset**
>
> The automatic displacement of the toolpath from the model edge equal to half the tool diameter. Ensures the cutting edge of the tool — not its centre — traces the model profile. Calculated and applied automatically by the CAM workbench.

Holding Tabs — Essential for Profile Cuts

When the Profile operation cuts all the way through the material, the cut-out piece becomes completely free from the surrounding stock. Without something holding it in place, the piece can shift, tip, or be thrown by the spinning bit — potentially damaging the piece, breaking the bit, or becoming a hazard. Holding tabs prevent this.

A tab is a small, uncut bridge of material left between the cut-out piece and the surrounding stock. It holds the piece in place during the final passes of the profile cut. After the job is complete and the machine has stopped, you use a sharp chisel or craft knife to cut through the tabs and release the piece.

Adding Tabs to a Profile Operation

1. Double-click your Profile operation in the Model tree to re-open it.
2. Look for the Dressup section, or go to CAM → Dressup → Tags. A Tags dialog opens.
3. Set Tab Width to 4 mm (the width of the bridge left uncut).
4. Set Tab Height to 3 mm (how tall the tab is — the tab starts this many mm above the final depth).
5. Set the number of tabs. For a rectangular badge (100 × 60 mm): 4 tabs, one on each side. For a circular coaster: 4–6 tabs evenly spaced.
6. Click Generate to auto-place the tabs. You can drag the tab positions in the 3D view to reposition them — avoid placing tabs at corners.
7. Click OK. The Profile toolpath now shows small gaps at each tab location.

> ⚠ **Important**
>
> Never run a Profile cut without tabs on any piece larger than about 30 × 30 mm.
>
> A free piece being driven by the cutting bit can shatter the bit, gouge the machine bed, and become a projectile.
>
> 4 tabs minimum for any rectangular or circular part. 6 tabs for anything larger than 200 mm in any dimension.

> 💡 **Tip**
>
> Position tabs on the longest, most accessible sides of the piece — not at corners, and not near any important design feature. You will sand or pare the tab marks smooth after cutting, so place them where sanding is easy and inconspicuous.

The CAM Simulator — Your Last Line of Defence Before Cutting

The CAM Simulator is a 3D animation of the complete cutting job. It shows the virtual cutting bit removing material from the virtual stock in the sequence the G-code will execute. Running the Simulator before every job is not optional — it is a required step in the pre-cut workflow.

The Simulator reveals problems that the 2D toolpath view cannot: tool paths that collide with clamps, operations in the wrong order, missing passes, unexpected rapid moves that would crash the tool into material. A five-minute Simulator check has prevented countless broken bits and ruined workpieces.

Running the Simulator

1. In the CAM workbench, go to CAM → CAM Simulator. The Simulator dialog opens.
2. The 3D view changes to show the stock as a solid block and the cutting bit at its starting position.

3. Click the Play button. The simulation runs at normal speed — use the Speed slider to run it faster.
4. Watch the simulation complete all operations in sequence. The material is visually removed as each operation runs.
5. After the simulation completes, inspect the result from multiple angles. Does it look like your intended design?

Colour	Move Type	What to Check
	Rapid move (G0)	The tool moves at maximum speed with no cutting. Must ALWAYS be above material. A yellow line that passes through your workpiece is a crash risk — stop and investigate.
	Cutting move (G1)	The tool moves at feed rate, cutting the material. These green lines should trace exactly the geometry you designed.
	Plunge move	The tool descends into the material at the plunge rate. These should only appear at the start of each cutting pass, never mid-way through a horizontal move.
	Retract / lift	The tool rises clear of the material. Should appear between operations and at the end of the job.

What to Look For in the Simulator

- All text is fully engraved — no missing letters or partial coverage.
- The border pocket (if present) is fully cleared to the correct depth.
- The profile cut traces the correct outer shape — the piece outline matches the design.
- Rapid moves (yellow) stay above the material surface — no yellow lines passing through the stock.
- The tabs are visible as small uncut bridges in the profile cut.
- The final depth of the profile cut is slightly below the stock bottom — confirming full cut-through.

> **Tip**
>
> Use the CAM Simulator's Accuracy setting to increase the resolution of the simulation if the result looks rough or faceted. Higher accuracy takes longer to compute but shows finer detail — useful for checking engraved text closely.

Post-Processing — Generating the G-code File

Post-processing converts FreeCAD's internal toolpath representation into a G-code file that the CNC 4040 Pro Max (via GRBL) can execute. Different CNC machines and controllers use slightly different G-code dialects — the post-processor is the translator that outputs the correct dialect for your specific machine. For the CNC 4040 Pro Max, the correct post-processor is GRBL.

Running the Post-Processor

1. Select your Job in the Model tree.
2. Go to CAM → Post Process. The Post Process dialog opens.
3. In the Post Processor drop-down, select grbl.
4. Click the folder icon next to the Output File field. Navigate to your project folder (e.g., CNC_Projects/Week_01/NameBadge/). Type a filename: NameBadge_v1.nc.
5. Click Process. FreeCAD generates the G-code file and saves it to the specified location.
6. A text window may open showing a preview of the G-code. Scroll through and verify that M3 (spindle on) appears before cutting moves, and M30 (program end) appears at the end.

> 📋 **Note**
>
> If the grbl post-processor does not appear in the drop-down, it may need to be located manually. Click the folder icon next to the Post Processor field (not the output file field) and navigate to FreeCAD's installation directory, then Mod → CAM → Path → Post → Scripts → grbl_post.py. Select it and it will appear in future sessions.

Verifying the G-code File

Before loading the file into UGS, open it in a plain text editor (Notepad on Windows, TextEdit on macOS, gedit on Linux) and scan through it. You are looking for:

- The file starts with G21 (millimetres), G90 (absolute mode), G17 (XY plane).
- M3 S_____ appears before the first cutting move (spindle on with speed).
- Tool change commands (T_ M6) appear between operations.
- M5 and M30 appear at the end (spindle off, program end).
- Z depth values are negative numbers for cutting moves (positive Z would be moving up, not cutting).

UGS Workflow — From File to Physical Cut

The complete UGS workflow was covered in Chapter 3. For reference, here is the compressed version — the six-step sequence you will follow for every job in this course, from Project 1 onwards. Use this as a mental checklist at the machine.

1. Connect & Home	Power on machine. Open UGS. Connect at 115200 baud. Status: Idle. Home machine ($H or Home button).
2. Install bit	Install correct bit (V-bit for engraving, end mill for profile). Tighten collet firmly.
3. Clamp material	Place material on bed. Clamp securely at 2+ points. Verify no clamp is in the cutting path.
4. Set X/Y/Z zero	Jog spindle to work origin. Zero X and Y. Lower Z to material surface using paper method. Zero Z. Raise Z 5 mm.
5. Load & visualise	File → Open in UGS → select .nc file. Inspect 3D visualiser — all operations visible, rapid moves above material.

6. Cut	PPE on. Send job. Supervise entire run. Pause for tool changes. Re-zero Z after each tool change.

PROJECT 1 — STEP-BY-STEP WALKTHROUGH

Personalised Name Badge

Complete step-by-step from blank file to physical cut piece

This is your first complete end-to-end CNC project. Follow every step in sequence. At the end you will have a physical engraved name badge that you designed yourself, cut on the CNC 4040 Pro Max. Do not skip steps — each one builds on the previous.

Design	Personalised name badge — your name or chosen text
Size	100 × 60 mm (adjust within range 80–120 × 50–80 mm)
Thickness	6 mm or 12 mm MDF
Operations	Engraving (V-bit), Profile (6 mm end mill)
Day	Day 5 of Week 1 — Project Day

Part A — Design in FreeCAD

1 **New document and Body**
Open FreeCAD. File → New. Part Design → Body. Save immediately as NameBadge.FCStd.

2 **Sketch the badge outline**
Part Design → Create a new Sketch → XY Plane → OK. Draw a centred rectangle: press R, click origin, drag and click. Constrain Horizontal Distance = 100, Vertical Distance = 60. Verify all green (0 DOF). Close sketch.

3 **Pad to material thickness**
Part Design → Pad. Set Length to your material thickness (6 or 12 mm). Click OK. Inspect the solid block.

4 **Add your name with ShapeString**
Workbench drop-down → Part. Part → Part ShapeString. Type your name. Set Height to 25 mm. Choose a bold font (Arial Bold, Impact). Set X to approximately -45 and Y to -12 to centre the text. Click OK.

5 **Verify the 3D model**
Switch back to Part Design. Orbit the 3D view. You should see: a flat rectangular solid with 3D text sitting on the top face. If the ShapeString appears buried or invisible, adjust its Z value in the Data panel (Properties → Data → Placement → Z) to match the Pad height.

> **Tip**
>
> Before continuing to CAM, save the file (Ctrl+S). Then orbit the model and verify it looks correct from multiple angles. A ShapeString at the wrong Z height is invisible in the default top-down view but obvious when you orbit.

Part B — CAM Setup

6	**Switch to CAM workbench** Workbench drop-down → CAM.
7	**Create the Job** CAM → Job. Select your Body model. Click OK. A Job appears in the tree.
8	**Configure Stock** Double-click Stock. Set: X = 102, Y = 62, Z = your material thickness (6 or 12 mm). Click OK.
9	**Add V-bit to Tool Library** CAM → Tools → Tool Library. Click +. Name: V-Bit 60deg. Type: V-Bit. Diameter: 3.175. Angle: 60. Click OK.
10	**Add End Mill to Tool Library** Click + again. Name: End Mill 6mm. Type: End Mill. Diameter: 6. Flutes: 2. Click OK. Close Tool Library.
11	**Operation 1 — V-bit Engraving** CAM → Engraving. Add base geometry: click the outline edges of your ShapeString letters (each letter needs its edges selected). Tool: V-Bit 60deg. Start Depth: 0. Final Depth: -0.8. Feed: 800. Plunge: 200. Click OK.
12	**Operation 2 — Profile cut** CAM → Profile. Add base geometry: click the outer rectangular edge of the badge. Side: Outside. Tool: End Mill 6mm. Final Depth: -(thickness + 0.5). Step Down: 1.5. Feed: 1000. Plunge: 300. Click OK.
13	**Add holding tabs** Double-click the Profile operation in the tree. CAM → Dressup → Tags. Tab Width: 4 mm. Tab Height: 3 mm. Number: 4. Generate. Position tabs on the four sides (not corners). Click OK.
14	**Run the Simulator** CAM → CAM Simulator. Play the simulation. Verify: text engraved fully, profile cut traces the rectangle, tabs visible, no yellow rapids through material. If satisfied, close Simulator.
15	**Post-process to GRBL** Select Job in the tree. CAM → Post Process. Post Processor: grbl. Output: NameBadge_v1.nc in your project folder. Click Process.

Part C — On the CNC 4040 Pro Max

16 **Complete the pre-cut checklist**
Go through all four sections of the Pre-Cut Safety Checklist (Chapter 2). Do not skip any item.

17 **Connect and home**
Open UGS. Connect to machine (115200 baud). Status: Idle. Home ($H). Status returns to Idle.

18 **Install V-bit**
Insert V-bit into ER11 collet. Tighten firmly with collet spanner.

19 **Clamp material**
Place MDF on machine bed. Clamp at all four corners — ensure no clamp is within the 100 × 60 mm cutting area.

20 **Set X/Y/Z zero**
Jog spindle to the centre of the material (work origin for a centred design). Zero X and Y. Lower Z carefully with paper method. Zero Z. Raise Z 5 mm.

21 **Load file and verify in UGS**
File → Open → NameBadge_v1.nc. Inspect 3D visualiser. Both operations visible. Rapid moves clear of material.

22 **Run engraving (Operation 1)**
PPE on. Click Send. The V-bit engraves your name. Watch closely for the first 30 seconds. The cut should be clean and consistent.

23 **Tool change to End Mill**
Spindle stops (M5). Jog Z up 30 mm. Remove V-bit. Insert 6mm end mill. Re-zero Z only (paper method). Resume in UGS.

24 **Profile cut (Operation 2)**
The end mill cuts the badge outline in multiple passes. Do not leave the machine. Watch for tab integrity on the final pass.

25 **Release and finish**
Spindle off. Raise Z. Remove clamps. Use chisel to cut the 4 tabs. Sand tab marks with 180-grit. Apply a clear coat finish if desired.

🏆 **Project 1 Complete**

Your personalised name badge is complete. You have just designed, programmed, and cut your first physical CNC piece from scratch. This is Bloom's Apply level — you used FreeCAD, CAM, and UGS to produce a specified outcome independently.

Take a photograph. Save your FreeCAD file and G-code file. Complete your project reflection sheet.

PROJECT 2 — STEP-BY-STEP WALKTHROUGH

Engraved Timber Coaster with Border

Circular coaster with recessed border pocket, engraved personal design, and profile cut-out

Project 2 builds directly on Project 1. You will add a new skill — the Pocket Shape operation — and run a three-operation job (Engraving → Pocket → Profile). The coaster introduces circular geometry and the annular sketch technique from Chapter 5.

Design	Circular coaster with recessed border ring and personal engraved element (name, monogram, or motif)
Size	100 mm diameter (adjust 90–120 mm)
Thickness	12–15 mm hardwood or MDF
Operations	3: Engraving (V-bit) → Pocket Shape border (End Mill 6mm) → Profile (End Mill 6mm)
Day	Day 10 of Week 2 — Project Day

Part A — Design

1 New file and Body
File → New. Save as Coaster.FCStd. Part Design → Body.

2 Sketch outer circle
Part Design → Create new Sketch → XY Plane. Press C. Click origin. Drag out — constrain Radius = 50 (100 mm diameter). Apply Coincident between circle centre and origin. Green. Close sketch.

3 Pad to thickness
Part Design → Pad. Length = 12 (or your material thickness). Click OK.

4 Sketch the border pocket (annular profile)
Click the top face. Part Design → Create new Sketch. Use external geometry to project the outer circle edge (Sketch → Sketcher geometries → Create external geometry, click the top circle edge — it appears purple). Draw a second circle, centre at origin, Radius = 40 (8 mm smaller than outer = 8 mm border width). Constrain Coincident to origin. Green. Close sketch.

5 Pocket the border ring
Part Design → Pocket. Depth = 2.5 mm. Click OK. A 2.5 mm recessed ring appears around the coaster perimeter.

6	**Add personal engraving** Switch to Part workbench. Part → Part ShapeString. Type your initial or a short word. Height: 20 mm. Position at centre (X: −10 to −15 approx., Y: −10). Click OK.

Part B — CAM (Three Operations)

7	**CAM Job and Stock** Switch to CAM. CAM → Job. Configure Stock: X = 105, Y = 105, Z = 12 (5 mm larger than diameter on each axis).
8	**Operation 1 — Engraving** CAM → Engraving. Select ShapeString edges. Tool: V-Bit 60deg. Final Depth: -0.8. Feed: 800. Plunge: 200. Click OK.
9	**Operation 2 — Pocket Shape (border ring)** CAM → Pocket Shape. Select the top face of the recessed border pocket (the ring-shaped recessed area). Tool: End Mill 6mm. Start Depth: 0. Final Depth: -2.5. Step Down: 1.5. Pattern: ZigZag. Step Over: 4 mm. Feed: 1000. Plunge: 300. Click OK. The pocket toolpath fills the ring area.
10	**Operation 3 — Profile cut** CAM → Profile. Select the outer circular edge. Side: Outside. Tool: End Mill 6mm. Final Depth: -12.5. Step Down: 1.5. Feed: 1000. Plunge: 300. Click OK. Add 4 tabs: width 4 mm, height 3 mm. Space evenly around the circle.
11	**Simulator check** CAM → CAM Simulator. Play the full 3-operation simulation. Verify: engraving complete, border ring fully cleared, circular profile cut with tabs visible. Check rapid moves are above material throughout.
12	**Post-process** CAM → Post Process. Post Processor: grbl. Save as Coaster_v1.nc.

Part C — On the CNC 4040 Pro Max

13	**Pre-cut checklist, connect, home** Complete pre-cut checklist. Connect UGS. Home machine.
14	**Install V-bit, clamp material, zero** V-bit in collet. Clamp square of material (110 × 110 mm minimum). Set X/Y/Z zero at the material centre (the sketch origin is at the centre for this circular design).
15	**Load Coaster_v1.nc in UGS — verify** Load file. Check visualiser — three operations should be visible: engraving lines, ring pocket fill, circular profile with gaps for tabs.
16	**Run Operation 1 — Engraving** Send job. Operation 1 (V-bit engraving) runs. Watch the first pass closely.

17 **Tool change — End Mill**
Spindle stops. Jog Z up. Remove V-bit. Insert 6 mm end mill. Re-zero Z only. Resume.

18 **Operations 2 & 3 run in sequence**
The end mill clears the border ring pocket (Operation 2), then cuts the circular profile (Operation 3). Supervise throughout.

19 **Release, sand, and finish**
Cut the 4 tabs. Sand all faces 120 → 180 → 240 grit. Apply food-safe finish (cutting board oil) if the coaster will be used with drinks.

🏆 **Project 2 Complete**

Project 2 introduces the Pocket Shape operation and your first three-operation job. You managed a tool change, configured a pocket clearing operation, and produced a multi-operation G-code file verified in the Simulator before cutting. This is the Apply level — independent execution of a complete workflow.

Week 2, Days 6–10 — Lesson Content Overview

Week 2 covers the CAM material in this chapter. Days 6–9 teach each component progressively; Day 10 is Project 2. The following table summarises the learning focus for each day.

Day 6 (Remember → Understand)	ShapeString and CAM workbench overview. Students add a ShapeString to their DoorSign model from Chapter 6, then explore the CAM workbench — Job creation, Stock, Tool Library. No operations yet. Exit ticket: name the four CAM pillars.
Day 7 (Understand)	CAM Engraving operation. Students set up a V-bit engraving of their ShapeString. Run the Simulator — observe toolpath lines. Adjust Final Depth and re-simulate. Discuss: how does depth affect the visual appearance of engraved text?
Day 8 (Understand → Apply)	CAM Profile operation and holding tabs. Students add a Profile operation to their DoorSign job. Add 4 tabs. Run the full Simulator showing both operations in sequence. Observe the tab gaps in the profile.
Day 9 (Apply)	Post-processing and UGS pre-cut workflow. Students post-process their DoorSign job to GRBL. Open the .nc file in a text editor and identify: G21, G90, M3, M5, M30, T1 M6, T2 M6. Load in UGS. Run a full air-cut (Z raised 20 mm). Supervised by instructor.
Day 10 (Apply — Project Day)	Project 2: Engraved Timber Coaster. Students complete the full project independently — design, CAM, machine. Instructor available for consultation only.

Chapter Summary

- The CAM workbench has four pillars: Job (the container), Stock (the virtual material block), Tools (cutting bit definitions), and Operations (the actual cutting instructions). Every CAM job uses all four.
- The Stock must match the physical material dimensions exactly. Z height is the most critical — it determines all depth calculations.
- The Tool Library stores named tool definitions. Accurate tool diameter is essential — the CAM workbench uses it to calculate profile offsets, pocket coverage, and engraving paths. Build the library once, reuse it across all projects.
- The four feed/speed settings are: Feed Rate (horizontal speed), Plunge Rate (vertical entry speed — always slower), Depth per Pass (incremental depth per cutting loop), and Spindle Speed (RPM). Starting values for MDF and timber are provided in the feeds table in this chapter.
- The Engraving operation follows 2D edges with the V-bit at a set depth. Select the ShapeString letter edges as base geometry. Final Depth of -0.8 mm is a reliable starting value for MDF.
- The Profile operation cuts along the outer or inner boundary of a profile at progressive depths. The toolpath is automatically offset by half the tool diameter. Always run Profile as the last operation in any job.
- Holding tabs are essential for all profile cut-outs larger than 30 × 30 mm. Set tab width to 4 mm, height to 3 mm, minimum 4 tabs per piece. Never profile without tabs.
- The CAM Simulator must be run before every job. Check all operations appear, rapid moves stay above material, and the result matches the design intent.
- Post-process to grbl post-processor. Verify the .nc file contains G21, G90, M3, M5, and M30 before loading in UGS.
- Project 1 (Name Badge) and Project 2 (Engraved Coaster) are the course's first two physical cut projects. Both are completed end-to-end using the complete FreeCAD → CAM → UGS → machine workflow.

Key Vocabulary — Chapter 7

Term	Definition
CAM workbench	FreeCAD's Computer-Aided Manufacturing workbench. Generates toolpaths and G-code from 3D models.
Job	The CAM container object that links the toolpaths to a 3D model body and holds all Stock, Tool, and Operation settings.
Stock	A virtual block representing the physical material on the CNC machine bed. Must match actual material dimensions in X, Y, and Z.
Tool Library	A database of cutting bit definitions stored within a FreeCAD CAM Job. Each tool definition specifies the bit's type, diameter, and cutting geometry.
Engraving operation	A CAM operation that directs the V-bit to follow 2D edges at a set depth. Used for text, decorative lines, and profile outlines.
Profile operation	A CAM operation that cuts along the outer or inner boundary of a profile to a set depth, with automatic tool-diameter offset. Used to cut out parts from stock.
Pocket Shape	A CAM operation that clears all material from an enclosed 2D area to a set depth. Used for recessed panels, border rings, and flat-bottomed cavities.

Holding tab	A small bridge of uncut material left by the Profile operation to hold the cut-out piece in place until the job is complete. Removed by hand after cutting.
Step Down	The depth increment for each cutting pass in a multi-pass operation. A 12 mm thick profile cut with 1.5 mm step down takes 8+ passes.
Feed rate	The horizontal cutting speed in mm/min. Set in the CAM operation and written as the F parameter in G-code.
Plunge rate	The vertical entry speed in mm/min. Always slower than feed rate — typically 20–40% of feed rate.
CAM Simulator	FreeCAD's built-in 3D animation that shows virtual material removal for all operations in a job. Must be run before every cutting session.
Post-processor	The software translator that converts FreeCAD's internal toolpath representation into a specific G-code dialect. Use grbl for the CNC 4040 Pro Max.
Profile offset	The automatic displacement of a Profile toolpath from the model edge, equal to half the tool diameter. Applied automatically — do not manually offset sketches.
Outside profile	A Profile operation where the tool cuts around the outer boundary of the profile — the standard mode for producing cut-out parts.
Inside profile	A Profile operation where the tool cuts along the inner boundary — used for cutting holes and interior profiles from the inside.

Looking Ahead — Part Three: Core Skills

Chapters 4–7 complete Part Two — the foundations. You can now design a 3D CNC-ready model in FreeCAD and cut it on the machine. Part Three (Chapters 8–11) introduces the skills that turn a competent beginner into a versatile designer: parametric spreadsheet-driven design, arrays and pattern repetition, SVG artwork import, and multi-body joinery. These are the skills that characterise professional CNC workflow.

PART THREE

Core Skills — Parametric & Pattern Design

Weeks 3–5 • Chapters 8 through 10 • Bloom's Levels: Understand → Apply

📖 **About Part Three**

Part Three covers Weeks 3 and 4 — Days 11 through 20. These are the weeks where you transition from following instructions to making design decisions. The skills introduced here — parametric design, pattern repetition, SVG artwork import, and joinery fundamentals — are the tools that distinguish a versatile CNC designer from someone who can only follow a recipe.

Part Three operates at the Understand and Apply levels of Bloom's Taxonomy. You will not just use these tools — you will explain why they work the way they do and choose between them intelligently. By the end of Day 20, you will have completed four complete CNC projects and accumulated a foundation of skills that the course's advanced weeks will build upon.

CHAPTER
8
of 24

PART THREE — CORE SKILLS

Parametric Design

The Spreadsheet workbench, cell aliases, driven dimensions, the Refresh cycle, and Project 3

Chapter Learning Objectives

By the end of this chapter, you will be able to:

- Create a Spreadsheet in FreeCAD and organise it with descriptive labels and values
- Assign named aliases to spreadsheet value cells
- Link sketch dimensional constraints to spreadsheet aliases using the =Spreadsheet.Alias syntax
- Link Pad depth, ShapeString height, and Pocket depth to spreadsheet values
- Update a complete 3D model by changing a single spreadsheet value and pressing Refresh
- Write compound formulas combining multiple spreadsheet values
- Diagnose and fix the three most common spreadsheet linking errors
- Explain which CAM settings update automatically after a parametric change and which need manual updating
- Apply a consistent G-code file versioning convention for parametric models

Parametric Design — From Fixed to Living Models

A fixed-dimension model is a photograph of a design. A parametric model is the design itself — a set of rules that can produce any size, any variant, in seconds.

In Chapter 1 you were introduced to the concept of parametric design through the shelf bracket analogy. You are now ready to implement it. Parametric design in FreeCAD means replacing the hard-coded numbers in your sketch constraints and feature settings with references to a spreadsheet. Instead of the width constraint reading '150', it reads '=Spreadsheet.Width'. The number 150 lives in the spreadsheet. The sketch reads it from there.

This sounds like it adds complexity. In reality, it removes it — but the benefit only becomes visible when you need to change something. The table below illustrates the difference in a concrete scenario: a client changes the size of an order after the design is already complete.

✕ Without parametric design	✓ With parametric design
Client requests badges in size A (100 × 60 mm). You design the badge with hard-coded dimensions. 'Great.'	Client requests badges in size A (100 × 60 mm). You design the badge with spreadsheet parameters.
Client then requests the same badge in size B (120 × 75 mm) and size C (80 × 50 mm) for different name lengths.	Client requests size B and C. You open the spreadsheet, change Width and Height, press Refresh. Done in 90 seconds.
You manually re-enter every constraint value for size B. Repeat for size C. Each redesign takes 20–30 minutes.	Export G-code for each size with version numbers. Three production-ready files in under five minutes.
One constraint is accidentally missed on size C. The border inset is wrong. The error is only discovered after cutting.	Every dimension updates together. The border inset (= Width − 2×BorderInset) recalculates automatically. No missed dimensions.
Total time for three size variants: 60–90 minutes of redesign work.	Total time for three size variants: under 5 minutes from size A to three exported G-code files.

The time savings become even more dramatic as models grow in complexity. A simple badge has perhaps 8 constrained dimensions. A finger joint box (Week 6) has 40+. A parametric wall clock (Week 9) has 80+. At that scale, a non-parametric approach is not just slow — it is genuinely error-prone. Parametric design is not a convenience; at that complexity level, it is a necessity.

The Spreadsheet Workbench

FreeCAD's Spreadsheet workbench provides a spreadsheet editor — similar in appearance to a simple version of Microsoft Excel or Google Sheets — directly inside your FreeCAD document. The spreadsheet object is stored as part of your .FCStd file, travels with your model, and is always available without opening any external application.

Creating a Spreadsheet

1. In your FreeCAD document, go to the Workbench drop-down and select Spreadsheet.
2. Go to Spreadsheet → Create Spreadsheet. A Spreadsheet object appears in the Model tree.
3. Double-click the Spreadsheet in the Model tree to open it. The spreadsheet editor appears in the main panel.
4. You will see a grid of cells — columns labelled A, B, C... and rows numbered 1, 2, 3...

> 📋 **Note**
>
> A FreeCAD document can contain multiple spreadsheet objects. For this course, use one spreadsheet per design file, named clearly (default name 'Spreadsheet' is fine for single-spreadsheet files). Keeping all parameters in one spreadsheet makes the model easier to maintain.

Navigating the Spreadsheet

The spreadsheet editor works like any basic spreadsheet: click a cell to select it, type to enter a value, press Tab to move right, Enter to move down. You can resize columns by dragging the column dividers. The toolbar above the spreadsheet provides basic formatting options — alignment, font size, borders — though you will rarely need them for a parameter spreadsheet.

Organising Your Spreadsheet — Best Practice

A well-organised parameter spreadsheet is easy to read, easy to maintain, and easy to use as a reference at the machine. A poorly organised one wastes time and causes confusion. Apply these conventions from the start and they will become automatic.

The Three-Column Layout

The recommended layout uses three columns:

- Column A — Label: a plain-English description of the parameter. For example: 'Badge Width'. This is for human reading only — FreeCAD does not use the label column.
- Column B — Value: the numerical value in millimetres (or degrees, or other units). This is the cell that FreeCAD reads. This is where the alias is set.
- Column C — Units/Notes: optional but useful. Enter 'mm', 'deg', or a note like 'adjust for material' to document what each parameter controls.

The spreadsheet below shows a complete parameter set for a simple sign, organised using the three-column layout:

	A	B	C
1	Parameter	Value	Units/Notes
2	SignWidth	150	mm
3	SignHeight	80	mm
4	Thickness	12	mm – set to actual material
5	BorderInset	10	mm – from each edge
6	TextHeight	28	mm
7	EngravingDepth	0.8	mm
8	HoleRadius	4	mm – keyring hole

💡 **Tip**

Row 1 is a header row (labels only). Start your parameters from Row 2. This keeps the layout clean and makes it obvious that Row 1 is documentation, not data.

Cell Aliases — The Key to Parametric Linking

An alias is a name you give to a spreadsheet cell. It is the bridge between the spreadsheet and the sketch.

Without aliases, you would have to reference spreadsheet cells by their grid location — B2, B3, B4. A sketch constraint linked to B2 is confusing: what does B2 mean? Is it the width or the height or the thickness? An alias replaces the obscure cell reference with a meaningful name: =Spreadsheet.SignWidth instead of =Spreadsheet.B2. The constraint is self-documenting — you can read the sketch and instantly understand what each dimension is linked to.

More importantly, aliases are stable. If you add a row to your spreadsheet and the value you care about shifts from B2 to B3, a cell reference link breaks. An alias link does not break — as long as the alias name stays the same, the link remains valid regardless of where the cell physically is in the spreadsheet.

📖 **Key Term: Alias**

A user-assigned name given to a spreadsheet cell, used to reference that cell in sketch constraints and feature properties. Aliases are case-sensitive and cannot contain spaces. Example: 'SignWidth' is a valid alias; 'Sign Width' is not.

Setting a Cell Alias

1. Click the cell you want to name — for example, B2 (which contains your Width value).
2. Go to Spreadsheet → Properties (or right-click the cell and choose Properties).
3. In the Properties dialog, find the Alias field.
4. Type the alias name — for example: SignWidth. Follow the naming rules: no spaces, no special characters, start with a letter. Capitalised words are conventional (SignWidth, not signwidth).
5. Click OK. The alias is now set. The cell's value (150) is now also accessible as Spreadsheet.SignWidth.
6. Repeat for every value cell. Your B column cells should each have a descriptive alias.

Alias Naming Rules

No spaces (use camelCase: SignWidth, not Sign Width). No special characters except underscore. Must start with a letter, not a number. Case-sensitive: SignWidth and signwidth are different aliases. Keep names descriptive but concise — they appear inside sketch constraint dialogs where space is limited.

Linking Sketch Constraints to Spreadsheet Values

Once your spreadsheet has aliases set, you can link sketch dimensional constraints to those aliases. This is the core action of parametric design — replacing a typed number in a constraint with a formula that reads from the spreadsheet.

The =Spreadsheet.Alias Syntax

In any FreeCAD field that accepts a formula — a sketch constraint dialog, a Pad length field, a Pocket depth field — you can click the small blue = icon (formula button) next to the value field to switch from number entry to formula entry. Once in formula mode, type your alias reference:

Syntax Reference

```
=Spreadsheet.SignWidth  // References the alias 'SignWidth' in the Spreadsheet object
=Spreadsheet.SignHeight  // References the alias 'SignHeight'
=Spreadsheet.Thickness  // References the material thickness alias
=Spreadsheet.SignWidth - 2 * Spreadsheet.BorderInset  // Compound formula: width minus twice the border inset
=Spreadsheet.TextHeight * 0.5  // Formula: half the text height
=Spreadsheet.HoleRadius * 2  // Formula: double the hole radius (for a diameter)
```

The formula is evaluated immediately — you can see the resulting number in the constraint dialog. If the formula produces an error (shown as red or 'unresolvable'), the alias name is likely misspelled or the Spreadsheet object has not been saved yet.

Step-by-Step: Linking a Sketch Width Constraint

1. Open a sketch that contains your rectangle. Double-click the sketch in the Model tree to enter the Sketcher.
2. Click the width dimension constraint on the rectangle (the small number label on the sketch). The constraint dialog opens.
3. In the value field, you will see the current hard-coded number (e.g., 150). Look for a small blue = button to the right of the field. Click it.
4. The field changes to formula entry mode. Delete the existing number.

5. Type: =Spreadsheet.SignWidth — exactly as written, using the exact capitalisation of your alias.
6. Press Enter or click OK. The constraint value changes to show the result (150 mm, if that is your alias value). The constraint label on the sketch may show a small equals sign (=) to indicate it is formula-driven.
7. Repeat for the height constraint: =Spreadsheet.SignHeight.
8. Close the sketch. Your sketch is now parametrically linked.

Try It Now

Open the simple sign or badge model you built in Chapters 5–6.

Switch to Spreadsheet workbench. Create a spreadsheet with at least 4 parameters: Width, Height, Thickness, BorderInset.

Enter values and set aliases for each.

Double-click the base sketch. Link the Width and Height constraints to your aliases.

Close the sketch. Go to Edit → Refresh (or press Ctrl+Shift+R).

Now go back to the Spreadsheet. Change SignWidth from 150 to 200. Press Enter.

Go to Edit → Refresh. Watch the sketch in the 3D view — the rectangle should expand to 200 mm.

This is parametric design working. You changed one number and the model updated.

Linking Pad, Pocket, and ShapeString to Spreadsheet Values

Parametric linking is not limited to sketch constraints. Any numeric field in Part Design that shows the blue = formula button can accept a spreadsheet reference. The most important fields to link are the Pad depth (material thickness), Pocket depth (engraving or recess depth), and ShapeString height (text size).

Linking Pad Depth

1. Double-click the Pad feature in the Model tree. The Pad dialog opens.
2. In the Length field, click the blue = formula button.
3. Type: =Spreadsheet.Thickness
4. Click OK. The Pad depth is now driven by your Thickness alias.
5. Test: change Thickness in the spreadsheet from 12 to 18. Refresh the model. The solid should become 18 mm thick.

Linking Pocket Depth

The Pocket depth is often a formula rather than a direct alias — the engraving depth might always be a fixed ratio of the material thickness, for example: =Spreadsheet.Thickness * 0.25

(25% of material thickness). This ensures that if you cut the design in thicker material, the recesses scale proportionally.

Linking ShapeString Height

Select the ShapeString in the Model tree. In the Data tab of the Properties panel, find the Size (or Height) field. Click the = button and type: =Spreadsheet.TextHeight. The text height now updates with the spreadsheet.

> **Tip**
>
> When linking ShapeString, also consider linking its X position. If you want the text to always appear at the left edge of the recessed panel, for example, you could set its X position to: =−Spreadsheet.SignWidth/2 + Spreadsheet.BorderInset + 5. This positions the text exactly where you want it regardless of the sign's width.

The Refresh Cycle — Keeping Everything in Sync

FreeCAD does not always update the model automatically when you change a spreadsheet value. You need to trigger the update manually. This is called the Refresh cycle, and it is the one workflow step that beginners most often forget, leading to apparent 'parametric design not working' problems that are actually just missed Refresh actions.

Step 1: Change the value	In the Spreadsheet, click the value cell you want to change. Type the new value. Press Enter.
Step 2: Refresh the model	Go to Edit → Refresh model (or press Ctrl+Shift+R, or click the Refresh button in the toolbar). FreeCAD re-evaluates all formulas and updates all features.
Step 3: Verify	Look at the 3D view. The model should reflect the new value. Check the Model tree — all features should be green (no yellow warnings).
Step 4: Check CAM (if applicable)	Switch to the CAM workbench. Inspect the Stock settings — they do not update automatically and must be manually adjusted. Re-run the CAM Simulator to verify the updated toolpaths.

> **Important**
>
> Changing a spreadsheet value without pressing Refresh means the model still shows the old dimensions.
>
> If your model appears unchanged after editing the spreadsheet, press Ctrl+Shift+R (Refresh) and wait for FreeCAD to recompute.
>
> Refresh may take several seconds for complex models — be patient and wait for the spinning cursor to stop.

Compound Formulas — Combining Spreadsheet Values

The real power of parametric design emerges when you write compound formulas — expressions that calculate derived values from your base parameters. You do not need to store every value explicitly in the spreadsheet. Derived values can be calculated on the fly using standard arithmetic.

FreeCAD's formula engine supports: addition (+), subtraction (−), multiplication (*), division (/), power (^), and common mathematical functions (sqrt(), abs(), sin(), cos()). Parentheses control order of operations as in standard arithmetic.

📝 Syntax Reference

```
=Spreadsheet.SignWidth - 2 * Spreadsheet.BorderInset  // Inner rectangle width — sign width minus border on both sides
=Spreadsheet.SignHeight / 2  // Half the sign height — useful for positioning elements vertically
=Spreadsheet.HoleRadius * 2  // Hole diameter — double the radius
=Spreadsheet.Thickness * 0.25  // Pocket depth as 25% of material thickness
=Spreadsheet.SignWidth - 20  // Fixed 10 mm margin on each side regardless of sign width
=sqrt(Spreadsheet.SignWidth ^ 2 + Spreadsheet.SignHeight ^ 2)  // Diagonal of the sign — for calculating bit travel estimates
```

Compound formulas are most valuable for derived dimensions that you would otherwise have to calculate manually and re-enter every time a base parameter changes. The border inset formula is the classic example: you want the inner rectangle to always be BorderInset millimetres from each edge. Without a compound formula, changing the sign width means manually recalculating and re-entering the inner rectangle width. With the formula, it updates automatically.

Try It Now

In your sign sketch, replace the inner border rectangle's width constraint with the compound formula: =Spreadsheet.SignWidth - 2 * Spreadsheet.BorderInset

Replace the height constraint with: =Spreadsheet.SignHeight - 2 * Spreadsheet.BorderInset

Now change SignWidth from 150 to 200 in the spreadsheet. Refresh.

The outer rectangle should be 200 mm wide. The inner border rectangle should be 180 mm wide (200 − 2×10).

The border inset distance (10 mm each side) is maintained automatically.

Change BorderInset from 10 to 15. Refresh. Both the outer sign and the inner border rectangle update consistently.

Common Parametric Linking Errors

Three errors account for the vast majority of problems students encounter when first working with parametric design. Knowing them in advance — what they look like and how to fix them — will save you significant frustration.

Error Message / Symptom	Cause	Fix
Sketch turns red after linking	*Alias name is misspelled or does not match the spreadsheet alias exactly. Capitalisation differences count — 'signwidth' and 'SignWidth' are different.*	Double-click the failed constraint. Check the formula text for typos. Compare it letter-by-letter to the alias name in the Spreadsheet Properties. Fix the typo.
Formula accepted but model does not update	*Refresh was not triggered after changing the spreadsheet value. FreeCAD does not auto-update on spreadsheet edits.*	Press Ctrl+Shift+R (Refresh model) or go to Edit → Refresh model. Wait for the model to recompute.
'Unresolvable' or 'no object named Spreadsheet' error	*The Spreadsheet object has not been saved, or its name in the model tree is different from 'Spreadsheet'. If you renamed the spreadsheet, the formula reference needs updating.*	Check the spreadsheet's name in the Model tree. If it is 'Spreadsheet001' rather than 'Spreadsheet', update all formulas to reference =Spreadsheet001.AliasName.
Circular reference error	*A formula references its own result (directly or indirectly). Example: Cell B2 contains =B3, and Cell B3 contains =B2. FreeCAD cannot resolve the circular dependency.*	Redesign the formula chain so values flow in one direction. Identify the cell causing the loop in the Spreadsheet's error message and break the cycle.
Constraint turns yellow (driven but not fully green)	*A driven constraint is marked as such but the sketch still has remaining DOF from other elements.*	The linked constraints are working correctly — the yellow colour just indicates they are formula-driven rather than hard-coded. Fully constrain the remaining white elements.

Parametric CAM — What Updates and What Does Not

When you update a parametric model through the spreadsheet, the Part Design model updates automatically after Refresh. The CAM workbench, however, is partially automatic — some things update, others require manual intervention. Knowing which is which prevents the common mistake of updating a model but cutting from outdated G-code.

Updates automatically (after model Refresh)	Toolpaths that reference model geometry — Engraving paths following ShapeString edges, Profile paths following body edges. These re-generate automatically when the underlying geometry changes.
Requires manual update — Stock dimensions	The Stock X, Y, Z dimensions are entered manually and do not respond to model changes. If your model width changes from 150 to 200 mm, you must also update the Stock X value from 152 to 202 mm manually.
Requires manual update — Operation depths	Final Depth values entered as fixed numbers do not update. To make depths parametric, type the formula directly into the depth field: =−Spreadsheet.EngravingDepth for engraving, =−(Spreadsheet.Thickness + 0.5) for a through-cut profile.
Requires re-simulation	After any model or CAM change, always re-run the CAM Simulator to verify the updated toolpaths before post-processing.
Requires re-post-processing	After any update — even if only the Stock changed — re-run the post-processor and save the new .nc file with a new version number.

G-code File Versioning

A parametric model naturally produces multiple G-code variants — different sizes, different thicknesses, different configurations. A disciplined file versioning convention prevents the nightmare scenario of cutting from the wrong file.

Base convention	ProjectName_MaterialThickness_Version.nc — Example: LuggageTag_12mm_v1.nc
Size variants	LuggageTag_Small_12mm_v1.nc, LuggageTag_Large_12mm_v1.nc
Version updates	Increment the version number on every post-process: v1, v2, v3. Never overwrite an earlier version. Keep all versions.
Date stamping	For production files: LuggageTag_2024-11-15_12mm_v3.nc. Makes file history chronologically sortable.
Never delete	Old .nc files are your audit trail. If a piece comes out wrong and you are not sure which G-code was used, old files allow you to investigate.

PROJECT 3 — STEP-BY-STEP WALKTHROUGH

Parametric Luggage Tag / Keyfob

A fully parametric flat tag — every dimension spreadsheet-driven. Must demonstrate live update before cutting.

Project 3 has a requirement not present in Projects 1 and 2: before the instructor grants permission to cut, you must demonstrate the parametric update to them directly. Change two spreadsheet values, press Refresh, show the model updates cleanly, re-export G-code. Only then do you cut. This demonstration confirms that your model is genuinely parametric — not merely labelled as such.

Design	Personalised luggage tag or keyfob with rounded rectangle, keyring hole, engraved text
Size	Student's choice within: 80–120 mm × 40–60 mm
Thickness	6 mm MDF (lighter for a luggage tag)
Operations	Engraving (V-bit), Profile (3mm or 6mm end mill)
Requirement	ALL sketch dimensions and Pad depth must reference the spreadsheet. Zero hard-coded values. Demonstrate update to instructor before cutting.
Day	Day 15 of Week 3 — Project Day

Part A — Spreadsheet Setup

1 **New file and spreadsheet**
File → New. Save as LuggageTag.FCStd. Switch to Spreadsheet workbench. Create Spreadsheet. Enter the following parameters:

	A	B	C
1	Parameter	Value	Notes
2	TagWidth	100	mm — outer width
3	TagHeight	45	mm — outer height
4	Thickness	6	mm — match material
5	CornerRadius	6	mm — rounded corner arc
6	HoleRadius	3.5	mm — keyring hole radius
7	HolePosFromEdge	8	mm — hole centre from left edge
8	TextHeight	16	mm
9	EngravingDepth	0.6	mm

2 **Set aliases**
Click each B-column value cell. Spreadsheet → Properties → Alias. Set: TagWidth, TagHeight, Thickness, CornerRadius, HoleRadius, HolePosFromEdge, TextHeight, EngravingDepth. Save the file.

Part B — Design in Part Design

3 **Create Body and base sketch**
Switch to Part Design. Body. New Sketch on XY plane. Draw a centred rectangle: Horizontal Distance = =Spreadsheet.TagWidth, Vertical Distance = =Spreadsheet.TagHeight.

4 **Add rounded corners**
In the Sketcher, use Sketch → Sketcher tools → Create fillet. Click each corner of the rectangle. Set radius = =Spreadsheet.CornerRadius for each fillet constraint.

5 **Add the keyring hole**
Draw a circle near the left end of the tag. Centre at X = −(=Spreadsheet.TagWidth/2 − Spreadsheet.HolePosFromEdge), Y = 0. Constrain Radius = =Spreadsheet.HoleRadius.

6 **Close sketch and verify**
All elements green, 0 DOF. Close sketch. Part Design → Pad, Length = =Spreadsheet.Thickness. Inspect the 3D solid — a rounded-corner flat tag with a keyring hole.

7 **Add ShapeString text**
Switch to Part. Part → ShapeString. Your name or identifier. Height = (edit the ShapeString's Height property in Data panel to) =Spreadsheet.TextHeight. Bold font. Position centred on the face.

Part C — Parametric Demonstration (Required Before Cutting)

8 **Demonstrate the parametric update to your instructor**
Open the spreadsheet. Change TagWidth from 100 to 120. Change TagHeight from 45 to 55. Press Ctrl+Shift+R (Refresh). Show the instructor that the tag outline, the rounded corners, the keyring hole position, and the ShapeString all update. Change values back to your desired cut size. Refresh again. Only after instructor sign-off do you proceed to CAM.

Part D — CAM and Cutting

9 **Create CAM Job and Stock**
Switch to CAM. CAM → Job. Configure Stock: X = =TagWidth + 4 (type approximate value), Y = TagHeight + 4, Z = 6. (Stock does not accept spreadsheet formulas — enter approximate values based on current spreadsheet values.)

10 **Operation 1 — Engraving**
CAM → Engraving. Select ShapeString edges. V-Bit 60deg. Final Depth = −=Spreadsheet.EngravingDepth (type −0.6 or link via formula). Feed: 700. Plunge: 180.

11 **Operation 2 — Profile**
CAM → Profile. Select outer tag edge. End Mill 3mm or 6mm. Final Depth = −(Thickness + 0.5) = −6.5. Step Down: 1.0 (for 3mm) or 1.5 (for 6mm). Tabs: 4, width 3mm, height 2mm. Feed: 700 (3mm) or 1000 (6mm). Plunge: 200 or 300.

12 **Simulator and post-process**
Run CAM Simulator. Verify engraving covers all text, profile traces the rounded rectangle, tabs present. Post-process: grbl. Save as LuggageTag_6mm_v1.nc.

13 **Pre-cut checklist, connect, home**
Complete checklist. Connect UGS. Home machine.

14 **Install V-bit, clamp 6mm MDF, set zero**
V-bit in collet. Clamp small piece of 6mm MDF. Set X/Y/Z zero at centre (centred rectangle = centred zero).

15 **Load file, verify in UGS, run**
Load LuggageTag_6mm_v1.nc. Verify 3D visualiser. Send job. Tool change to end mill after engraving. Profile cut with tabs.

16 **Release, clean, finish**
Cut tabs. Sand 120 → 240 grit. Thread a keyring through the hole.

🏆 **Project 3 Complete — Parametric Demonstrated**

Project 3 is the first project that requires a live demonstration of parametric design before cutting. This requirement is not bureaucratic — it verifies that your model is genuinely driven by the spreadsheet and that you understand the update workflow. A model that only 'claims' to be parametric but still has hard-coded values will fail the demonstration.

Week 3, Days 11–15 — Lesson Content Overview

Day 11 (Understand)	Parametric concept and the Spreadsheet workbench. Students create their first spreadsheet with 6+ parameters and set aliases for every value. No linking yet — just the spreadsheet setup. Exit ticket: explain in one sentence what an alias does and why it is better than a cell reference.
Day 12 (Understand → Apply)	Linking sketch constraints to spreadsheet values. Guided exercise: link all constraints in a simple rectangle sketch to the spreadsheet. Test the link by changing a value and refreshing. Class discussion: what breaks if you misspell an alias name? Explore the error deliberately.
Day 13 (Apply)	Linking Pad depth, ShapeString height, and Pocket depth. Guided exercise: extend the linked model from Day 12 to include parametric depth values. Write two compound formulas. Independent exercise: produce 3 size variants of the same design using only spreadsheet changes.
Day 14 (Apply)	Parametric CAM workflow. Students update a CAM job after a model parameter change — what needs manual updating, what updates automatically. Post-process 3 variants with correct version-numbered filenames. Independent: verify that each .nc file produces a correctly sized visualisation in UGS.
Day 15 (Apply — Project Day)	Project 3: Parametric Luggage Tag. Full independent project. Parametric demonstration to instructor required before any cutting is permitted.

Chapter Summary

- Parametric design replaces hard-coded dimensions with spreadsheet references. A single spreadsheet value change propagates through the entire model automatically after Refresh. This is the difference between a dead drawing and a living model.
- The Spreadsheet workbench is built into FreeCAD. Create a Spreadsheet object, organise it with three columns (Label, Value, Notes), and set aliases on every value cell.
- Aliases give cells descriptive names (SignWidth, Thickness) rather than obscure cell references (B2). They are set via Spreadsheet → Properties → Alias. Aliases are case-sensitive and cannot contain spaces.
- Link sketch constraints using the formula syntax =Spreadsheet.AliasName, activated by clicking the blue = button in the constraint dialog. Pad depth, Pocket depth, and ShapeString height can all be linked the same way.
- Compound formulas combine multiple aliases: =Spreadsheet.SignWidth − 2 * Spreadsheet.BorderInset automatically maintains a consistent border margin regardless of sign size.
- The Refresh cycle is: change value → Ctrl+Shift+R (Refresh) → verify model updated. Always refresh before simulating, before post-processing, and before cutting.
- After a parametric model update, Stock dimensions and operation depth values with hard-coded numbers must be manually updated in the CAM workbench. Toolpath geometry updates automatically.
- Version every G-code file: ProjectName_Thickness_v1.nc. Increment the version number on every post-process. Never overwrite an earlier version.
- Project 3 requires a live parametric demonstration to the instructor before cutting. Change two parameters, Refresh, verify model updates, re-export G-code. Only then cut.

Key Vocabulary — Chapter 8

Term	Definition
Parametric design	A design approach where model geometry is controlled by named variables (parameters) in a spreadsheet, rather than fixed dimensions. Changing a parameter updates the entire model.
Spreadsheet workbench	FreeCAD's built-in spreadsheet editor, stored inside the .FCStd file. Used to create and manage parameter spreadsheets that drive model dimensions.
Cell alias	A user-assigned name given to a spreadsheet cell. Aliases replace obscure cell references (B2) with descriptive names (SignWidth) in formula links.
Formula syntax	The text format used to link a FreeCAD field to a spreadsheet alias: =Spreadsheet.AliasName — or for the default spreadsheet: =Spreadsheet.Width.
Driven constraint	A sketch constraint whose value is determined by a formula rather than a typed number. Displayed with a small = indicator. Turns yellow-green rather than plain green.
Compound formula	A formula that combines multiple spreadsheet values using arithmetic: =Spreadsheet.Width − 2 * Spreadsheet.BorderInset. Calculates derived values from base parameters.

Refresh cycle	The two-step process to update a parametric model: (1) change a spreadsheet value, (2) press Ctrl+Shift+R or go to Edit → Refresh model. Required after every spreadsheet edit.
Circular reference	An error where a formula references its own result, directly or indirectly. FreeCAD cannot resolve circular references and displays an error. Fix by breaking the dependency chain.
Dead drawing	Informal term for a non-parametric model or drawing where all dimensions are hard-coded fixed numbers, requiring manual re-entry for every variation.
Live model	Informal term for a fully parametric model where dimensions are driven by spreadsheet values that can be updated globally with a single change and Refresh.
G-code versioning	The practice of naming G-code files with version numbers (v1, v2, v3) to distinguish between different exports of the same design at different parameter values.
Parametric CAM	A CAM workflow where some operation depths and geometry reference spreadsheet values, allowing the CAM job to update alongside the model when parameters change.

Looking Ahead — Chapter 9: Arrays, Patterns & Repetitive Geometry

Chapter 9 introduces Part Design Linear Pattern and Polar Pattern — the tools that let you repeat a feature dozens of times across a model without drawing it once. Combined with Draft Array for tiling objects across a sheet and the CAM Pocket Shape operation with Adaptive strategy, Chapter 9 covers everything needed for Projects 4 (Decorative Patterned Tile) and 5.

CHAPTER 9 of 24

PART THREE — CORE SKILLS

Arrays, Patterns & Repetitive Geometry

Linear Pattern, Polar Pattern, Draft Array, material nesting, and the CAM Pocket Shape operation

Chapter Learning Objectives

By the end of this chapter, you will be able to:

- Apply Part Design Linear Pattern to repeat a feature in one or two directions
- Calculate the spacing between pattern occurrences from the Length and Occurrences values
- Apply Part Design Polar Pattern to distribute features in a circular or arc arrangement
- Use construction circles to position pattern features at a precise radius
- Use the Draft workbench Array, Move, and Scale tools to tile objects across a stock sheet
- Compare Draft Array and Part Design Linear Pattern and choose the appropriate tool for a given task
- Plan a material nesting layout to minimise stock waste
- Create a CAM Pocket Shape operation using ZigZag, Adaptive, and Offset clearing strategies
- Configure step-over, step-down, and finish pass settings for pocket operations

The Power of Repetition in CNC Design

Drawing one feature takes five minutes. Drawing forty identical features takes two hundred minutes — unless you know how to use patterns.

CNC machines excel at repetition. A machine that has cut one perfect hole can cut a thousand perfect holes in the same time it would take a human to mark out and drill ten by hand. But to take advantage of this, you need to be able to specify repetition efficiently in FreeCAD — not by drawing forty features by hand, but by drawing one and instructing FreeCAD to replicate it mathematically.

Part Design's Linear Pattern and Polar Pattern tools do exactly this for model features. The Draft workbench's Array tool does the same for entire objects — including imported shapes and multi-part nesting layouts. Together, these tools are the foundation of efficient CNC design for any project involving repeated elements: holes, pockets, teeth, slots, hour markers, honeycomb cells, and more.

Linear Pattern — Repeating Features Along an Axis

Part Design Linear Pattern takes an existing feature — a Pad, a Pocket, a Chamfer — and repeats it a specified number of times along a specified axis. The copies are evenly spaced. You set the total length the array spans and the number of copies, and FreeCAD calculates the spacing automatically.

Linear Pattern is a Part Design tool, which means it works on Part Design features — Pads, Pockets, Fillets, etc. It cannot directly tile entire bodies or imported objects. For those tasks, the Draft Array tool is used instead (covered later in this chapter).

> 📖 **Key Term: Linear Pattern**
>
> A Part Design tool that repeats an existing feature (Pad, Pocket, Chamfer, etc.) a specified number of times along a specified axis, with equal spacing calculated from the total length and occurrence count.

Creating a Linear Pattern — Step by Step

1. Create the feature you want to repeat — for example, a circular Pocket representing one hole.
2. In the Model tree, click on that feature to select it.
3. Go to Part Design → Linear Pattern. The Linear Pattern dialog opens.
4. Set Direction: choose X-axis, Y-axis, Z-axis, or a named edge from your model as the repeat direction.
5. Set Length: the total distance from the first occurrence to the last. NOT the spacing between features — the total span.
6. Set Occurrences: how many times the feature appears in total (including the original).
7. FreeCAD previews the result in the 3D view as you adjust values.
8. Click OK. The pattern is created and appears in the Model tree as a Linear Pattern feature.

The Spacing Formula — Understanding Length and Occurrences

The relationship between Length, Occurrences, and Spacing is the most important mathematical concept in this chapter. Beginners frequently confuse 'Length' with 'spacing' and end up with the first and last occurrences at the correct positions but everything else wrong. Understanding the formula prevents this.

> 📐 **Linear Pattern Spacing Formula**
>
> `Spacing = Length ÷ (Occurrences - 1)`
>
> Length is the total distance from the FIRST occurrence to the LAST. Occurrences is the total count including the original. The gap between adjacent features equals the total length divided by one less than the occurrence count.

Length	Occurrences	Spacing = Length ÷ (Occurrences − 1)	Real-world example
100 mm	5	`100 ÷ 4 = 25 mm`	*Five holes evenly spaced across 100 mm — 25 mm apart*
200 mm	9	`200 ÷ 8 = 25 mm`	*Nine holes across 200 mm — same 25 mm spacing, larger array*
150 mm	6	`150 ÷ 5 = 30 mm`	*Six pockets across 150 mm — 30 mm spacing*
80 mm	3	`80 ÷ 2 = 40 mm`	*Three features across 80 mm — 40 mm apart*

Try It Now

Create a 200 × 100 mm pad body.

Add a circular pocket (radius 5 mm) near the left edge of the top face.

Select the Pocket. Go to Part Design → Linear Pattern.

Set Direction: X-axis. Length: 180 mm (leaving 10 mm margin each side). Occurrences: 5.

Before clicking OK, calculate the expected spacing: 180 ÷ 4 = 45 mm.

Verify in the 3D view preview that the holes are 45 mm apart. Click OK.

Now change Occurrences to 9. Expected spacing: 180 ÷ 8 = 22.5 mm. Does the preview match?

2D Arrays — Two-Direction Linear Pattern

A 2D array is created by applying a second direction to an existing Linear Pattern. This tiles the feature across both the X and Y directions, creating a grid of repeated features. In the Linear Pattern dialog, a second row of direction, length, and occurrences controls the second axis — or you can apply a second Linear Pattern on top of the first.

1. Create a single feature and apply a Linear Pattern in the X direction (e.g., 5 occurrences across 120 mm).
2. With the Linear Pattern selected in the Model tree, go to Part Design → Linear Pattern again.
3. This time, set Direction to Y-axis. Set Length and Occurrences for the Y direction (e.g., 3 occurrences across 60 mm).
4. Alternatively, in FreeCAD v1.1.0: in the Linear Pattern dialog, look for 'Add Direction' to add a second axis within the same operation.
5. Click OK. A 5×3 grid of 15 features appears.

> **Tip**
>
> For a 2D array, always calculate both spacings before applying. For a 5×3 grid where features must be 30 mm apart in X and 25 mm apart in Y: set X Length = 30×4 = 120 mm with 5 occurrences, and Y Length = 25×2 = 50 mm with 3 occurrences. Writing these calculations down before touching FreeCAD saves repeated trial-and-error.

Polar Pattern — Circular Symmetry

Polar Pattern distributes a feature around a rotation axis. Where Linear Pattern tiles in a straight line, Polar Pattern tiles around a circle. The result is a rotationally symmetric arrangement — twelve hour markers equally spaced around a clock face, six bolt holes on a flange, eight slots on a decorative gear-like disc.

Polar Pattern is activated the same way as Linear Pattern: select the feature, go to Part Design → Polar Pattern. The dialog has three main settings: Axis (the rotation axis), Angle (the total arc to distribute features across), and Occurrences (how many copies, including the original).

Full Circle vs Partial Arc

Full circle (360°)	Distributes features evenly around the full circumference. Use this for clock faces (12 markers, 360°), symmetric flanges, and any design with complete rotational symmetry. Spacing per occurrence = 360° ÷ Occurrences.
Partial arc (<360°)	Distributes features across less than the full circle. Use for fan arrangements, partial sunbursts, and designs that need symmetry on one side only. Example: 5 features across 120° creates a fan with 30° between each feature (120° ÷ 4).
180° (half circle)	Produces features mirrored about the axis — effectively a symmetric arrangement about a central line. Useful when you need the same feature on both sides of a design without full circular symmetry.

Setting the Correct Axis

For flat CNC parts designed on the XY plane and viewed from above, the Polar Pattern rotation axis is always the Z-axis. The Z-axis passes through the origin vertically — rotating around it produces the circular distribution you see from the top view. If your features are distributing in unexpected directions, check that the axis is set to Z-axis, not X or Y.

Try It Now

Create a 200 mm diameter circular pad (a flat disc).

Add a small rectangular Pocket (4 × 12 mm) on the top face, positioned near the edge of the disc at the 12 o'clock position.

Select the Pocket. Part Design → Polar Pattern.

Set Axis: Z-axis. Angle: 360°. Occurrences: 12.

Click OK. Twelve equally spaced rectangular recesses appear around the disc — a clock face hour marker pattern.

Now try Angle: 180°, Occurrences: 7. Observe the half-circle fan arrangement.

Try Angle: 90°, Occurrences: 4. A quarter-circle fan.

Experiment freely — observe how Angle and Occurrences interact.

Using Construction Circles to Position Polar Pattern Features

By default, a Polar Pattern rotates features around the Body's origin. If you want features at a specific radius from the centre, draw the original feature at that radius before applying the pattern. Construction circles provide the most accurate and repeatable way to position features at a precise distance from the centre.

1. In the sketch where you draw the original feature, also draw a construction circle (toggle to construction mode before drawing, or draw and then toggle): diameter = 2 × the desired radius.
2. Constrain the centre of your feature (e.g., the centre of the mounting hole circle, or the midpoint of the rectangular recess) to lie on the construction circle using Constrain Point on Object.
3. All features that are Polar-Patterned from this original will also be at the same radius, because they are copies of the original's exact geometry.

Note

The construction circle technique is used explicitly in Projects 9 (clock face) and 10 (wall clock with polar-arrayed hour markers). Mastering it now will make those later projects significantly smoother.

The Draft Workbench — Arrays, Move, and Scale

The Draft workbench is FreeCAD's 2D/3D manipulation workbench. Unlike Part Design Linear Pattern (which only works on Part Design features), Draft tools work on any FreeCAD object — including imported SVG shapes, ShapeStrings, entire Part Design bodies, and converted mesh objects. Draft is the right tool when you need to tile, move, or scale objects that are not Part Design features.

Draft Array — Three Types

The Draft Array tool creates copies of a selected object in three possible arrangements:

Rectangular Array	Creates a grid of copies — rows and columns. You specify the spacing in X and Y directions and the count in each direction. This is the primary tool for nesting multiple parts on a stock sheet.
Polar Array	Creates copies in a circular arrangement around a specified centre point. Similar to Polar Pattern but works on complete objects rather than individual features.
Path Array	Distributes copies of an object along a curved path (a spline, polyline, or edge). Used for advanced decorative arrangements. Less common in standard CNC workflows.

Using Draft Rectangular Array for Material Nesting

When you need to cut multiple copies of the same part from one sheet of material, Draft Array lets you position them all at once. This is called nesting — arranging parts efficiently to minimise material waste. While professional nesting software optimises layouts automatically, Draft Array allows manual nesting that is sufficient for most course projects.

1. Complete your part design (Body or converted Part Shape).
2. Switch to Draft workbench. Select the part object.
3. Go to Draft → Array.
4. In the Array dialog, select Rectangular array type.
5. Set the X spacing (distance between part origins in the X direction). If your part is 100 mm wide and you want 5 mm between parts: X spacing = 105 mm.
6. Set the Y spacing (distance between part origins in Y). For a 60 mm tall part with 5 mm gap: Y spacing = 65 mm.
7. Set the Number in X (columns) and Number in Y (rows) to fill your stock sheet.
8. Click OK. Copies of the part tile across the defined area.
9. In the CAM workbench, select all tiled part profiles as base geometry for a single Profile operation. All copies will cut in one job.

Draft Move and Draft Scale

Draft Move	Moves a selected object by a specified X, Y, Z delta (offset). More precise than dragging in the 3D view. Go to Draft → Move. Select the object, specify the base point and target point (or enter a delta vector).
Draft Scale	Scales a selected object by a specified factor. Select the object, go to Draft → Scale. Enter a uniform scale factor (e.g., 0.5 for half size, 2.0 for double size) or non-uniform X/Y/Z scale factors. Note: scaling a Part Design body may not update associated CAM operations — use parametric design for production-quality resizing.
Draft Rotate	Rotates an object around a specified point. Select the object, go to Draft → Rotate. Specify the rotation centre point and the angle. Useful for angling parts in a nesting layout.

Draft Array vs Part Design Pattern — When to Use Each

Part Design Linear / Polar Pattern	Draft Rectangular / Polar Array
Works on Part Design features (Pad, Pocket, Chamfer, etc.)	Works on any FreeCAD object (bodies, imported shapes, ShapeStrings, mesh objects)
Remains part of the original body — integrated into the model history	Creates separate array objects — not integrated into the body's feature tree
Updates when the original feature changes (fully parametric)	Does not automatically update if the source object changes after array creation
Best for: repeating features within a single part (holes, teeth, markers)	Best for: tiling multiple complete parts on a stock sheet, importing and tiling SVG shapes
No control over spacing individual copies differently — all equal	Each copy can be manually positioned if needed (break the array and move individually)

> 💡 **Tip**
>
> The practical rule of thumb: if you are repeating a feature inside one part, use Part Design Linear or Polar Pattern. If you are arranging multiple complete parts or imported objects across a stock sheet, use Draft Array.

CAM Pocket Shape — Clearing Areas Efficiently

The Pocket Shape operation clears all material from an enclosed 2D area to a set depth. Unlike the Engraving operation (which follows edges) or the Profile operation (which cuts along a boundary), Pocket Shape fills in an entire enclosed region. The most important decision in setting up a Pocket Shape is choosing the clearing strategy — the pattern of tool movements used to remove the material.

Clearing Strategies

FreeCAD's CAM workbench offers three main clearing strategies for Pocket Shape operations. Each has different characteristics in terms of cut quality, tool wear, and cycle time. The table below compares them:

	ZigZag	Adaptive (Trochoidal)	Offset
How it works	Alternating back-and-forth straight passes across the pocket	Arcing tool path that limits engagement angle — never full-width cuts	Concentric passes shrinking inward from the pocket walls
Best for	Shallow wide pockets, soft materials, rapid material removal	Deep pockets, hardwood, production volume, long tool life	Clean pocket walls, decorative pockets where wall quality matters
Tool wear	Higher — full-width engagement at each pass reversal	Lower — constant chip load extends tool life significantly	Medium

Cut time	Fastest for shallow	Faster than ZigZag for deep — higher feed rate offsets the longer path	Slower — many passes
Surface quality	Good floor, marks at reversal points	Excellent floor and walls	Excellent walls, good floor
When to use in this course	Projects 1–4, shallow border pockets	Projects 5+ for any pocket deeper than 5 mm or in hardwood	Optional for decorative display pieces where wall appearance is critical

Configuring a Pocket Shape Operation

1. Switch to CAM workbench. Go to CAM → Pocket Shape.
2. In the Base Geometry section, click Add and select the floor face of the pocket (the face at the bottom of the recessed area in the 3D model, or the top face of the body if you want a pocket that starts from the top).
3. In the Tool section, select the end mill to use (6 mm end mill for most pockets in this course).
4. Set Depth settings: Start Depth = 0 (top of material), Final Depth = the pocket depth (negative value: -3 for a 3 mm deep pocket).
5. Set Step Down: how deep each pass goes. For MDF with a 6 mm end mill: 1.5 mm per pass. For hardwood: 1.0 mm.
6. In the Operation section, set Pattern (clearing strategy): ZigZag, Adaptive, or Offset.
7. Set Step Over: the horizontal distance the tool moves between adjacent passes, expressed as a percentage of tool diameter or as an absolute value. 60–70% of diameter is standard.
8. Optional: enable Finish Pass at the end — a final very-shallow pass (0.1–0.2 mm) at reduced feed rate to improve floor and wall quality.
9. Click OK. The pocket toolpath appears on the model.

Step-Over and Step-Down in Detail

Step-over and step-down are the two spatial parameters that most affect the balance between speed and quality in pocket clearing. Understanding their roles helps you make deliberate choices rather than accepting defaults.

Step-Down (depth per pass)	How deep the tool plunges per cutting loop. Smaller = more passes = more time but less force per pass. Larger = fewer passes = faster but higher cutting forces and risk of bit breakage. Rule of thumb: step-down ≤ 0.5 × tool diameter for hardwood, ≤ 0.75 × diameter for MDF.
Step-Over (horizontal engagement)	The distance between adjacent tool passes expressed as a fraction of the tool diameter. 50% = overlap with previous pass by half the tool diameter = very clean but slow. 75% = less overlap = faster but may leave witness marks between passes. 60–70% is the practical sweet spot for MDF.
Adaptive step-over	In Adaptive strategy, step-over is expressed as an engagement angle limit rather than a fixed distance. FreeCAD automatically adjusts the tool path to maintain constant engagement. Typical setting: 50% of diameter as the 'step-over' target.

Finish Pass depth	When enabled, the finish pass cuts a thin layer (0.1–0.2 mm) on the final pass at a slower feed rate. It removes the tool marks from the roughing passes and leaves a smooth floor. Enable for display-quality pieces; skip for functional parts where aesthetics are secondary.

> **Tip**
>
> For your first Pocket Shape experiments, use ZigZag strategy with 60% step-over and 1.5 mm step-down. These settings work reliably for MDF and are predictable in the Simulator. Once you are comfortable with pockets, switch to Adaptive for any pocket deeper than 5 mm — you will notice the machine sounds smoother and tools last longer.

PROJECT 4 — STEP-BY-STEP WALKTHROUGH

Decorative Patterned Tile

A square tile with at least one Linear or Polar Pattern element, a border pocket, and full CAM setup with Adaptive strategy

Project 4 is the first genuinely decorative project in the course. There are requirements (a pattern element, a border pocket, a profile cut) but the design is yours. The goal is to produce something you would actually want to keep or give as a gift — this personal investment in the design makes the technical learning more durable.

Design	Decorative square tile — personal pattern using Linear and/or Polar Pattern
Size	150 × 150 mm (student choice within 120–200 mm)
Thickness	12 mm MDF or hardwood
Operations	Pocket Shape (Adaptive, for pattern pockets and border) + Profile (end mill)
Requirement	At least ONE element must be created using Linear Pattern or Polar Pattern (not manually drawn)
Day	Day 20 of Week 4 — Project Day

Part A — Design

1 **New file, spreadsheet, and Body**
New file, save as PatternedTile.FCStd. Create Spreadsheet with parameters: TileSize=150, Thickness=12, BorderInset=12, PatternPocketDepth=5, BorderPocketDepth=3. Set aliases. Create Body.

2 **Sketch and Pad the tile base**
Part Design → Sketch on XY. Centred rectangle: =Spreadsheet.TileSize × =Spreadsheet.TileSize. Close sketch. Pad: =Spreadsheet.Thickness.

3 **Design your pattern element — one feature**
On the top face, create a sketch. Draw one pocket element: a circle, a square, a hexagon, a slot — your choice. Constrain it carefully. Position it somewhere on the face (near the centre is fine for the first element). Close sketch. Apply Pocket: depth = =Spreadsheet.PatternPocketDepth.

4 **Apply Linear or Polar Pattern**
Select the Pocket in the Model tree. Apply Linear Pattern (for rows/columns) or Polar Pattern (for circular arrangement). Set the parameters to distribute across the tile face.

> 💡 **Tip**
>
> For a linear grid: if your tile is 150 mm and you want a 4×4 grid of 8 mm circles with 15 mm spacing, you need 4 occurrences across 45 mm (15×3) in each direction. Start your first circle at X=−22.5, Y=−22.5 so the grid is centred on the origin.

5 **Add the border pocket**
Create a new sketch on the top face. Project outer edges. Draw an inner rectangle inset by =Spreadsheet.BorderInset on each side. Close sketch. Pocket: depth =Spreadsheet.BorderPocketDepth.

6 **Optional: add chamfer**
Apply Part Design Chamfer (1 mm) to the top four rim edges for a polished look.

Part B — CAM

7 **CAM Job and Stock**
Switch to CAM. Create Job. Configure Stock: X = TileSize + 4, Y = TileSize + 4, Z = Thickness.

8 **Operation 1 — Pocket Shape for pattern elements**
CAM → Pocket Shape. Select all pattern pocket faces. End Mill 6mm. Strategy: Adaptive. Step-over: 60%. Step-down: 1.5 mm. Final Depth: −5 (or match your PatternPocketDepth). Feed: 1100. Plunge: 320. Click OK.

9 **Operation 2 — Pocket Shape for border**
CAM → Pocket Shape. Select the border pocket face. End Mill 6mm. Strategy: ZigZag (border is shallow — ZigZag is fine). Step-over: 65%. Step-down: 1.5 mm. Final Depth: −3. Feed: 1100. Plunge: 320. Click OK.

10 **Operation 3 — Profile**
CAM → Profile. Select outer square edge. End Mill 6mm. Outside. Final Depth: −12.5. Step Down: 1.5. Tabs: 4, width 4mm, height 3mm. Feed: 1000. Plunge: 300. Click OK.

11 **Simulator and post-process**
Run CAM Simulator. All three operations must appear. Pattern pockets fully cleared with Adaptive toolpath. Border pocket cleared. Profile with tabs. Post-process grbl. Save as PatternedTile_12mm_v1.nc.

Part C — Machine

12 **Pre-cut checklist, connect, home**
Complete all four sections of pre-cut checklist. Connect UGS. Home machine.

13 **6mm end mill in collet, clamp material, zero**
All three operations use the end mill — no tool change needed for this project. Clamp 155×155mm+ stock. Set X/Y/Z zero at material centre.

14 **Load, verify, run**
Load PatternedTile_12mm_v1.nc. Verify 3D visualiser — three operations visible. Send job. Supervise all operations.

15 **Release and finish**
Cut 4 tabs. Sand 120 → 180 → 240 grit. Apply varnish or Danish oil. Photograph for portfolio.

🏆 **Project 4 Complete**

Project 4 demonstrates that CNC is not just about functional parts — it produces genuinely beautiful decorative objects. The pattern you designed is original. The techniques you used (Linear or Polar Pattern, Adaptive pocket clearing) are the same ones professional CNC designers use daily. Well done.

Week 4, Days 16–20 — Lesson Content Overview

Day 16 (Apply)	Linear Pattern fundamentals. Students create a single pocket, then apply Linear Pattern with 5 occurrences. Calculate spacing before applying — verify the result matches the calculation. Extend to a 2D array (3×4 grid). Independent exercise: recreate a given hole pattern (provided as a dimensioned drawing) using the minimum number of Linear Pattern operations.
Day 17 (Apply)	Polar Pattern. Students create a clock-face-like disc with 12 equally spaced markers using Polar Pattern (360°, 12 occurrences). Then create a fan arrangement (Polar Pattern, 120°, 5 occurrences). Discuss: how would you use Polar Pattern to create a gear-like toothed outline? (Preview of Week 7)
Day 18 (Apply → Analyse)	Draft workbench — Array, Move, Scale. Import a provided SVG file. Use Draft Array to tile it 3×2 across a stock sheet. Calculate material waste percentage for the current layout. Rearrange the array (change spacing) to reduce waste. Compare: which layout wasted less? Introduce the Draft Array vs Part Design Pattern comparison.
Day 19 (Analyse)	CAM Pocket Shape — strategy comparison. Students set up the same pocket three times with ZigZag, Adaptive, and Offset strategies. Compare the Simulator output for each — toolpath appearance, estimated run time (shown in UGS after loading the .nc file). Discuss: which would you choose for (a) a shallow decorative pocket in MDF, (b) a 15 mm deep hardwood pocket, (c) a pocket where wall quality is critical?
Day 20 (Apply — Project Day)	Project 4: Decorative Patterned Tile. Full independent project.

Chapter Summary

- Linear Pattern repeats a Part Design feature along an axis. The key formula is: Spacing = Length ÷ (Occurrences − 1). Length is the total span from first to last, not the gap between adjacent features.
- 2D arrays combine two Linear Pattern directions to create a grid. Calculate both spacings before applying to avoid trial-and-error.
- Polar Pattern distributes a feature around a rotation axis. For flat CNC parts on the XY plane, the axis is always Z. Full circle = 360°. Partial arc = any angle < 360°. Construction circles position features at a precise radius before applying the pattern.
- Draft Array works on any FreeCAD object — bodies, imported shapes, ShapeStrings. Use it for tiling multiple complete parts on a stock sheet. Draft Move and Draft Scale provide precise repositioning and resizing of objects.
- Use Part Design Linear/Polar Pattern for features within a single part. Use Draft Array for multiple parts or imported geometry across a stock sheet.
- CAM Pocket Shape has three clearing strategies: ZigZag (fast, simple, best for shallow pockets), Adaptive (constant chip load, best for deep/hard material), and Offset (best for clean walls). Use Adaptive for any pocket deeper than 5 mm.
- Step-down controls the depth increment per pass (≤ 0.5× tool diameter for hardwood). Step-over controls horizontal engagement (60–70% of tool diameter is the standard sweet spot). A finish pass at reduced feed improves surface quality.

Key Vocabulary — Chapter 9

Term	Definition
Linear Pattern	A Part Design tool that repeats an existing feature a specified number of times along a specified axis. Spacing = Length ÷ (Occurrences − 1).
Occurrences	The total number of copies in a Linear or Polar Pattern, including the original feature. A pattern with Occurrences = 5 produces five copies total (the original plus four duplicates).
Length (pattern)	In Linear Pattern: the total distance from the first occurrence to the last occurrence. Not the spacing between adjacent copies.
2D array	A Linear Pattern applied in two directions simultaneously, producing a grid of features. Configured by setting two directions within the Linear Pattern dialog.
Polar Pattern	A Part Design tool that distributes a feature in a circular arrangement around a rotation axis. Configured with Axis, Angle (total arc), and Occurrences.
Draft Array	A Draft workbench tool that creates copies of any FreeCAD object in a rectangular, polar, or path arrangement. Works on complete objects, not just individual Part Design features.
Material nesting	The process of arranging multiple parts on a stock sheet to minimise material waste. Draft Array is the primary tool for manual nesting in this course.
Pocket Shape	A CAM operation that clears all material within an enclosed 2D area to a specified depth. Uses a clearing strategy (ZigZag, Adaptive, or Offset) to plan the tool's path through the area.
ZigZag strategy	A Pocket Shape clearing pattern using alternating back-and-forth straight passes. Fast and simple; best for shallow pockets in soft material.

Adaptive strategy	A Pocket Shape clearing pattern that limits the tool's engagement angle to maintain a constant chip load. Higher feed rates, extended tool life, less heat; best for deep pockets and hardwood.
Offset strategy	A Pocket Shape clearing pattern using concentric passes shrinking inward from the pocket walls. Produces excellent wall quality; used for high-finish decorative pockets.
Step-down	The depth increment for each cutting pass in a multi-pass CAM operation. Should not exceed 0.5× tool diameter for hardwood, 0.75× for MDF.
Step-over	The horizontal distance between adjacent passes in a Pocket Shape operation. Typically 60–70% of tool diameter for standard clearing.
Finish pass	An optional final pass in a Pocket Shape operation at very shallow depth and reduced feed rate. Removes roughing marks and produces a smooth floor and walls.
Construction circle	A circle drawn in the Sketcher and set to construction mode, used as a reference to position features at a precise radius from the origin before applying Polar Pattern.

Looking Ahead — Chapter 10: Importing Artwork & Organic Curves

Chapter 10 introduces B-splines for freeform curve drawing and SVG file import through the Draft workbench. You will learn to source, prepare, and import vector artwork into FreeCAD, combine it with native geometry, and create a CAM Engraving job that traces the imported paths. By the end of Chapter 10 you will have completed Project 5 — the Artistic Knotwork or Custom Artwork Panel.

PART THREE

Core Skills — Parametric & Pattern Design

Weeks 3–5 • Chapters 8 through 10 • Bloom's Levels: Understand → Apply

📖 **About Part Three**

Part Three covers Weeks 3 through 5. The skills introduced here — parametric design, pattern repetition, SVG artwork import, and joinery fundamentals — are the tools that distinguish a versatile CNC designer from someone who can only follow a recipe. Part Three operates at the Understand and Apply levels of Bloom's Taxonomy. You will not just use these tools — you will explain why they work the way they do and choose between them intelligently.

CHAPTER
10
of 24

PART THREE — CORE SKILLS

Importing Artwork

SVG files, B-splines, Draft Import, scaling imported geometry, combining native and imported paths for CAM

Chapter Learning Objectives

By the end of this chapter, you will be able to:

- Explain the difference between a B-spline and a polyline and choose the correct approach for any given CNC design
- Draw a closed B-spline in the Sketcher using control points and shape it to match a reference form
- Apply B-spline constraints to fix endpoint positions and ensure tangent joins between curves
- Prepare an SVG file in Inkscape for import: convert text to paths, simplify nodes, and remove fills
- Import an SVG file into FreeCAD using Draft Import and inspect the imported objects in the Model tree
- Identify which SVG content imports correctly and troubleshoot common import failures
- Scale, move, and rotate imported Draft objects to match required dimensions and position
- Convert Draft Wire and Draft BSpline objects to Part Shapes suitable for CAM operations
- Combine imported SVG geometry with native FreeCAD sketch geometry in a single multi-operation CAM job
- Resolve Z-height, closure, and node density problems that commonly affect imported geometry
- Post-process and verify a combined-source CAM job in UGS

A round bit follows a path. The quality of what it produces is determined entirely by the quality of that path — whether it comes from a rectangle you drew in two minutes or from a Celtic knotwork design prepared in Inkscape. The machine does not know the difference. You do.

The first nine chapters of this course taught you to construct geometry natively inside FreeCAD — rectangles, circles, pockets, patterns, text. These tools are fast, precise, and parametrically linkable. They are the right tools for structural shapes, functional geometry, and anything that needs to be dimensionally controlled.

But not every design starts inside FreeCAD. Some of the most visually compelling CNC work begins as hand-drawn artwork, as vector graphics sourced from a design library, or as a logo prepared in a dedicated illustration tool. Chapter 10 teaches you how to bring that external artwork into FreeCAD and use it alongside your native geometry. It also introduces the B-spline — FreeCAD's tool for drawing smooth organic curves directly in the Sketcher without needing any external software at all.

By the end of this chapter you will be able to take virtually any source of 2D artwork — drawn by hand, imported from the web, created in Inkscape, or built with B-splines — and turn it into a clean engraving toolpath on the CNC machine.

B-Splines — Smooth Freeform Curves in the Sketcher

Most of the geometry drawn in this course has been made from straight lines and circular arcs — predictable, easy to constrain, and easy to dimension. Straight lines and arcs are the correct approach for the vast majority of CNC work. But some design elements — a leaf shape, a fish silhouette, a flowing decorative motif — cannot be described cleanly by lines and arcs. These shapes call for a different type of curve: the B-spline.

> 📖 **Key Term: B-Spline**
>
> A smooth mathematical curve defined by a series of control points. The curve is attracted towards each control point but does not necessarily pass through them (except at the endpoints, if constrained). B-splines produce smooth, continuous curves ideal for CNC toolpaths because they generate smooth machine motion rather than a series of tiny linear segments.

How B-Splines Differ from Other Curve Types

Approach	How It Works	Best For
Polyline	Series of straight line segments connected end to end. Each segment is a true straight line.	Polygons, sharp-cornered profiles, any shape where edges are genuinely straight
Arc by centre	A true circular arc — constant radius, smooth curve.	Rounded corners, fillets, holes, any curve with constant radius
B-Spline	Smooth curve defined by control points — the curve flows toward each point without touching it.	Organic shapes, artistic motifs, any curve whose radius varies continuously

The critical advantage of B-splines for CNC work is the smoothness of the resulting toolpath. When the CAM software reads a B-spline edge, it generates G-code with arc moves (G2/G3 commands) rather than hundreds of tiny G1 linear segments. The machine follows a truly smooth path, which produces cleaner cut surfaces, reduces vibration, and extends tool life. A polyline approximation of a curve generates a rougher toolpath with visible facets — particularly in harder materials.

Drawing a B-Spline in the Sketcher

1. Activate the tool — Sketch → Sketcher geometries → Create B-Spline.
2. Click in the sketch view to place control points. Each click adds one point. The preview curve updates in real time.

3. Place 5–8 control points for your shape. Rough positions are fine — you will refine by dragging.
4. Close the curve by clicking on the first control point. A closed B-spline is essential for Pad and Pocket operations.
5. Right-click or press Escape to deactivate the tool.
6. In selection mode, drag individual control points to pull the curve into the desired shape.

> **Tip**
>
> Control points outside the body of the shape push the curve outward. Control points on or inside the body pull the curve inward. This magnetic analogy helps you intuitively predict how moving a point will change the curve.

B-Spline Degree

Every B-spline has a degree — a mathematical property controlling curve smoothness. The default degree 3 is the standard for virtually all CNC design work.

Degree 1	The curve is a polyline — passes exactly through every control point with straight segments.
Degree 2	A smooth curve that passes closer to control points. Moderate smoothness.
Degree 3 (default)	Smooth curve with excellent shape control. The standard for CNC design work.
Degree 5+	Very smooth curves, less affected by individual control points. Useful for highly organic forms.

Constraining B-Splines

Constrain endpoint position	Fix at least one endpoint with Coincident or a dimensional constraint to anchor the B-spline in the sketch. Without this, the entire curve can drift.
Constrain endpoint tangency	Where a B-spline meets a straight line or another B-spline, a Tangent constraint ensures a smooth join with no kink in the CAM toolpath.
Constrain bounding box	Add construction lines and constrain control points to lie within a defined bounding rectangle to give dimensional control over the overall envelope.

Note

A B-spline does not need to be fully constrained (all green) the way a rectangle or circle must be. The important constraints are: endpoint positions and tangency at joins. Focus on anchoring the shape to the coordinate system — unconstrained internal control points are acceptable.

Try It Now

Open a new Sketch on the XY plane. Draw a closed B-spline with 7 control points roughly resembling a leaf — wide in the middle, pointed at both ends.

Drag the control points one at a time. Moving a point on the 'equator' widens or narrows the body. Moving a point near the tip sharpens or rounds it.

Select the leftmost endpoint and apply a Fix constraint (Block) to anchor it. Now drag other control points — the anchored endpoint stays fixed.

Add a second constraint: fix the right endpoint 80 mm to the right. The leaf's overall length is now defined even though the internal shape remains freeform.

B-Splines vs Polylines — When to Use Which

B-Spline vs Polyline Decision Guide

Use a B-spline when:

- The design is organic — curves that continuously change direction
- The finished piece will be viewed closely and surface smoothness matters
- You are creating artwork for engraving where the flowing curve is the design
- You sourced the design from an illustration and want to reproduce its feel

Use a polyline when:

- The design has genuinely straight edges or circular arcs
- The design is functional — joinery, mounting holes, structural profiles
- Precision dimensioning is required — you need to specify exact sizes
- You are designing parts that assemble with others where tolerances matter

Important

For the structural outline of a CNC part — the profile that the machine cuts along — always use geometric shapes with constraints. Reserve B-splines for decorative elements: engraving paths, inlay designs, artistic borders. Using a B-spline for a functional part's outer profile makes dimensional verification difficult.

SVG Files — Vector Art for CNC

Scalable Vector Graphics (SVG) is the dominant format for 2D vector artwork on the web and in design software. It stores geometric shapes as mathematical descriptions — paths, curves, and lines — rather than as grids of coloured pixels. Because SVG content is geometric, it can be scaled to any size without quality loss and can be interpreted as toolpaths by FreeCAD's CAM workbench.

📖 **Key Term: SVG (Scalable Vector Graphics)**

A vector image format that stores artwork as mathematical path descriptions — lines, curves, and shapes defined by coordinates — rather than pixel grids. SVG files are resolution-independent and can be edited in Inkscape, Adobe Illustrator, or any modern web browser. For CNC work, only the path geometry is relevant — fills, colours, gradients, and effects are invisible to FreeCAD.

What an SVG File Contains — What Imports and What Does Not

SVG Content Type	Imports to FreeCAD?	CNC Use
Closed paths (outlines, shapes)	✅ Yes — Draft Wire / Draft BSpline	*Profile cuts, engraving outlines, pocket boundaries*
Open paths (lines, curves)	✅ Yes — Draft Wire	*Engraving paths, decorative line work*
Text objects	❌ No — convert to paths first	*N/A until converted in Inkscape*
Fills (colour areas)	❌ Ignored — outlines only	*N/A*
Gradients and effects	❌ Ignored by FreeCAD importer	*N/A*
Raster images embedded in SVG	❌ Cannot import — needs path	*N/A*
Groups	⚠️ May import partially — ungroup	*Best to ungroup in Inkscape first*

💡 **Tip**

The single most important rule for SVG import into FreeCAD: only path outlines matter. Fill colours and visual styles are completely irrelevant to the CNC machine. When evaluating an SVG for CNC use, ignore how it looks on screen and focus entirely on whether the path outlines are clean, closed (where required), and free of excess nodes.

Inkscape — Preparing SVG Files for Import

Inkscape is a free, open-source vector graphics editor available at inkscape.org. It is the standard tool for preparing SVG artwork for FreeCAD import. You need to know four specific operations that address the most common import problems.

Operation 1 — Convert Text to Paths

If your SVG contains text, that text is stored as a text object — not as a geometric path. FreeCAD cannot import text objects. Convert the text to a path in Inkscape before saving.

1. Open the SVG in Inkscape.
2. Select all text: Ctrl+A, or click the text object directly.
3. Convert: Path → Object to Path (Shift+Ctrl+C).

4. Verify: selected items should now show as 'Path' not 'Text' in the status bar.
5. Save as Plain SVG (File → Save a Copy → Plain SVG format).

> ⚠ **Important**
> Once text is converted to paths it cannot be edited as text again. Always keep a backup copy of the original SVG with text objects intact before converting.

Operation 2 — Simplify Nodes

Complex SVG artwork — particularly files created by tracing a raster image — can contain far more node points than necessary. Excessive nodes slow down FreeCAD's CAM calculation and produce jagged toolpaths with thousands of tiny linear segments.

1. Select the path you want to simplify. Press N to open the Node editor.
2. Press Ctrl+A to select all nodes on the path.
3. Path → Simplify (Ctrl+L). Inkscape reduces the node count while maintaining the shape.
4. Repeat if needed. Check the node count in the status bar — aim for fewer than 100 nodes per path for CNC work.
5. Zoom in to verify the simplified shape still looks acceptable before saving.

Operation 3 — Ungroup All Objects

SVG artwork often contains grouped objects. FreeCAD handles groups inconsistently — some import correctly, others partially, and some fail entirely. Ungrouping before import is the safest approach: press Ctrl+A to select all, then Object → Ungroup (Shift+Ctrl+G). Repeat several times until the status bar shows 'path' rather than 'group' for all selected objects.

Operation 4 — Save as Plain SVG

Inkscape's default save format is 'Inkscape SVG', which adds Inkscape-specific metadata that can cause import issues. Before importing into FreeCAD, always save a copy as Plain SVG: File → Save a Copy → change file type to Plain SVG. Use a descriptive name that makes it clear this is the prepared version — for example, 'Knotwork_ForImport.svg'.

Draft Import — Bringing SVG into FreeCAD

FreeCAD imports SVG files through the Draft workbench's Import tool. The imported objects appear as Draft objects in the Model tree — Draft Wire (for straight-edged paths) or Draft BSpline (for curved paths). These are temporary forms that must be converted before they can be used by the CAM workbench.

1. Switch to the Draft workbench.
2. Go to File → Import. In the file dialog, navigate to your Plain SVG file. Ensure the file type filter includes SVG and click Open.

3. FreeCAD will ask whether to import as SVG or as Shapes. Select 'Import as Shapes'. This imports the path geometry as usable 3D objects.
4. After import, new objects appear in the Model tree — typically named Wire, Wire001, BSpline, BSpline001, etc. Each corresponds to one path in the SVG file.
5. The imported shapes appear in the 3D view, typically lying flat in the XY plane. They may be very small or very large depending on the SVG file's unit settings — rescaling is covered in the next section.
6. Count the imported objects in the Model tree and compare to the number of paths you prepared in Inkscape. If they do not match, at least one path failed to import.

Import Troubleshooting

Symptom	Likely Cause	Fix
Nothing imports — error	SVG saved as Inkscape SVG with incompatible elements, or file contains only raster content	Re-save as Plain SVG in Inkscape. Remove any embedded rasters. Verify the SVG contains at least one path element.
Text objects did not import	Text objects are not paths — FreeCAD cannot import them	Convert text to path in Inkscape: Path → Object to Path. Re-export and re-import.
Only some paths imported	Some paths are open or have zero area	In Inkscape, check all paths are properly open or closed. Remove zero-area paths.
Shapes appear tiny (1–5 mm)	SVG saved in pixels — FreeCAD interprets 1 pixel as 0.3528 mm (96 dpi)	Normal result — rescale using Draft Scale after import.
Shapes appear very large (500+ mm)	SVG saved in mm at A4 or larger document size	Rescale using Draft Scale to target size.
Curved paths import as line segments	Too many nodes, or path type was Polyline not BSpline	Apply Path → Simplify in Inkscape before import. Use smooth node handles.
Shapes at wrong Z height	Draft Import sometimes places objects at Z > 0	Select all imported objects. Use Draft Move with Z delta = −(current Z) to return to Z=0.

> 📋 **Note**
> The most reliable way to verify a complete import is to count paths in Inkscape before exporting and compare that count to the number of objects in the FreeCAD Model tree. If the counts do not match, open the SVG in Inkscape, inspect each path individually, and re-prepare.

Scaling, Positioning, and Rotating Imported Objects

Imported SVG geometry almost never arrives at the correct size and position. SVG files use a mix of units (pixels, millimetres, points), and the relationship between SVG units and FreeCAD's millimetre coordinate system varies depending on how the file was saved. Rescaling is almost always required.

Measuring and Calculating the Scale Factor

Before rescaling, measure the current dimensions of the imported object. Select it in the Model tree and check the Properties panel → Data tab for the bounding box values, or use Draft → Dimension to measure directly in the 3D view. Then calculate:

📐 **Scale Factor Formula**

Scale Factor = Target Size ÷ Current Size

Example: imported shape is 523 mm wide. Target size is 150 mm. Scale factor = 150 ÷ 523 = 0.287

If you know the SVG was saved in pixels at 96 dpi (web standard): SVG pixels × (25.4 ÷ 96) = mm. A 500-pixel-wide SVG imports at approximately 132 mm.

Applying Draft Scale

1. Select all imported Draft objects in the Model tree.
2. Go to Draft → Scale. The Scale dialog opens.
3. Enter your calculated scale factor in the X and Y fields. Set Z to 1.0 — never scale the Z axis for flat imported shapes.
4. Click 'Pick' and click the origin (0,0,0) as the reference point. This keeps the object near its current position.
5. Click OK. Verify by measuring again.

Draft Move — Positioning at the Correct Location

After scaling, the object may be anywhere in the coordinate space. Use Draft Move to reposition it precisely. Go to Draft → Move. Click your reference point on the imported shape, then enter the target coordinates (typically 0,0,0) in the dialog. The object moves to the new position.

💡 **Tip**

For decorative CNC work, centring the imported artwork on the origin is often most convenient. Measure the object's width and height, then use Draft Move with a target point of (−width/2, −height/2, 0). The object's centre will be exactly at the origin, making it easy to overlay with a native body also centred there.

Draft Rotate

If an imported SVG arrives rotated, use Draft → Rotate. Select the imported objects, click the centre of rotation in the 3D view, then enter the rotation angle in degrees. Positive values rotate anticlockwise, negative values rotate clockwise.

Converting Draft Objects to Part Shapes — Draft Upgrade

This is the step most beginners miss — and the one that makes imported geometry usable in CAM. FreeCAD's CAM workbench requires Part-level geometry. Draft objects (Draft Wire, Draft BSpline) are not Part objects, and the CAM workbench cannot see their edges for toolpath generation.

> 📖 **Key Term: Draft Upgrade**
>
> A Draft workbench operation that converts Draft objects (Wire, BSpline) into Part workbench objects (Part Wire, Part Face, Part Shell). This conversion is required before imported SVG geometry can be used in CAM operations. Found at Draft → Upgrade.

1. Select all Draft Wire and Draft BSpline objects in the Model tree that you want to use in CAM.
2. Go to Draft → Upgrade. FreeCAD processes the selected objects.
3. New objects appear in the Model tree — typically named Wire, Wire001, Face, Face001. These are Part-level objects.
4. Hide the original Draft objects (Space bar). You now work exclusively with the converted Part objects.
5. Switch to the CAM workbench. When creating an Engraving operation, you should now be able to select the edges of the converted Part objects as base geometry.

Input Path Type	Draft Upgrade Output
Closed path (loop)	Part Face — a flat filled region. Usable in CAM Pocket Shape and Engraving.
Open path (line with two endpoints)	Part Wire — a single edge. Usable in CAM Engraving operations.
Multiple closed paths, same plane	Part Shell or compound. If CAM fails, convert each path individually.

> ⚠ **Important**
>
> If Draft Upgrade fails or produces an unexpected result, the most common cause is a path that appears closed but has a tiny gap between start and end points — common in SVG files traced from raster images. In Inkscape, open the Node editor (N), select both the start and end nodes, and use Path → Join nodes (Shift+J) to force the path closed. Re-export and re-import.

Combining Imported and Native Geometry in CAM

The most useful application of SVG import is combining external artwork with native FreeCAD geometry in a single CNC job. The structural shape is parametric FreeCAD geometry. The decorative element is imported. The CAM job brings both together.

The structural shape — the overall panel profile, pockets, and functional features — is designed natively in FreeCAD where it can be fully parametric and dimensionally controlled. The decorative element — the engraved knotwork, the custom logo, the B-spline leaf — is imported from an SVG file or drawn as a B-spline. The CAM job brings both together into a single toolpath sequence.

The Recommended Workflow — Follow in Order

Phase 1 — Native design	Build the structural shape in Part Design: sketch, constrain, Pad, Pocket. This is the physical body of the piece.
Phase 2 — Import artwork	Import the SVG, scale to the correct size, move to the correct position over the native body.
Phase 3 — Convert	Apply Draft Upgrade to convert all imported Draft objects to Part objects.
Phase 4 — Z-height check	CRITICAL: Verify all imported Part objects are at Z=0. Objects at any other Z height produce incorrect toolpaths.
Phase 5 — CAM Job	Create the CAM Job referencing the native body. Add all operations: Engraving for imported paths, Pocket for native pockets, Profile for the outer edge.
Phase 6 — Simulate	Run the CAM Simulator. Verify every imported path appears in the engraving toolpath. Check depths and tabs.
Phase 7 — Post-process	Post-process to GRBL, load in UGS, verify the visualiser shows all expected toolpaths before cutting.

The Z-Height Problem — Critical Check After Every Import

The most common CAM failure with imported geometry is a Z-height mismatch. SVG files are 2D documents — they contain no Z-axis information. FreeCAD places imported objects at Z=0 by default, but various operations including Draft Upgrade can sometimes change the Z position. Even Z=0.001 mm causes the CAM Engraving operation to behave incorrectly, because it sees the engraving edge as fractionally above the top face of the model.

After converting imported objects, always perform a Z-height check: select each imported Part object, go to the Properties panel → Data tab, and check Placement → Position → z. It must be exactly 0. If it is not, use Draft → Move with a Z delta equal to the negative of the current Z value. For example, if z = 0.15, move by (0, 0, −0.15).

Common Combined-Geometry CAM Issues

Problem	Symptom	Solution
Imported path not closed	Engraving only covers part of the outline — tool lifts mid-path	In Inkscape, open Node editor, select start and end nodes, use Path → Join nodes. Re-export, re-import.
Too many nodes	CAM simulation runs very slowly or FreeCAD becomes unresponsive	Apply Path → Simplify in Inkscape before import. Aim for fewer than 100 nodes per path.
Imported edge not selectable in CAM	Clicking the shape in the Engraving dialog does not select it	Draft Upgrade was not performed, or produced a compound object. Re-upgrade each path individually.
Engraving path at wrong depth	Simulator shows engraving at correct depth on some paths, Z=0 rapid on others	Z-height mismatch. Check all imported objects are at exactly Z=0 after Draft Upgrade.
Profile ignores imported shapes	The Profile CAM operation does not see the imported outline	Profile operations only work with Part Design Body edges. Use the native body outline for Profile cuts.

PROJECT 5 — STEP-BY-STEP WALKTHROUGH

Artistic Knotwork or Custom Artwork Panel

An engraved decorative panel combining imported SVG artwork with native FreeCAD geometry — Week 5 Project Day

Project 5 is the first project that requires you to source and prepare your own design material before project day. During Days 21–24, find and prepare an SVG file you want to engrave. Good sources include SVGRepo.com, publicly available Celtic knotwork collections, animal silhouettes, botanical illustrations, or decorative border designs. Avoid complex multi-layered illustrations — a design with 2–5 distinct paths is ideal for a first combined-geometry project.

Design	Decorative panel combining at least one imported SVG element with at least one natively designed feature
Size	Student's choice: 150 × 150 mm or 200 × 100 mm — set in Spreadsheet
Material	15–18 mm hardwood (pine, cedar) — the extra depth makes relief engraving more visually striking
Operations	V-bit Engraving (imported paths + native text), Pocket (border recess), Profile (outer panel)
Key req.	At least one element from an SVG import AND at least one natively drawn structural feature
Day	Day 25 of Week 5 — Project Day

Part A — Spreadsheet and Native Body

1

New file and Spreadsheet

File → New. Save as ArtworkPanel.FCStd. Switch to Spreadsheet workbench. Create Spreadsheet with aliases: PanelWidth = 150, PanelHeight = 150, Thickness = 15, BorderInset = 12, BorderDepth = 3, EngravingDepth = 0.8.

2

Create the panel body

Switch to Part Design. New Body → New Sketch on XY plane. Draw a centred rectangle: Width = =Spreadsheet.PanelWidth, Height = =Spreadsheet.PanelHeight. Fully constrain. Close sketch. Pad → =Spreadsheet.Thickness.

3

Add the border pocket

New sketch on the top face of the panel. Draw two concentric rectangles — outer matching the panel edge, inner inset by =Spreadsheet.BorderInset on each side. Constrain fully. Close sketch. Pocket → Depth = =Spreadsheet.BorderDepth. This creates the recessed border frame.

4

Optional native text

If adding native text: Part → ShapeString. Enter text, select a bold sans-serif font. Height linked to spreadsheet. Position centred in the recessed panel area.

Part B — Import, Scale, and Convert

5

Import SVG

Switch to Draft workbench. File → Import → navigate to your prepared Plain SVG → Import as Shapes. Inspect the Model tree — verify the expected number of path objects appeared.

6

Measure current size

Select an imported Wire object. Note current dimensions from the Properties panel or use Draft Dimension.

7

Calculate and apply scale factor

Target size = approximately 120 mm (leaving 12 mm border each side of 150 mm panel). Scale factor = 120 ÷ (current measured width). Draft → Scale. Enter scale factor for X and Y, set Z = 1.0. Reference point: origin.

8

Draft Move to centre

Select all imported objects. Draft → Move. Position the artwork centred over the panel. If both panel and SVG were centred on the origin, target (0, 0, 0).

9

Z-height check

Select each imported object. Verify Data → Placement → Position → z = 0.000. If any object is not at exactly Z=0, correct it with Draft Move (Z delta = negative of current Z value).

10

Draft Upgrade

Select all imported Draft objects. Draft → Upgrade. New Part objects appear in the Model tree. Hide the original Draft objects — work exclusively with the converted Part objects from this point.

Part C — CAM Setup

11 **Create CAM Job**
CAM → Job. Select the native panel body as the model. Configure Stock: X = PanelWidth + 4, Y = PanelHeight + 4, Z = Thickness.

12 **Tool Library**
Add two tools: 60° V-bit (for engraving all imported paths and text) and 6mm end mill (for border pocket and profile). Enter feed rates and step-down appropriate for hardwood.

13 **Operation 1 — V-bit Engraving**
CAM → Engraving. Select all imported Part edges and any native ShapeString edges. V-Bit 60°. Final Depth = −0.8 mm. Feed: 600 mm/min. Plunge: 150 mm/min.

14 **Operation 2 — Pocket Shape**
CAM → Pocket Shape. Select the border pocket face. 6mm end mill. Final Depth = −3 mm. Strategy: Adaptive. Step-over: 65%. Step-down: 1.0 mm. Finish pass: 0.1 mm.

15 **Operation 3 — Profile**
CAM → Profile. Select the outer panel edge. 6mm end mill. Final Depth = −(Thickness + 0.5) = −15.5 mm. Step-down: 1.5 mm. Outside. Tabs: 4, width 4mm, height 3mm.

16 **Verify operation order**
Operations in the CAM tree must be: 1 = Engraving (V-bit) → 2 = Pocket (6mm EM) → 3 = Profile (6mm EM). Never run Profile before all pocket and engraving operations are complete.

Part D — Simulate, Post-Process, and Cut

17 **Run CAM Simulator**
CAM → Simulator. Verify: all imported paths appear in the engraving toolpath, the border pocket is fully cleared, profile includes tabs, depths correct. Look for any engraving moves showing as rapid moves (yellow) — if present, a Z-height error exists. Exit CAM, correct the Z position of the affected object, re-run.

18 **Post-process**
CAM → Post Processor → GRBL. Save as ArtworkPanel_15mm_v1.nc.

19 **Load and verify in UGS**
Open UGS. Connect. File → Open → select .nc file. Open the 3D Visualiser. Verify three distinct toolpath zones are visible: fine engraving paths, pocket clearing paths, and outer profile path.

20 **Pre-cut checklist and zero**
Complete the pre-cut checklist. Set X/Y/Z zero. V-bit installed for the first operation (engraving).

21 **Run the job**

Send the job. After the engraving operation, UGS will pause for tool change. Install 6mm end mill, re-zero Z from the stock surface, resume for pocket and profile operations.

22 **Evaluate**

Collect the finished piece. Is the imported artwork clearly engraved? Does the border recess have clean walls and flat floor? Does the profile edge cut cleanly? Note anything you would do differently.

🏆 **Project 5 Complete**

Project 5 is the first time the machine cuts artwork that did not originate inside FreeCAD. If the engraving follows the imported paths cleanly, every step of the import and conversion workflow was correct — from Inkscape preparation through Draft Upgrade to Z-height verification. That chain of steps is now part of your permanent workflow.

Week 5, Days 21–25 — Lesson Content Overview

Day 21 (Apply)	B-Splines in Sketcher. Students draw a closed B-spline leaf shape, experiment with control point placement, add and remove knots for local detail control, and apply endpoint and tangency constraints. Comparison exercise: draw the same organic shape as a polyline vs B-spline and compare the CAM toolpath quality in the Simulator.
Day 22 (Analyse)	SVG files and Draft Import. Students import a provided SVG with a deliberate problem (text not converted), observe the partial import, fix the issue in Inkscape, re-import successfully. Discussion: what makes an SVG 'CNC-ready'? Students source their own SVG for Project 5 and complete Inkscape preparation.
Day 23 (Analyse)	Scaling, positioning, and Draft Upgrade. Students import the prepared class SVG, calculate the scale factor, apply Draft Scale and Draft Move, perform the Z-height check, and apply Draft Upgrade. The session ends with a combined model (native body + converted Part objects) ready for CAM.
Day 24 (Analyse)	Combined-source CAM. Students set up the full 3-operation job: Engraving (imported edges) + Pocket (native recess) + Profile (native edge). Deliberately introduce a Z-height error on one object — observe the Simulator symptom and apply the fix. Post-process and verify in UGS.
Day 25 (Apply — Project Day)	Project 5: Artistic Knotwork or Custom Artwork Panel. Full independent project using the student's own SVG source material. See walkthrough above.

Chapter Summary

- B-splines are smooth mathematical curves defined by control points. They are ideal for organic CNC artwork — leaf shapes, flowing motifs, decorative borders — where varying-radius curves are required. B-splines produce arc-based G-code that gives smoother machine motion than polyline approximations.
- SVG files store artwork as path descriptions, not pixels. For CNC work only path outlines matter — fills, colours, gradients, and effects are invisible to FreeCAD's importer. Closed paths become Draft Wire or Draft BSpline objects; text and raster content cannot be imported.
- Inkscape preparation is always required before importing SVG files. The four essential steps are: convert text to paths, simplify nodes (under 100 per path for CNC), ungroup all objects, and save as Plain SVG (not Inkscape SVG).
- Draft Import (File → Import in the Draft workbench) brings SVG paths into FreeCAD as Draft Wire or Draft BSpline objects. These are temporary forms — they must be converted before the CAM workbench can use them.
- Draft Scale, Move, and Rotate are used to bring imported objects to the correct size, position, and orientation. The scale factor formula is: Scale Factor = Target Size ÷ Current Size.
- Draft Upgrade converts Draft objects to Part-level objects (Part Wire, Part Face) that the CAM workbench can read. This step is required for every import workflow — without it, imported shapes cannot be selected as CAM base geometry.
- The Z-height check is the most critical step after Draft Upgrade. All imported Part objects must be at exactly Z=0. Even a 0.001 mm Z offset causes incorrect toolpaths. Check Placement → Position → z in the Properties panel for every imported object.
- Combined CAM jobs treat native body geometry (Profile, Pocket) and imported Part geometry (Engraving) as separate sources in the same Job. Operation order is always: Engraving → Pocket → Profile. Profile is always the final operation.

Key Vocabulary — Chapter 10

Term	Definition
B-Spline	A smooth mathematical curve defined by control points. The curve flows towards each point without necessarily touching it. Produces smooth arc-based G-code in CAM, ideal for organic artwork and decorative engraving paths.
Control point	A positional handle that defines the shape of a B-spline. Dragging a control point pulls the curve towards it. Control points outside the curve body push it outward; points inside pull it inward.
B-Spline degree	A mathematical property controlling the smoothness and shape influence of a B-spline. Degree 3 (the default) is the standard for CNC design work. Higher degree = smoother but harder to control.
SVG	Scalable Vector Graphics — a vector image format storing artwork as mathematical path descriptions. Resolution-independent and directly importable into FreeCAD for use as CAM geometry.
Inkscape	A free, open-source vector graphics editor used to prepare SVG files for FreeCAD import. Available at inkscape.org. Required for: converting text to paths, simplifying nodes, ungrouping, and saving as Plain SVG.

Plain SVG	An SVG file saved without Inkscape's proprietary metadata extensions. Required for reliable import into FreeCAD. Save via File → Save a Copy → Plain SVG format.
Draft Import	The FreeCAD tool (accessed via File → Import in the Draft workbench) used to bring SVG path geometry into FreeCAD as Draft Wire and Draft BSpline objects.
Draft Wire	A Draft workbench object created when a straight-edged or polyline SVG path is imported. Must be converted to a Part object via Draft Upgrade before use in CAM.
Draft BSpline	A Draft workbench object created when a curved SVG path is imported. Must be converted to a Part object via Draft Upgrade before use in CAM.
Scale factor	The multiplier applied to imported objects to bring them to the correct size. Calculated as: Target Size ÷ Current Size. Applied using Draft → Scale.
Draft Upgrade	The Draft workbench operation that converts Draft Wire and Draft BSpline objects into Part-level objects (Part Wire, Part Face). Required before imported geometry can be used in the CAM workbench.
Z-height check	A verification step performed after Draft Upgrade. All imported Part objects must be at exactly Z=0. Any Z offset — even 0.001 mm — causes incorrect toolpaths in the CAM Engraving operation.
Combined geometry	A FreeCAD design and CAM job that uses both native Part Design body geometry (for structural features) and imported Part objects (for decorative engraving) within the same file and CAM Job.
Node (SVG)	A control point on an SVG path. Complex files can contain hundreds of unnecessary nodes that slow CAM calculation. Simplified with Path → Simplify in Inkscape before import.

Looking Ahead — Chapter 11: Multi-Body Design & CNC Joinery Fundamentals

Chapter 11 moves into Part Four and introduces multi-body FreeCAD files, tolerance design, dogbone fillets, and the CNC joinery skills needed for the Two-Panel Finger Joint Box project. The chapter introduces the clearance formula that ensures finger joints actually fit together, and covers the pockets-before-profiles CAM rule that prevents parts from shifting mid-job.

PART FOUR

Intermediate Skills — Curves, Joinery & Addons

Weeks 5–7 • Chapters 11 through 14 • Bloom's Levels: Apply → Analyse → Evaluate

📖 **About Part Four**

Part Four moves you from competent beginner to versatile intermediate designer. These three weeks introduce the skills that separate someone who can follow instructions from someone who can design independently: organic curve drawing, SVG artwork import, CNC joinery with correct tolerances, gear and mechanical part generation, and Boolean solid operations.

The Bloom's levels climb steeply here. You are not just using tools — you are comparing strategies, analysing why joints fail, and making justified design decisions. By the end of Day 35 (Week 7) you will have produced a working gear pair, a correctly fitting finger joint box, and an artistically engraved decorative panel.

CHAPTER
11
of 24

PART FOUR — INTERMEDIATE SKILLS

Multi-Body Design & CNC Joinery Fundamentals

Managing multiple bodies, tolerance design, dogbone fillets, multi-part CAM, and the Finger Joint Box

Chapter Learning Objectives

By the end of this chapter, you will be able to:

- Create and manage multiple Part Design Bodies within a single FreeCAD file
- Rename, show, and hide individual bodies for efficient model tree management
- Share a single Spreadsheet across multiple bodies using the Clearance cell strategy
- Explain the difference between nominal dimension and actual machined dimension
- Design joints with press, snug, hand, and sliding fit using the clearance formula
- Draw dogbone fillets at inside corners of slot profiles and calculate the correct diameter
- Describe the T-bone fillet as an alternative to dogbone for specific geometries
- Set up a multi-part CAM job that nests two panels on one stock sheet
- Apply the pockets-before-profiles rule to a multi-operation job
- Cut, assemble, and evaluate the fit quality of a two-panel finger joint box

Why Joinery Is the Ultimate CNC Test

Engraving tests your ability to produce good toolpaths. Joinery tests your entire design workflow — because a joint either fits or it does not, and if it does not fit, there is a specific reason you can find, diagnose, and fix.

CNC joinery — the use of a CNC router to cut interlocking joints like finger joints, box joints, and mortise-and-tenon joints — is one of the most rewarding and technically demanding things you can do on the CNC 4040 Pro Max. It rewards precision in every step of the workflow: precise sketching, correct clearance values, accurate dogbone fillets, correct operation ordering, and well-executed machine zeroing.

It also reveals errors with immediate physical feedback. A joint that is 0.3 mm too tight will not assemble. A joint that is 0.5 mm too loose will rattle. There is no ambiguity. This directness makes joinery one of the most valuable learning contexts in the course — every mistake has a clear symptom, and diagnosing the symptom teaches you something permanent.

This chapter covers all the tools you need for CNC joinery: multi-body files for managing several panels in one FreeCAD file, tolerance design for controlling fit, dogbone fillets for correcting the round-bit/square-corner problem, and the CAM and machine workflows specific to multi-part joinery jobs.

Multi-Body Files — One File, Many Parts

A multi-body FreeCAD file contains more than one Part Design Body within a single .FCStd document. This is the standard approach for any joinery project — all panels, pieces, and components live in one file, share one Spreadsheet, and can be updated together with a single Refresh. Keeping joinery components in separate files creates synchronisation problems: if you change the finger width in one file, you must remember to update every other file. With multi-body design, you change it once.

Creating a Second Body

1. Open your FreeCAD document. You should already have a Body containing your first panel.
2. Right-click in the Model tree on an empty area (not on any feature). Select Add a Body, or go to Part Design → Body.
3. A second Body appears in the Model tree alongside the first. Both bodies are inside the same Document.
4. The new Body is automatically 'active' — any new sketches or features will be added to it. Click the Body you want to be active to make it active (it shows in bold in the Model tree in v1.1.0).

Renaming Bodies for Clarity

As soon as you create a second body, rename both of them. Working with 'Body' and 'Body001' is confusing; working with 'FrontPanel' and 'SidePanel' is not.

1. Right-click a Body in the Model tree.
2. Select Rename from the context menu.
3. Type a descriptive name: FrontPanel, SidePanel, LidPanel, BasePlate — whatever describes the part clearly.
4. Press Enter. The name updates immediately.

Showing, Hiding, and Working on Individual Bodies

Hiding a Body (Space bar)	Click the Body in the Model tree, then press Space bar (or click the eye icon). The body disappears from the 3D view. This is essential when working on one panel — hiding the others reduces visual clutter and makes it easier to select faces and edges on the active part.
Showing a Body	Press Space bar again (or click the eye icon) to make the body visible again. Show all bodies when doing a visual fit check.
Making a Body active	Double-click a Body in the Model tree to make it active. All new sketches and features will be added to the active body until you activate another.
Visual fit check	Show all bodies. Use the Properties panel → Data → Placement to move bodies relative to each other for a visual check. Set the Placement X, Y, Z values to position the panels as they would sit when assembled. This is not a functional assembly — it is a visual sanity check before cutting.

Sharing a Spreadsheet Across Multiple Bodies

The most powerful aspect of multi-body design is that all bodies in the same FreeCAD file share access to the same Spreadsheet object. This means one set of parameters drives every panel in the project simultaneously.

A joinery spreadsheet for a simple box might look like this:

BoxWidth	150 mm — outer width of the box
BoxHeight	100 mm — outer height of the box
BoxDepth	80 mm — outer depth of the box
PanelThickness	9 mm — material thickness
FingerWidth	9 mm — width of each finger (ideally matches material thickness)
Clearance	0.15 mm — the gap added to slots to allow tab insertion
FingerDepth	9 mm — how deep the fingers extend (= PanelThickness, for full-depth joints)

The FrontPanel Body sketches reference =Spreadsheet.BoxWidth and =Spreadsheet.PanelThickness. The SidePanel sketches reference =Spreadsheet.BoxDepth and =Spreadsheet.PanelThickness. The Clearance alias is used in every joint slot calculation. Change PanelThickness from 9 to 12 — one Refresh updates every panel, every joint, every dimension consistently.

> **Tip**
> Set up the Spreadsheet first and populate every alias before creating any Body sketches. Linking to an alias that does not yet exist causes an error. Build your parameter foundation before your geometry.

Tolerance Design — Engineering the Right Fit

A CNC machine can cut to 0.1 mm accuracy. A good joint design uses that accuracy deliberately — building the correct amount of clearance into every mating surface.

Tolerance is the designed gap between two mating surfaces. In manual woodworking, this gap is achieved by fitting and trimming until the joint is right. In CNC woodworking, the gap is calculated in advance and built into the model. The machine cuts it exactly right — every time, for every piece — if the tolerance value is correct.

The gap is always applied to the slot (the hole), never to the tab (the protruding piece). The tab is cut at nominal size — the exact dimension you specify. The slot is cut at nominal + 2×Clearance — slightly larger on both sides to allow the tab to enter.

The Clearance Formula

```
Slot Width  =  Tab Width + (2 × Clearance)
```

The slot is wider than the tab by one Clearance value on each side (two sides total). This gives the tab room to enter the slot without requiring force. The Clearance value is stored in the Spreadsheet and applied to all slot dimensions. The tab is always the nominal size; the slot is always nominal + 2×Clearance.

Example: Tab Width = 9 mm, Clearance = 0.15 mm

```
Slot Width = 9 + (2 × 0.15) = 9.30 mm
```

The Five Fit Types — Choosing the Right Clearance

The correct clearance depends on the application. A box joint that will be glued needs to be snug (the glue fills any remaining gap). A joint that must be repeatedly disassembled needs to slide freely. The table below describes all five fit types and their appropriate clearance values.

Fit Type	Clearance per Side	Description & CNC Use
Press fit	**0.00–0.10 mm**	Two parts lock together permanently under hand pressure — cannot be separated without tools. No gap visible.
Snug fit	**0.10–0.20 mm**	Parts slide together with gentle hand pressure and hold securely without adhesive. The ideal for finger joints that will be glued.
Hand fit	**0.20–0.30 mm**	Parts slide together easily by hand with slight play. Suitable for joints that must be assembled and disassembled repeatedly.
Sliding fit	**0.30–0.50 mm**	Visible gap — parts slide together freely. Suitable for adjustable components or where the joint will be filled with glue for strength.
Loose fit	**>0.50 mm**	Too much gap — the joint wobbles. Usually an error in the clearance setting or a machine calibration issue.

Starting Clearance Values for CNC 4040 Pro Max in MDF/Timber

Every machine is slightly different. The values below are reliable starting points, but you should always cut a test joint in scrap material before committing to your final stock. Adjust the Clearance alias in the spreadsheet based on the test result.

MDF (9–12 mm) — first attempt	Clearance = 0.15 mm (snug fit). Test the joint before cutting final panels.
Pine / softwood — first attempt	Clearance = 0.20 mm. Softwood fibres compress slightly under pressure — needs slightly more clearance than MDF.
Hardwood — first attempt	Clearance = 0.25 mm. Hardwood does not compress. A tighter joint will not go together at all.
Plywood / birch ply	Clearance = 0.15–0.20 mm. The ply face grain may close the joint slightly — start at 0.20 mm.

If test joint is too tight	Increase Clearance by 0.05–0.10 mm. Refresh model. Re-export G-code. Cut another test.
If test joint is too loose	Decrease Clearance by 0.05 mm. Refresh. Re-export. Cut test. (Rarely needed — tight is more common than loose.)

📖 **Key Term: Nominal dimension**

The designed, theoretical size of a feature — the number you type into your sketch constraint. The actual machined dimension will vary slightly from nominal due to tool deflection, feed rate, and material properties. Tolerance design accounts for this variation.

Try It Now

In a new sketch, draw two rectangles side by side representing a tab (9 mm wide) and a slot ($9 + 2\times0.15 = 9.30$ mm wide).

Close the sketch. Do not extrude — just observe the dimensions.

Change the Clearance in your spreadsheet from 0.15 to 0.30. Refresh.

The slot width updates to 9.60 mm. The tab remains 9 mm.

This visual exercise makes the clearance formula tangible before applying it to real joinery geometry.

Dogbone Fillets — Solving the Round-Bit Problem

A round bit cannot cut a square inside corner. It always leaves a curved radius. Dogbone fillets solve this — they are small circular cutouts placed at inside corners to give the mating part room to seat fully.

This is one of the most practically important concepts in CNC joinery. It is also the one most beginners overlook until they find their carefully designed joints do not assemble — because the tabs physically cannot enter the slots due to the radius left by the bit in each corner.

❌
Slot (viewed from above)

```
┌─────────┐
│         │   ← Slot walls
└──⌒──┘    │  ← Round corner!
           │    from bit
```

The mating tab cannot seat fully because the rounded corner prevents full contact.

☑
Slot with dogbone (viewed from above)

```
┌─────────┐
└o───────o┘   ← Circular
  overcuts at      relief cuts
```

The mating tab now seats fully because the circular relief clears the corner.

How Dogbones Work

A dogbone fillet is a circle placed at each inside corner of a slot profile. The circle's diameter equals the cutting bit's diameter. When the CNC machines the slot, it cuts the slot profile normally — and at each corner, it also cuts the circular fillet. The result is a slot with small semicircular extensions at each inside corner. These extensions (the 'bones' of the dogbone) provide enough clearance for a square-cornered tab to seat fully.

The name 'dogbone' comes from the shape of the modified corner when viewed from above: a rectangular slot with a circular bulge at each end corner looks like a cartoon dog's bone.

> 📖 **Key Term: Dogbone fillet**
> A small circular cutout placed at each inside corner of a CNC-cut slot or joint profile, with diameter equal to the cutting bit diameter. It compensates for the bit's radius that would otherwise prevent a square-cornered mating part from seating fully.

Drawing a Dogbone Fillet in the Sketcher

1. In the Sketcher, draw your slot profile as normal — a rectangle or other profile with straight sides.
2. Identify each inside corner — the corners where the mating tab must seat against the material. For a finger joint slot, this is typically the two bottom corners of each notch.
3. At each inside corner, draw a circle. The circle's diameter should equal your cutting bit's diameter. For a 6 mm end mill: circle diameter = 6 mm (radius = 3 mm).
4. Constrain the circle's centre to the inside corner point using Constrain Coincident. The circle is now exactly centred on the corner.
5. The circle overlaps slightly with the slot rectangle. This is correct — it represents the over-cut area.
6. Close the sketch. When pocketed or profiled, the machine will cut the circular relief at each corner.

Calculating the Correct Dogbone Circle Size

Rule	Dogbone circle diameter = the diameter of the cutting bit you are using for the profile/pocket cut.
Example: 6 mm end mill	Dogbone circles: diameter = 6 mm (radius = 3 mm). Constrain Radius = 3 for each circle.
Example: 3 mm end mill	Dogbone circles: diameter = 3 mm (radius = 1.5 mm). Smaller bit = smaller dogbones = less visible.
Why bit diameter?	The bit leaves a radius of bit_diameter/2 at each corner. A dogbone of the same diameter over-cuts exactly far enough to clear the corner without over-cutting further than necessary.
Minimum recommended bit	For 9 mm finger joints, the 6 mm end mill is appropriate. For 6 mm finger joints, a 3 mm end mill with 3 mm dogbones is needed.

T-Bone Fillets — An Alternative Approach

A T-bone fillet is a rectangular notch (rather than a circle) cut at each inside corner. Where a dogbone circle is centred exactly on the corner point, a T-bone notch extends perpendicularly from the corner along one of the adjacent walls.

T-bones are less common than dogbones but have one advantage: they are easier to sand smooth after assembly because the rectangular notch is in-line with the slot wall rather than protruding into the corner. They are visible from the side of the assembled joint, which some designers consider less attractive. Use dogbones for most joinery — switch to T-bones when post-assembly finishing visibility is a concern.

> **Tip**
>
> If you find dogbone fillet drawing repetitive, install the Dogbone FreeCAD addon (Tools → Addon Manager → search 'Dogbone'). This addon automatically places correctly-sized dogbone circles at every inside corner of a selected sketch profile with one click. It is covered briefly in Chapter 12 (Addons).

Multi-Part CAM — Cutting Several Panels in One Job

When your FreeCAD file contains multiple bodies, you can include all of them in a single CAM Job and cut every panel from one stock sheet in a single machine session. This is how efficient joinery production works: one setup, one set of zero-point measurements, and all panels cut in one continuous job.

Setting Up a Multi-Body CAM Job

1. Switch to the CAM workbench. Go to CAM → Job.
2. In the Model field, you can add multiple bodies. Click the drop-down and select one body. Then click Add to add additional bodies.

3. Alternatively, if FreeCAD does not show an Add option, create the Job with one body first, then add the second by right-clicking the Job in the Model tree → Add Model.
4. Set the Stock dimensions large enough to accommodate all panels side by side with at least 5 mm clearance between parts and 10 mm from each edge.
5. In the Properties panel of each Body, use the Placement X, Y values to position the bodies within the stock area — place them side by side, not overlapping.

Operation Ordering — The Pockets-Before-Profiles Rule

In any multi-operation job involving both pockets (or recesses) and profiles (cut-outs), the correct sequence is always: pockets first, profiles last. This rule is not arbitrary — it is a fundamental safety and quality principle with a clear physical justification.

> 📐 **The Pockets-Before-Profiles Rule**
>
> Always complete all pocket and engraving operations before running any profile cut that frees a piece from the stock. A profile cut severs the part from the surrounding material. Once severed, the part can shift — and a shifted part will be cut incorrectly by any subsequent operation. Profiles are always the final operation in any job.

In a multi-panel job, this means: run all pocket operations on Panel A first, then all pockets on Panel B, then the profile cut on Panel A, then the profile cut on Panel B. Do not interleave pockets and profiles for the same panel, and never run a profile before all pockets on that panel are complete.

Tabs for Multi-Part Jobs

Every profile cut in a multi-part job requires its own set of tabs. Configure tabs independently for each Panel's profile operation — do not share tabs between different parts. The number and position of tabs may differ between panels depending on size and shape.

> ● **Safety Rule**
>
> NEVER profile-cut a panel while other panels still need pockets or engraving.
>
> If a panel shifts after profile cutting (even while held by tabs), the remaining operations on that panel are no longer correctly positioned.
>
> Always check: 'Are all pockets complete on this panel before I profile cut it?' The answer must be YES before any profile operation runs.

PROJECT 6 — STEP-BY-STEP WALKTHROUGH

Two-Panel Finger Joint Box

Design, cut, and assemble a two-panel finger joint using correct clearance, dogbones, and multi-part CAM

Project 6 is the first joinery project — and the most technically demanding project so far. It tests every skill from the previous five weeks: parametric design, Sketcher constraints, Part Design features, correct tolerance application, dogbone fillets, multi-body management, and multi-part CAM. The finished pieces must physically assemble. If they do not, you diagnose why, adjust the Clearance value, re-export G-code, and cut again.

Design	Two interlocking panels (FrontPanel + SidePanel) — front and side of a simple open-topped box
Material	9 mm MDF or birch plywood, two panels each approx. 130 × 100 mm
Thickness	9 mm — both panels the same thickness (required for matching fingers)
Bit	6 mm upcut end mill for all operations
Finger width	9 mm — equals material thickness for clean, equal-width fingers
Clearance	0.15 mm (adjust after test cut)
Operations	Profile cut only (the finger notches ARE the profile shape — no separate pocket)
Day	Day 30 of Week 6 — Project Day

Understanding Finger Joint Geometry

A finger joint has alternating tabs (protruding fingers) and slots (recesses). For CNC cutting, the finger width ideally equals the material thickness — for 9 mm material, use 9 mm wide fingers. This creates a visually regular pattern and ensures each finger is exactly one 'tab-width' wide, which makes calculating the joint positions straightforward.

The alternating rule: Panel A starts with a tab at the top. Panel B starts with a slot at the top. They are mirror images of each other in the joint area. Panel A's tabs fit into Panel B's slots and vice versa.

Part A — Spreadsheet Setup

1 — **Create file and Spreadsheet**
New file, save as FingerJointBox.FCStd. Create Spreadsheet with: PanelHeight=100, PanelThickness=9, FingerWidth=9, Clearance=0.15, NumFingers=5.

2 — **Set all aliases**
PanelHeight, PanelThickness, FingerWidth, Clearance, NumFingers.

Part B — FrontPanel Design

3

Create FrontPanel Body
Part Design → Body. Rename it FrontPanel.

4

Sketch the front panel outline with finger joints on the right edge
On XY plane, draw the panel outline. Width = =Spreadsheet.PanelThickness (the panel is thin — the 'width' you see from the front is the thickness). Height = =Spreadsheet.PanelHeight. This is a 9×100 mm side profile of the panel. Now draw the finger tabs on the right edge: alternating protrusions and recesses, each =Spreadsheet.FingerWidth wide. The panel starts with a tab at the top. Draw 5 fingers total (NumFingers). Constrain every element using spreadsheet aliases where possible.

Tip

The easiest approach: draw the plain panel rectangle, then add rectangular protrusions (tabs) and recesses (slots) on the right edge. Constrain each tab/slot width to =Spreadsheet.FingerWidth. Position the first tab at Y = =Spreadsheet.PanelHeight – Spreadsheet.FingerWidth (top edge), then place subsequent alternating tabs/slots downward at FingerWidth intervals.

5

Add dogbone fillets at all inside corners of the slots
At each inside corner of each slot on the right edge, add a circle diameter = 6 mm (= bit diameter). Apply Coincident to each circle centre at the inside corner. All inside corners should have a dogbone circle.

6

Close sketch and Pad
Fully constrain the sketch (all green). Close sketch. Part Design → Pad. Length = width of the box front (e.g., =Spreadsheet.PanelHeight or a BoxWidth alias — this is the physical length of the front panel). For a simple test joint, Pad to 60 mm.

Part C — SidePanel Design

7

Create SidePanel Body
Part Design → Body. Rename SidePanel. Make it the active body.

8

Sketch SidePanel — mirror of FrontPanel joints
The SidePanel joint profile is the INVERSE of the FrontPanel: where FrontPanel has a tab, SidePanel has a slot (and vice versa). SidePanel starts with a slot at the top edge. The SLOT width = =Spreadsheet.FingerWidth + 2*Spreadsheet.Clearance (= 9 + 0.30 = 9.30 mm). The TAB width = =Spreadsheet.FingerWidth (= 9.00 mm, nominal — no clearance on tabs). Add dogbone fillets at all inside corners of the SidePanel slots too.

9

Pad the SidePanel
Pad to 60 mm (same as FrontPanel length for this test joint).

Part D — Visual Fit Check

10

Position panels for visual assembly check
Show both bodies. In the Properties panel → Data → Placement of SidePanel, set X = 0, Y = 0. Set FrontPanel Placement to position it as it would sit when assembled (tabs entering slots from the side). Rotate one panel 90° using Placement Angle if needed. Verify visually that tabs and slots appear to align.

> **Tip**
>
> The visual fit check does not verify the clearance — it only checks geometry alignment. A correctly aligned model that still does not assemble physically is a clearance problem, not a geometry problem.

Part E — CAM Setup

11

Create CAM Job with both bodies
CAM → Job. Add FrontPanel. Add SidePanel. Configure Stock: large enough for both panels side by side (e.g., 130 × 220 mm × 9 mm). Position both panels within the stock with 10 mm between them.

12

Add 6mm End Mill to Tool Library
Name: EndMill6mm. Diameter: 6. Flutes: 2.

13

Profile Operation — FrontPanel
CAM → Profile. Select FrontPanel outer edges (the entire panel profile including tabs). Side: Outside. Final Depth: −9.5. Step Down: 1.5 mm. Tabs: 4, width 4mm, height 3mm. Feed: 900. Plunge: 250.

14

Profile Operation — SidePanel
Repeat the same profile setup for SidePanel. Note: the SidePanel slots already have the clearance built into the sketch — the profile cut follows the slot geometry as drawn, which is already 0.30 mm wider than the FrontPanel tabs.

15

Simulator check — critical for joinery
Run CAM Simulator. Verify: (1) both panels appear in simulation, (2) the finger profiles are clearly visible with correct tab/slot alternation, (3) dogbone circles appear at slot corners, (4) tabs are present on all profile cuts, (5) no rapid moves pass through either panel.

16

Post-process
grbl post-processor. Save as FingerJointBox_9mm_v1.nc.

Part F — Cutting and Assembly

17

Pre-cut checklist — all four sections
This is a complex job. Take extra time on the checklist. Verify: both panels will fit within the clamping area with no clamp in the cutting path of either panel.

18 **6mm end mill, clamp material, set zero**
Clamp a piece of 9mm MDF or ply large enough for both panels. Set zero at the centre of the stock (both panels are positioned relative to the same origin).

19 **Load and verify in UGS**
Load FingerJointBox_9mm_v1.nc. In the UGS 3D visualiser, verify both panel outlines are visible, including the finger profiles. Can you see the tab/slot alternation pattern in each panel?

20 **Run the job — supervise throughout**
Send the job. Watch the first 30 seconds of each profile cut closely — the finger profiles require precise tracking. Do not leave the machine.

21 **Release tabs and test fit**
Spindle off, Z raised. Remove clamps. Chisel both panels free from tabs. Do NOT sand or modify the joint surfaces yet. Attempt assembly immediately.

🔧 **Fit Assessment Guide**

Assembly Assessment

Assess the fit by trying to push the panels together by hand:

- The joint does not move at all — too tight. Increase Clearance by 0.10 mm. Re-export and cut test joint.
- The joint needs firm hand pressure to close fully — snug fit. Excellent. This is the target for glued joints.
- The joint slides together easily with slight resistance — hand fit. Good for unglued assemblies.
- The joint is loose and moves freely — sliding fit. Decrease Clearance by 0.05 mm if you want snug. Fine as-is for glueable joints where the glue fills the gap.
- The joint falls apart — too loose. Decrease Clearance by 0.10 mm.

22 **Adjust, re-export, and re-cut if needed**
If the fit is not acceptable: update Clearance in the spreadsheet, Refresh, re-export with version number v2, cut a small test joint from scrap first, then cut the final panels.

🏆 **Project 6 — The Real Test**

Project 6 is the proving ground. A joint that fits is a complete success of the entire CNC workflow from sketch to machine. A joint that does not fit is not a failure — it is information: your clearance value needs adjusting by a known amount. Adjust it, recut, and assemble. This iteration is exactly how professional CNC operators calibrate new machines and new materials.

Week 6, Days 26–30 — Lesson Content Overview

Day 26 (Analyse)	Multi-body files and shared Spreadsheets. Students create a 2-body file, rename bodies, practice show/hide, and connect both bodies to one Spreadsheet. Change one Spreadsheet value — verify both bodies update. Discussion: why is this preferable to two separate files?
Day 27 (Analyse)	Tolerance design in depth. Instructor demonstrates: cut two identical slot profiles — one with 0 clearance, one with 0.15 mm, one with 0.30 mm. Students assemble all three and characterise the fit. Class builds a calibration table for their specific machine and material combination.
Day 28 (Analyse)	Dogbone fillets. Students draw a slot profile without dogbones, then add dogbones, and visually compare. Instructor demonstrates (using a printed model) how a tab physically cannot enter a slot without dogbones. Students practice the full dogbone drawing workflow in the Sketcher.
Day 29 (Analyse → Evaluate)	Multi-part CAM and operation ordering. Students set up a 2-body CAM job. Experiment with operation order — intentionally run a profile before a pocket and observe the problem (demonstrated, not necessarily cut). Apply pockets-before-profiles rule. Simulator verification for the complete job.
Day 30 (Analyse — Project Day)	Project 6: Two-Panel Finger Joint Box. Full independent project. Assessment includes physical assembly quality.

Chapter Summary

- A multi-body FreeCAD file contains multiple Part Design Bodies in a single document. All bodies share access to the same Spreadsheet, making parametric joinery straightforward to manage. Rename, show, and hide bodies using the Model tree for efficient workflow.
- The clearance formula for CNC joinery is: Slot Width = Tab Width + 2 × Clearance. The tab is always at nominal size. The slot is always widened by one Clearance value on each side. Store Clearance as a Spreadsheet alias and link all slot dimensions to it.
- Five fit types: Press (0.00–0.10 mm), Snug (0.10–0.20 mm), Hand (0.20–0.30 mm), Sliding (0.30–0.50 mm), and Loose (>0.50 mm). For finger joints that will be glued, target snug fit. Always cut a test joint in scrap before final material.
- Dogbone fillets solve the round-bit/square-corner problem. A dogbone is a circle placed at each inside corner of a slot, with diameter equal to the cutting bit's diameter. Without dogbones, the bit radius prevents the mating tab from seating fully.
- T-bone fillets are rectangular alternatives to dogbones. Use them when post-assembly side-view appearance is important. Dogbones are preferred for most CNC joinery.
- Multi-part CAM: all bodies can share one CAM Job. Position them within the stock using Placement. Apply the pockets-before-profiles rule: all pockets on a panel must complete before its profile cut runs.
- Project 6 assembles in real life. A joint that does not fit is not a failure — it is calibration data. Adjust Clearance, increment the G-code version number, and cut again until the fit is correct.

Key Vocabulary — Chapter 11

Term	Definition
Multi-body file	A FreeCAD .FCStd document containing more than one Part Design Body. Used for joinery projects where multiple panels share a common Spreadsheet and coordinate system.
Nominal dimension	The theoretical designed size of a feature before any manufacturing variation is considered. The tab is cut at the nominal dimension; the slot is cut at nominal + clearance.
Clearance	The designed gap between two mating surfaces that allows them to assemble. For CNC joinery: Slot Width = Tab Width + 2 × Clearance (one Clearance per side).
Press fit	A joint clearance of 0.00–0.10 mm. Parts lock permanently under hand pressure. Used for permanent joints that will not be disassembled.
Snug fit	A joint clearance of 0.10–0.20 mm. Parts slide together with gentle firm pressure. The target fit for glued finger joints.
Sliding fit	A joint clearance of 0.30–0.50 mm. Parts slide together freely. Used for adjustable or repeatedly-assembled connections.
Dogbone fillet	A circular cutout at each inside corner of a slot profile, diameter equal to the cutting bit diameter. Compensates for the bit radius that would otherwise prevent a square tab from seating.
T-bone fillet	A rectangular notch at inside corners of a slot, extending along one adjacent wall. An alternative to dogbone with better in-line visual appearance when viewed from the joint side.
Finger joint	A CNC joinery method using alternating tabs and slots along panel edges to create an interlocking right-angle connection. Tab width typically equals material thickness.
Alternating rule	In finger joints: if Panel A starts with a tab, Panel B must start with a slot. The two panels are mirror images of each other in the joint region.
Pockets-before-profiles rule	The CAM operation ordering principle: all pocket and engraving operations must complete before any profile cut that frees a piece. A freed piece can shift, making subsequent operations inaccurate.
Test joint	A small sample cut in scrap material, identical to the intended joint, used to verify and calibrate the clearance value before committing to final stock.
Active Body	The Part Design Body currently selected for editing. All new sketches and features are added to the active body. Change the active body by double-clicking another body in the Model tree.
Visual fit check	Positioning multiple bodies in the FreeCAD 3D view using Placement values to simulate the assembled state. Verifies geometry alignment before cutting — not a substitute for a physical test joint.

Looking Ahead — Chapter 12: FreeCAD Addons

Chapter 12 covers the FreeCAD Addon Manager and the installation and use of key community addons — in particular the FCGear workbench for mathematically correct involute gear generation. Addons extend FreeCAD's capabilities significantly and are an essential part of the intermediate designer's toolkit.

PART FOUR

Intermediate Skills — Curves, Joinery & Addons

Weeks 5–7 • Chapters 11 through 14 • Bloom's Levels: Apply → Analyse → Evaluate

📖 **About Part Four**

Part Four covers Weeks 5 through 7 — Days 21 through 35. These are the weeks where FreeCAD begins to feel genuinely powerful rather than merely functional. The skills introduced here — importing external artwork, designing precision joinery, and extending FreeCAD with community addons — represent the transition from learning the software to wielding it with judgement. Part Four operates at the Apply, Analyse, and Evaluate levels of Bloom's Taxonomy. By the end of Week 7, you will have designed and cut a working mechanical gear pair — one of the most technically demanding and visually rewarding projects in this course.

CHAPTER 12 of 24

PART FOUR — INTERMEDIATE SKILLS

FreeCAD Addons

Expanding FreeCAD with the Addon Manager, FCGear, and community workbenches for CNC

Chapter Learning Objectives

By the end of this chapter, you will be able to:

- Navigate to the FreeCAD Addon Manager and explain what it is and where addons come from
- Distinguish between workbench addons and macro addons
- Install the FCGear workbench using the Addon Manager and verify it is working
- Describe the key addons relevant to CNC work: Curves, Fasteners, Dogbone, SheetMetal
- Apply a structured risk-assessment approach before installing any third-party addon
- Update, disable, and remove addons using the Addon Manager
- Explain why restarting FreeCAD after addon installation is required

FreeCAD ships with powerful tools. The FreeCAD community ships with even more. The Addon Manager is the bridge between the two.

The core FreeCAD installation — the software you have been using since Week 1 — contains everything you need to design, model, and CAM-process flat CNC parts. It is comprehensive, professionally capable, and free. But FreeCAD is also an open platform: developers around the world build additional workbenches, tools, and macros that extend FreeCAD far beyond its factory defaults.

These community extensions are called addons. They live in a publicly accessible repository, are catalogued inside FreeCAD's Addon Manager, and can be installed in minutes. Some addons are essential for specific workflows — FCGear, which you will use in Week 7 to design the gear pair, is one of the most polished and widely used FreeCAD addons in existence. Others are niche but invaluable when you need them.

This chapter covers everything you need to know about working with addons: where they come from, how to evaluate them, how to install and manage them, and which ones matter most for CNC work. It is a short chapter by design — the addon system is straightforward once you understand the underlying concepts — and it sets the stage for Chapter 13, which puts FCGear to work designing a full spur gear pair.

What Are FreeCAD Addons?

FreeCAD's architecture separates core functionality — the engine, the data model, the rendering system — from the tools that sit on top of it. Everything you interact with in FreeCAD is technically a workbench: the Sketcher, Part Design, CAM, Draft, Spreadsheet — all of them are workbenches built using the same API that any developer can access. This means there is no fundamental difference between a built-in workbench and a community-developed one. Both use the same underlying system. The only practical differences are distribution method and maintenance responsibility.

> 📖 **Key Term: Addon**
>
> An extension to FreeCAD developed by the community and distributed through the official FreeCAD addon repository. Addons include workbenches (which add entire new tool sets), macros (small scripts that automate specific tasks), and preference packs (visual theme and settings bundles). Addons are not part of the core FreeCAD installation but can be installed, updated, disabled, and removed through the built-in Addon Manager.

Two Types of Addons

Workbench Addons	Add a complete new workbench with its own toolbar, menu items, and tools. After installation and restart, the workbench appears in the workbench drop-down alongside the built-in workbenches. FCGear is a workbench addon. Examples: FCGear, Curves, Fasteners, SheetMetal, Assembly4.
Macro Addons	Small scripts that automate a single task without adding a full workbench. Macros are accessed via Tools → Macros. They appear in a macro list and are run individually. The Dogbone macro adds dogbone fillets to a selected sketch automatically — a useful shortcut for the joinery work from Chapter 11. Examples: Dogbone, HoleFeature, Timber, BoundBox.

Where Addons Come From

All addons available through the Addon Manager are hosted in the official FreeCAD addon repository on GitHub at github.com/FreeCAD/FreeCAD-addons. The repository is a curated list of addon source locations — each addon lives in its own GitHub repository maintained by its developer, and the central list points to them. The FreeCAD core team reviews addons for inclusion in the official list but does not write or maintain most of them.

This means the quality, update frequency, and compatibility of addons varies. A well-maintained addon like FCGear has been updated to work with every recent FreeCAD version, has extensive documentation, and has thousands of active users reporting bugs. A less-maintained addon might not have been updated in two years and may produce unpredictable results with FreeCAD v1.1.0. Knowing how to evaluate an addon before installing it is an important skill — covered in detail later in this chapter.

The Addon Manager

The Addon Manager is FreeCAD's built-in tool for browsing, installing, updating, and managing addons. Access it from the main menu: Tools → Addon Manager. It does not require internet access to browse the cached list, but does require a connection to download and install.

Opening and Navigating the Addon Manager

1. Go to Tools → Addon Manager. The Addon Manager window opens. The first time, FreeCAD downloads the current addon list from GitHub — this may take 10–30 seconds.
2. The left panel shows all available addons. Use the tabs to filter: All, Workbenches, Macros, Preference Packs. The search bar at the top filters by name.
3. Click any addon to see its description, version information, maintainer, and GitHub link in the right panel. Check the 'Tested with FreeCAD version' field to verify compatibility with v1.1.0.
4. Addons show a coloured status indicator: green = installed and up to date, orange = installed but update available, grey = not installed.
5. Click Install/Update. FreeCAD downloads the addon from its GitHub repository.
6. After installing any workbench addon, FreeCAD will prompt you to restart. Always restart immediately — skip the restart and the workbench will not appear in the drop-down.

> ⚠ **Important**
> Always restart FreeCAD completely after installing a workbench addon. Click the 'Restart Now' button in the Addon Manager dialog. Skipping the restart means the new workbench will not appear in the drop-down list and you may experience unpredictable FreeCAD behaviour.

Updating Addons

Addon developers release updates to fix bugs, add features, and maintain compatibility with new FreeCAD versions. The Addon Manager shows orange indicators for installed addons with available updates. To update: open Tools → Addon Manager, click 'Check for updates' in the top-right area, then click any orange-flagged addon and select 'Update'. Restart FreeCAD after updating workbench addons.

> 💡 **Tip**
> Check for addon updates at the start of a new course module. An update to FCGear, for example, might fix a gear geometry issue that would otherwise affect your Week 7 project. Two minutes in the Addon Manager at the start of each week is good practice.

Disabling and Removing Addons

You can temporarily disable an addon without uninstalling it — useful for troubleshooting if you suspect an addon is causing FreeCAD instability. Select the addon in the Addon Manager and click 'Disable'. The addon will no longer load on startup but its files remain. Re-enable by clicking 'Enable'.

To permanently remove an addon, select it and click 'Remove'. FreeCAD deletes the addon files. A restart is required for removal to take full effect. Plan accordingly before removing an addon you have used in saved files — those documents will still open, but the addon's features will no longer be available.

Risk Management — Evaluating Addons Before Installing

Installing software from the internet always carries risk. FreeCAD's Addon Manager mitigates this significantly by limiting the official list to addons hosted on GitHub with public source code — you can inspect exactly what will be installed before installing it. However, risk management still requires active judgement on your part. Use the framework below before installing any addon you are not already familiar with.

Evaluation Question	How to Check	What a Good Answer Looks Like
Is the addon in the official Addon Manager list?	Search for it in Tools → Addon Manager	*Yes — it appears in the list. Installing from outside the Addon Manager (downloading random .zip files) carries significantly higher risk.*
When was it last updated?	Click the addon — check 'Last commit' date, or visit the GitHub link	*Updated within the last 12 months. An addon not updated in 2+ years may have compatibility problems with FreeCAD v1.1.0.*
What FreeCAD versions has it been tested with?	Check 'Tested with' field in Addon Manager, or the README on GitHub	*The README explicitly mentions FreeCAD 0.21, 1.0, or 1.1. Older-only references are a yellow flag.*
How many users does it have?	Check GitHub Stars and Forks count, or search the FreeCAD Forum	*Hundreds of stars and active forum discussion indicate a healthy user base that surfaces serious bugs quickly.*
Is there documentation?	Look for a README, a wiki page, or a FreeCAD Forum thread	*A dedicated documentation page or active Forum thread. An addon with no documentation is harder to troubleshoot.*
Does the FreeCAD community recommend it?	Search the FreeCAD Forum (forum.freecad.org) for the addon name	*Multiple positive forum threads, use in official tutorials, or recommendation in FreeCAD documentation.*

> 📋 **Note**
>
> For this course, all required addons have been pre-evaluated. FCGear is a well-maintained, widely-used addon that is safe to install. The other recommended addons listed in this chapter — Curves, Fasteners, Dogbone, SheetMetal — are similarly established. You are not required to install anything beyond FCGear for the Week 7 project.

Recovering from an Addon That Crashes FreeCAD on Startup

If an addon causes a crash on startup — preventing FreeCAD from launching at all — you can manually disable it by navigating to FreeCAD's addon folder and removing or renaming the addon's folder. Paths by platform:

Windows	C:\Users\[username]\AppData\Roaming\FreeCAD\Mod\
macOS	~/Library/Application Support/FreeCAD/Mod/
Linux	~/.FreeCAD/Mod/

Remove the offending folder, restart FreeCAD normally, then use the Addon Manager to formally uninstall it. If FreeCAD behaves unpredictably after installing a new addon, use the Disable function to isolate whether that addon is the cause before taking more drastic steps.

Installing FCGear — Step by Step

FCGear is the addon you will use in Week 7 to generate spur gears for Project 7. It is one of the most capable and best-documented addons in the FreeCAD ecosystem. Installation takes less than five minutes and requires an internet connection.

1. Open Addon Manager — Tools → Addon Manager. Wait for the addon list to load.
2. Search for 'FCGear' in the search bar. The FCGear workbench should appear immediately in the filtered list.
3. Click FCGear. Read the description panel — note the current version number and the 'Last updated' date. Verify it mentions FreeCAD compatibility.
4. Click 'Install'. FreeCAD downloads the FCGear files from GitHub. This typically takes 15–60 seconds.
5. When installation completes, FreeCAD prompts you to restart. Click 'Restart now'.
6. After restart, open the Workbench drop-down. Scroll through the list — 'FCGear' should appear as a new entry.
7. Select FCGear from the workbench drop-down. The toolbar changes to show FCGear-specific tools: InvoluteGear, CycloidGear, BevelGear, and others. Installation confirmed.

Try It Now

Go to Tools → Addon Manager. Search for 'FCGear'. Click the result and read the full description — note the maintainer name, the last update date, and the GitHub URL.

Click Install. Watch the progress bar. When the restart prompt appears, click 'Restart now'.

After FreeCAD restarts, open the workbench selector. Confirm FCGear appears in the list.

Select FCGear and look at the toolbar — you should see at least four gear types available. You do not need to generate any gears yet — that is Chapter 13's work. Confirming the workbench is present and accessible is the complete goal of this step.

If FCGear Does Not Appear After Restart

Did you restart?	Close FreeCAD completely (the entire application, not just the document) and reopen it. The workbench drop-down requires a full restart.
Did the download complete?	Open Tools → Addon Manager and check FCGear's status. If it shows orange rather than green, the installation may have been partial. Click Update.
Check the Mod folder	Navigate to the FreeCAD Mod folder for your OS (see paths above). A folder named 'FCGear' should exist and contain Python (.py) files. If missing, re-install from the Addon Manager.
Internet connection issue	If GitHub was temporarily unreachable during installation, the download may have been incomplete. Try reinstalling when connectivity is confirmed.

Other Recommended Addons for CNC Work

While only FCGear is required for this course, the following addons are worth knowing about. They address specific CNC design challenges you will encounter as your skills develop. Each is listed with its purpose and an indication of when it becomes relevant.

Addon	Type	Purpose	When Relevant
FCGear	*Workbench*	Generates parametric gear profiles: involute spur gears, bevel gears, cycloid gears, worm gears, and timing belt pulleys. Resulting solid bodies work directly in CAM.	*Required — Week 7. Used for all mechanical component work beyond this course.*
Curves	*Workbench*	Advanced curve and surface tools: interpolated curves through specific points, blend surfaces between edges, geodesic curves. Powerful for freeform 3D surface design.	*Optional — useful when 3D CAM and organic surface work becomes more advanced.*
Fasteners	*Workbench*	Inserts standard hardware models — screws, bolts, nuts, washers, threaded inserts — all based on ISO and DIN standards with correct dimensions.	*Optional — useful when designing CNC-cut parts that incorporate purchased hardware.*
Dogbone	*Macro*	Automatically adds dogbone fillets to all inside corners of a selected sketch. Replaces the manual circle-at-corner method from Chapter 11 with a single operation.	*Optional — useful for all joinery work. Dramatically speeds up dogbone workflow for complex joint designs.*
SheetMetal	*Workbench*	Parametric sheet metal design: bends, flanges, hems, and flat pattern unfolding. The DXF flat pattern export is useful for nesting in CNC and laser cutting contexts.	*Optional — primarily relevant for laser cutting or flat-pack furniture with precise fold lines.*

Assembly4	*Workbench*	Full assembly constraints: coincident, distance, angle, and axial joints between separate Part Design bodies. Allows kinematic simulation of moving assemblies.	*Optional — useful for assemblies where fit and motion need verification before cutting.*

Tip

Install addons one at a time and test FreeCAD stability between installations. Installing multiple addons simultaneously makes it difficult to identify which one caused a problem if issues arise. FCGear first — verify it works — then optionally explore the others.

Using Addons Responsibly

FreeCAD addons are developed by volunteers — most addon maintainers are working professionals or enthusiasts who maintain their addons in their own time without payment. When you benefit from an addon, there are several ways to contribute to the community that made it possible.

Report bugs clearly	If an addon produces an incorrect result or crashes FreeCAD, report it on the addon's GitHub Issues page or the FreeCAD Forum. A clear bug report — FreeCAD version, addon version, exact steps to reproduce, screenshot — is genuinely valuable to maintainers.
Acknowledge in your work	If you produce course work, client work, or tutorial content using a community addon, acknowledge it. Naming FCGear in a project description or tutorial gives deserved credit and costs nothing.
Vote on the FreeCAD Forum	Forum posts about addons benefit from engagement. If an addon solved a problem for you, a brief thank-you post on the Forum or a GitHub Star takes 30 seconds and encourages the maintainer to continue.
Respect licences	Most addon source code is licensed under LGPL or similar open-source licences. If you build on or redistribute addon code, check the licence and comply with its attribution requirements.

Week 7, Day 31 — Lesson Content Overview

Chapter 12 covers Day 31 of Week 7. Days 32–35 are covered in Chapter 13 (Gears & Mechanical Parts), which puts FCGear immediately to work.

Day 31 (Analyse)	FreeCAD Addons and Addon Manager. Students install FCGear using the Addon Manager, verify it appears in the workbench selector, and explore the Addon Manager independently to identify other addons relevant to their own CNC interests. Class discussion: what risks exist when installing third-party software? What checks would you run before installing an unknown addon? Students build a class 'recommended addons shortlist' — each nominates one addon with a one-sentence justification.

Chapter Summary

- Addons extend FreeCAD without modifying the core installation. They add entire new workbenches (large tool sets) or single-task macros through a standardised system managed from inside FreeCAD via Tools → Addon Manager.
- The Addon Manager is the safe, official way to install addons. It connects to the curated GitHub repository and provides install, update, disable, and remove functionality in one interface. Never install addons from unofficial .zip files.
- A restart is required after installing or updating any workbench addon. Without a full restart, the new workbench will not appear in the workbench selector.
- Evaluate before installing: check the update date, FreeCAD version compatibility, user community size (GitHub Stars/Forks), and available documentation. Addons not updated in 2+ years may have compatibility issues with v1.1.0.
- FCGear is the required addon for this course. It generates parametric involute gears and other mechanical profiles as solid bodies suitable for Part Design, Boolean operations, and CAM toolpaths. Chapter 13 puts it to work immediately.
- If an addon causes FreeCAD instability, use the Disable function to isolate whether it is the cause. If it causes a crash on startup, remove the addon's folder manually from the Mod directory, then formally uninstall from the Addon Manager.
- Other notable CNC addons: Curves (advanced 3D surfaces), Fasteners (standard hardware models), Dogbone (automatic dogbone fillets — speeds up joinery work), SheetMetal (flat-pattern DXF), Assembly4 (kinematic assembly constraints).
- Addon developers are volunteers. Report bugs clearly, acknowledge addon use in your work, and respect the open-source licences under which addons are distributed.

Key Vocabulary — Chapter 12

Term	Definition
Addon	A community-developed extension to FreeCAD distributed through the official addon repository. Includes workbenches, macros, and preference packs. Not part of the core installation — managed through the Addon Manager.
Addon Manager	FreeCAD's built-in tool for browsing, installing, updating, disabling, and removing addons. Accessed via Tools → Addon Manager. Connects to the official FreeCAD addon repository on GitHub.
Workbench addon	An addon that adds a complete new workbench with its own toolbar, menu items, and tool set. Requires a full FreeCAD restart after installation before the workbench appears in the selector.

Macro addon	An addon that automates a single task without adding a full workbench. Macros are accessed via Tools → Macros and run individually. No restart required after installation.
FCGear	A community workbench addon that generates parametric gear profiles: involute spur gears, bevel gears, cycloid gears, and worm gears. The required addon for Week 7 (Project 7 — Spur Gear Pair Display).
GitHub	A web-based platform for hosting and sharing source code. The official FreeCAD addon repository and all individual addon repositories are hosted on GitHub. The Addon Manager downloads directly from GitHub.
Mod folder	The directory on your computer where FreeCAD stores installed addon files. Path varies by OS: Windows: AppData\Roaming\FreeCAD\Mod\, macOS: ~/Library/Application Support/FreeCAD/Mod/, Linux: ~/.FreeCAD/Mod/. Useful for manual recovery if an addon prevents FreeCAD from starting.
Status indicator	The colour icon shown next to each addon in the Addon Manager: green = installed and current, orange = installed but update available, grey = not installed.
Restart required	After installing or updating any workbench addon, FreeCAD must be fully closed and reopened before the new workbench appears in the workbench selector. The Addon Manager shows a 'Restart now' button.
Risk assessment	A structured evaluation of an addon before installation — checking update date, FreeCAD version compatibility, user community size, and documentation quality — to determine whether the addon is safe and stable to install.

Looking Ahead — Chapter 13: Gears & Mechanical Parts — FCGear Workbench

Chapter 13 puts FCGear immediately to work. You will generate a mathematically correct involute spur gear pair, calculate pitch diameter and gear ratios, add centre bores using Part Design, apply Part Boolean operations, and configure the roughing and finishing CAM strategy for precision gear tooth profiles — culminating in Project 7, the Spur Gear Pair Display.

CHAPTER
13
of 24

PART FOUR — INTERMEDIATE SKILLS

Gears & Mechanical Parts — FCGear Workbench

Involute gear design, FCGear parameters, Boolean operations, and the roughing + finishing CAM strategy

Chapter Learning Objectives

By the end of this chapter, you will be able to:

- Explain the involute gear profile in plain English and why it is used
- Generate a spur gear using FCGear InvoluteGear and configure its key parameters
- Calculate Pitch Diameter from Module and Teeth
- Design a mating gear pair for a specified gear ratio
- Apply the Module matching rule when designing gears that must mesh
- Add a centre bore and keyway to a gear using Part Design features
- Apply Part Boolean Cut, Fuse, and Common operations
- Distinguish between Part Boolean operations and Part Design Pocket
- Configure a roughing pass and a finishing pass for precision profile CAM
- Set up the profile offset for a roughing allowance

Why Gears?

Nothing demonstrates the precision of CNC machining more clearly than a set of gears that mesh and turn smoothly. Gears are mathematically exact, visually compelling, and immediately functional.

A wooden gear pair that actually works — that spins smoothly, transmits rotation, and produces a predictable speed change — is one of the most satisfying things you can make on a CNC machine. It is also one of the most technically demanding: the tooth profile must be mathematically correct, the cutting must be precise, and the two gears must share the same fundamental dimension (Module) or they will not mesh at all.

This chapter introduces the FCGear addon workbench, which generates mathematically correct involute gear geometry automatically from a small set of parameters. You do not need to understand involute calculus to produce a working gear — but you do need to understand the key parameters. Once you do, gear design becomes as straightforward as setting values in a dialog box.

The chapter also covers Part Boolean operations — Cut, Fuse, and Common — which are essential not just for gears but for any compound 3D form that cannot be built using Part Design's Pad and Pocket alone. Boolean operations give you the ability to combine and subtract any two solid shapes, regardless of how complex they are.

Involute Gear Fundamentals — Plain English

An involute is the curve traced by the end of a string being unwound from a cylinder. This particular curve has a remarkable property: when two gears with involute tooth profiles mesh, the contact point between their teeth moves smoothly along a straight line, producing constant, smooth force transmission regardless of minor variations in the centre distance between the gears.

Why does this matter for CNC? Because involute gears are forgiving of small manufacturing errors. If your gear centre distance is 0.5 mm off due to machine positioning, the gears will still mesh and turn smoothly — the involute profile compensates. A non-involute tooth profile would bind or slip under the same conditions.

You will never draw an involute curve by hand in FreeCAD. The FCGear addon calculates the involute geometry mathematically and generates the complete tooth profile automatically. Your job is to understand the parameters that control the gear size and shape.

> 📖 **Key Term: Involute profile**
>
> The mathematically defined tooth shape used in standard spur gears. Involute geometry ensures smooth, constant force transmission during meshing and tolerates small variations in gear centre distance. The standard for virtually all industrial gearing.

FCGear Workbench — Installation and Interface

FCGear is a community addon workbench that extends FreeCAD with gear-generation tools. It must be installed through the Addon Manager before use. If you completed the Chapter 12 (Addons) exercises, FCGear should already be installed. If not, go to Tools → Addon Manager, search 'FCGear', install it, and restart FreeCAD.

Accessing FCGear

1. After installation, FCGear appears in the Workbench drop-down menu.
2. Switch to the FCGear workbench.
3. The menu bar now includes a Gear menu with options for different gear types.
4. For this course: Gear → InvoluteGear creates a standard spur gear.

FCGear InvoluteGear — All Parameters Explained

When you select Gear → InvoluteGear, a gear object is created immediately using default values. All parameters are modified in the Properties panel (Data tab) rather than in a dialog box. Click the gear in the Model tree to select it, then look at the Data tab — you will see all the gear parameters listed there, ready to edit.

The Key Parameters

Module (m)	The most fundamental gear parameter. Module controls tooth size — a larger Module means physically larger, wider teeth. Module = Pitch Diameter ÷ Number of Teeth. Standard values: 1, 1.25, 1.5, 2, 2.5, 3, 4, 5, 6. For wood CNC: use Module 3 or larger.
Teeth (z)	The number of teeth on the gear. More teeth = larger gear (for a given Module). Minimum recommended for wood: 12 teeth (fewer teeth produce weak, fragile geometry). Standard designs: 15–40 teeth for functional gears.
Height	The thickness of the gear in mm — how thick the gear is in the Z direction. For display gears: 12–18 mm. For functional gears that transmit real load: 15–25 mm.
Pressure Angle (α)	The angle of force transmission at the tooth contact point. Standard value: 20°. Do not change this unless you have a specific technical reason. Both gears in a meshing pair must have the same pressure angle.
Backlash	A small gap intentionally built into each tooth space to prevent jamming. Standard: 0.0 for display gears (press fit meshing). Set to 0.05–0.1 mm for functional gears that must turn freely under load.
Simple	A toggle in some FCGear versions that simplifies the tooth profile for faster rendering. Keep this off (false) for cutting — only use it for concept visualisation when the full involute calculation is slow.

The Most Important Parameter: Module

Module is the single most important parameter in gear design. It controls three things simultaneously: the physical size of each tooth, the pitch diameter of the gear, and which other gears it can mesh with. Two gears can only mesh with each other if they share the same Module value — a Module 3 gear cannot mesh with a Module 4 gear, even if all other dimensions happen to line up.

> ⚙ **Gear Rule**
>
> The Module Matching Rule: Two gears only mesh correctly if their Module values are identical. This is the fundamental compatibility rule in gearing. Before designing any gear pair, decide on the Module first. All subsequent parameter choices flow from it.

Why Module 3+ for Wood?

Small-Module gears have narrow, delicate teeth that are easily broken in timber. Module 1 (the size used in watchmaking) has teeth just 1 mm wide at the tip — completely impractical for a wood CNC router at any reasonable feed rate. Module 2 is marginal. Module 3 produces teeth approximately 9.4 mm tall and 9.4 mm wide at the pitch circle, which is the minimum for reliable wood gears. Module 4–5 gears are preferred when strength matters.

Pitch Diameter and Gear Ratios — The Core Calculations

Pitch Diameter Formula

```
Pitch Diameter (mm)  =  Module × Teeth
```

The Pitch Diameter is the effective working diameter of the gear — the circle along which meshing contact occurs. Two meshing gears' pitch circles are tangent to each other. The centre distance between two meshing gears = (PD1 + PD2) ÷ 2.

Gear Ratio Formula

```
Gear Ratio  =  Driven Teeth ÷ Driver Teeth
```

The gear ratio determines the speed relationship between the two gears. A 2:1 ratio means the driven gear turns at half the speed of the driver. To achieve 2:1, the driven gear needs exactly twice as many teeth as the driver — both with the same Module.

The table below is a practical reference. Find your desired Module and Tooth count — the intersection shows the Pitch Diameter in millimetres. Use this when planning a gear pair to check whether the gears will fit within your stock size and the CNC 4040 Pro Max's work area.

M \ T	T=12	T=16	T=20	T=24	T=30	T=40
M=2	24mm	32mm	40mm	48mm	60mm	80mm
M=3	36mm	48mm	60mm	72mm	90mm	120mm
M=4	48mm	64mm	80mm	96mm	120mm	160mm
M=5	60mm	80mm	100mm	120mm	150mm	200mm
M=6	72mm	96mm	120mm	144mm	180mm	240mm

Rows highlighted in red (Module 1–2) are too small for reliable wood CNC cutting. Rows in grey (Module 3+) are suitable.

Worked Example — Designing a 2:1 Gear Pair

You want a gear pair where the driven gear turns at half the speed of the driver. Module = 3. Driver has 15 teeth. Required calculations:

Driver teeth	15 (chosen)
Driven teeth	15 × 2 = 30 (for 2:1 ratio)

Driver Pitch Diameter	Module × Teeth = 3 × 15 = 45 mm
Driven Pitch Diameter	3 × 30 = 90 mm
Centre distance	(45 + 90) ÷ 2 = 67.5 mm
Actual gear ratio	30 ÷ 15 = 2.0 (exactly 2:1)
Verification	If driver turns at 100 RPM → driven turns at 50 RPM. Correct.

> 💡 **Tip**
> Always verify the Pitch Diameter of both gears against your stock material and machine work area before cutting. A Module 5 gear with 40 teeth has a Pitch Diameter of 200 mm — this just barely fits in the CNC 4040 Pro Max's 400 × 400 mm work area, but only if you plan the layout carefully.

Adding Features to Gears — Bores, Keyways, and Lightening Holes

The FCGear object is a solid cylinder with gear teeth cut into its perimeter. By itself it has no shaft hole. Adding a centre bore, keyway, and other features requires Part Design operations applied to the gear object — or Part Boolean operations, which we cover in the next section.

Adding a Centre Bore

The centre bore is the hole through which the shaft passes. For a display gear mounted on a wooden dowel, the bore diameter should match the dowel (typically 6, 8, or 10 mm for display work). For a functional gear on a metal shaft, the bore matches the shaft diameter precisely.

1. With the FCGear object in your Model tree, switch to the Part Design workbench.
2. Click the face of the gear (top flat face).
3. Go to Part Design → Create a new Sketch.
4. Draw a circle at the origin. Constrain Diameter to your desired bore size (e.g., 8 mm for an 8 mm dowel).
5. Close the sketch. Apply Part Design → Pocket → Through All.
6. The bore passes completely through the gear thickness.

Adding a Keyway

A keyway is a rectangular notch cut along the bore wall. It accepts a matching key on the shaft, preventing the gear from spinning on the shaft independently. For display gears, a keyway is optional but adds to the visual appeal and mechanical authenticity.

1. Create a new sketch on the same gear face used for the bore.
2. Draw a small rectangle: width = approximately 1/4 of the bore diameter, depth = 3–5 mm. Centre it at the top of the bore circle (12 o'clock position).
3. Apply Pocket → Through All.
4. The keyway slot is cut from the bore wall through the full gear thickness.

Part Boolean Operations — Cut, Fuse, and Common

Part Boolean operations work on complete solid bodies. Unlike Part Design Pocket (which removes material defined by a 2D sketch from an existing body), Boolean operations work by combining or subtracting two independently existing 3D solids. The second solid acts as the cutter or combiner — and it can be any shape: a rotated box, an extruded profile, another gear, a complex imported mesh.

Boolean operations are accessed from the Part workbench — switch to Part before using them. They require two separate objects: the base solid (the one being operated on) and the tool solid (the shape that does the cutting or merging). The order of selection matters for Boolean Cut — select the base first, then the tool.

Operation	Symbol	What it Produces	CNC Use Cases
Part Cut	A − B	Subtracts the second selected solid from the first. The volume occupied by B is removed from A. B itself disappears from the result.	*Cutting a keyway slot (any shape) from a gear body. Removing an angled feature that is impossible with Part Design Pocket. Creating a French cleat slot in a panel.*
Part Fuse	A ∪ B	Combines two separate solids into one unified solid. The boundary between the two original solids is dissolved — the result is a single merged body.	*Combining a gear body with a hub extension. Merging two separate bracket arms into one part. Adding a boss to an existing plate.*
Part Common	A ∩ B	Keeps only the volume that is shared by both A and B. Everything outside the overlap is removed. Both originals disappear; only the intersection remains.	*Finding the actual overlap between two bodies for inspection. Generating the cross-section of a complex intersection. Rarely used in standard CNC workflows but powerful for analysis.*

Part Boolean vs Part Design Pocket — When to Use Which

Use Part Design Pocket when...	The material to remove is defined by a 2D sketch on an existing face. The pocket is vertical (perpendicular to the sketch plane). The feature needs to be parametric and editable. This covers 90% of CNC operations.
Use Part Boolean Cut when...	The shape to remove is a 3D object rather than a 2D extruded profile. The cut is at an angle that cannot be achieved with a simple sketch. You need to subtract a rotated, moved, or otherwise positioned 3D solid from another solid.
Real example: gear keyway	A simple rectangular keyway can be done with Part Design Pocket (draw rectangle on face, pocket through). A helical slot or angled keyway requires Boolean Cut with a matching 3D shape.

Real example: French cleat	A 45° angled slot in a panel cannot be cut with Part Design Pocket (Pocket is always perpendicular to the sketch). It requires Boolean Cut with a rotated rectangular box — exactly as shown in the tool rack project.

Try It Now

In a new FreeCAD document, create a simple box: Pad a 60×60mm square to 20mm height.

Switch to Part workbench. Part → Primitives → Box. Set Length=15, Width=15, Height=25.

Position this small box overlapping one corner of the large box (set its Placement X=50, Y=50, Z=0).

Now select the large box first, then Ctrl+click the small box.

Go to Part → Boolean → Cut.

A corner notch is removed from the large box — the shape of the small box.

Now try Part → Boolean → Fuse with two overlapping boxes — observe them merge into one body.

Notice: Boolean operations consume both input objects. The originals are gone.

Precision CAM — Roughing and Finishing Passes

One pass can remove material quickly, or one pass can cut accurately. The roughing + finishing strategy does both — by splitting the work into two passes with different priorities.

For most CNC parts in this course, a single profile pass is sufficient. For precision parts — gears, toleranced fits, display pieces that must look perfect — a single pass is often not enough. The roughing pass removes material quickly, leaving a thin skin of material on the profile. The finishing pass removes just that last skin slowly and precisely, producing a clean, dimensionally accurate surface.

This is the professional standard for precision CNC machining across industries from aerospace to woodworking. The extra time the finishing pass adds is always repaid in surface quality and dimensional accuracy.

	Roughing Pass	Finishing Pass
Purpose	Remove bulk material quickly	Cut to exact final dimension with clean surface
Tool offset	+ 0.2–0.3 mm from final profile (leaves material for finishing)	0 mm (cuts exactly on the model profile)
Feed rate	Standard or higher — speed prioritised	20–40% slower — surface quality prioritised

Step down	Standard per-material value	Same or full depth in one pass (finishing is the last pass)
Passes	Multiple passes through material	Typically one final pass
Result	Rough surface, close to dimension	Smooth surface, accurate final dimension
When used	Always — before finishing	For precision parts (gears, toleranced fits, display pieces)

Configuring the Profile Offset for Roughing

In FreeCAD CAM, the roughing pass is set up as a Profile operation with a positive offset value. The offset moves the toolpath away from the model's nominal edge — leaving a thin layer of material. A finishing pass then runs with zero offset, cutting exactly to the model edge.

1. Create the first Profile operation. In the operation settings, find the Extra Offset field (or Offset Value in some FreeCAD versions).
2. Set Extra Offset to 0.25 mm (leaves a 0.25 mm skin of material after roughing).
3. Set Feed Rate to standard roughing speed (e.g., 900 mm/min for hardwood gears).
4. Set Step Down to your standard value (1.0 mm for hardwood).
5. Click OK. This is the roughing pass.
6. Create a second Profile operation targeting the same geometry.
7. Set Extra Offset to 0 mm (cuts exactly to the model edge).
8. Set Feed Rate to 60–70% of roughing speed (e.g., 600 mm/min). Slower = better surface.
9. Set Step Down to the full profile depth (one pass on the finishing run).
10. Click OK. This is the finishing pass.

> **Tip**
>
> Run the roughing pass first in the Simulator. Observe that it leaves a thin outline around the gear teeth — the 0.25 mm skin. Then run the finishing pass and observe it trace the exact tooth profile, removing that skin. The visual difference in simulation quality is immediately apparent.

PROJECT 7 — STEP-BY-STEP WALKTHROUGH

Spur Gear Pair Display

Design and cut a 2:1 meshing spur gear pair with centre bores, mounted on a simple display frame

Project 7 is the most mechanically satisfying project in the course. Two wooden gears that mesh and turn smoothly. The process tests your ability to apply FCGear parameters correctly, add Part Design features to addon-generated objects, configure a roughing + finishing CAM strategy, and manage a multi-part job with 5+ operations.

Design	Two spur gears (driver: 15T, driven: 30T) + a simple rectangular display frame
Module	3 — for robust timber teeth
Material	15–18 mm pine or hardwood (MDF teeth are fragile — use solid timber)
Bit	3 mm end mill for gear tooth profiles (roughing + finishing), 6 mm end mill for bores and frame
CAM strategy	Roughing pass (offset +0.25 mm) + Finishing pass (offset 0) for both gear profiles
Day	Day 35 of Week 7 — Project Day

Part A — Gear Design

1 **New file — save as SpurGearPair.FCStd**
File → New. Save. Switch to FCGear workbench.

2 **Create the Driver Gear (15 teeth)**
Gear → InvoluteGear. In the Data tab (Properties panel), set: Module = 3, Teeth = 15, Height = 15, Pressure Angle = 20, Backlash = 0. Rename the gear object 'DriverGear' in the Model tree.

3 **Verify Driver Gear dimensions**
Pitch Diameter should be 3×15 = 45 mm. Check this by looking at the gear in the 3D view — the teeth should extend beyond 45mm (the addendum circle is larger than the pitch circle), which is correct.

4 **Create the Driven Gear (30 teeth)**
Gear → InvoluteGear again. Set: Module = 3, Teeth = 30, Height = 15, Pressure Angle = 20, Backlash = 0. Rename 'DrivenGear'. Pitch Diameter = 3×30 = 90 mm.

5 **Add centre bores (Part Design)**
Switch to Part Design. Click the top face of DriverGear. New sketch. Circle at origin, diameter 8 mm. Close. Part Design → Pocket → Through All. Repeat for DrivenGear with the same bore size.

6 **Optional: add keyways**
On each gear's top face, new sketch. Small rectangle (2×3 mm) at 12 o'clock on the bore. Pocket → Through All.

7 **Design the display frame**
Create a new Body named 'DisplayFrame'. Sketch: a rectangle with two holes positioned at the correct centre distance. Centre distance = (45+90)÷2 = 67.5 mm. Two circles at X=0, Y=0 and X=67.5, Y=0, both diameter 8.5 mm (slightly larger than bore — the dowels pass through these). Frame outer dimensions: 150 × 60 mm. Pad to 10 mm.

Part B — CAM Setup

8 **Create CAM Job — all three bodies**
CAM → Job. Add DriverGear, DrivenGear, DisplayFrame. Stock: 250 × 120 × 18 mm. Position driver and driven gear side by side, with frame below them.

9 **Add tools: 3mm end mill (T1) and 6mm end mill (T2)**
Tool Library: T1 = EndMill3mm, Diameter 3. T2 = EndMill6mm, Diameter 6.

10 **Operations 1 & 2 — Driver gear roughing + finishing**
Using T1 (3mm end mill): Op 1 = Profile, DriverGear outer edge, Extra Offset +0.25 mm, Feed 800 mm/min, Step Down 0.8 mm. Op 2 = Profile, same edge, Extra Offset 0, Feed 550 mm/min, full depth in one pass. Add tabs (4 tabs, 3mm wide, 2mm high) to Op 1 only — the finishing pass removes the same tabs at the same positions.

11 **Operations 3 & 4 — Driven gear roughing + finishing**
Same setup as Operations 1 & 2 for DrivenGear (larger gear — more time, same strategy).

12 **Operation 5 — Bore pockets (both gears)**
CAM → Pocket Shape or Drilling. Select both gear bore faces. T2 (6mm end mill) or drill bit. Through All. Feed 400, Plunge 150.

13 **Operation 6 — Display frame profile**
CAM → Profile. DisplayFrame outer edge + two hole profiles. T2. Through depth. Tabs: 4.

14 **Simulator — all 6 operations**
Run the full CAM Simulator. This job takes time to simulate — use the Speed slider. Verify: tooth profiles are cleanly cut, bores present, frame cut out. Look carefully at the gear teeth — in the roughing pass simulation, a thin skin should remain. In the finishing pass, it should be removed precisely.

15 **Post-process**
grbl. Save as SpurGearPair_15mm_v1.nc.

Part C — Cutting

16	**Pre-cut checklist, connect, home** This is the most complex job so far — take extra time with the pre-cut checklist.
17	**Install 3mm end mill, clamp stock, set zero** 3mm end mill for the gear operations. Clamp 250×120mm timber, minimum 15mm thick. Set X/Y/Z zero at the centre of the stock.
18	**Run Operations 1–4 (roughing + finishing, both gears)** Send job. Supervise closely during gear tooth cutting — the 3mm bit is small and can deflect or break if feed rate is too aggressive. If you hear unusual chatter, use UGS feed rate override to reduce to 80%.
19	**Tool change to 6mm end mill** Swap to 6mm end mill. Re-zero Z only. Resume job for Operations 5–6 (bores and frame).
20	**Release, test mesh, finish** Cut all tabs. Release all three pieces. Test: place driver and driven gear with their centres at 67.5 mm apart. Do they mesh and turn smoothly? Sand tooth faces lightly with 220 grit. Apply Danish oil for a rich finish. Mount on the display frame using 8 mm dowels.

🏆 **Project 7 Complete — Gears That Actually Mesh**

Two wooden gears that mesh and spin smoothly — this is what CNC precision looks like in practice. The roughing + finishing strategy, the correct Module, the precise bore placement at exactly the calculated centre distance: every step of this project contributed to the final result.

If the gears bind (teeth catch): check the Module values are identical, and check the centre distance. If teeth break during cutting: increase Module, reduce feed rate, or switch to a harder timber.

Week 7, Days 31–35 — Lesson Content Overview

Day 31 (Analyse)	FCGear addon review and installation (if not done in Chapter 12). Students generate their first gear, explore all parameters, and calculate pitch diameters for three different Module/Teeth combinations. Key task: predict which two gears from a list could mesh (identify matching Module values). Discussion: what makes Module the fundamental compatibility criterion?
Day 32 (Analyse)	Gear ratio design. Students design a 3:1 gear pair (choosing their own Module and driver tooth count). Calculate all dimensions: pitch diameters, centre distance, RPM at given input. Independent: design a compound gear train (driver → intermediate → driven) that achieves a 4:1 total ratio using two gear pairs.

Day 33 (Evaluate)	Boolean operations. Students run the three-operation Try It Now (Cut, Fuse, Common) from this chapter. Then apply Boolean Cut to a practical problem: create the centre bore for their gear using Boolean Cut with a cylinder rather than Part Design Pocket — compare the two approaches. Discuss: what could Boolean do that Pocket cannot?
Day 34 (Evaluate)	Roughing + finishing CAM strategy. Students set up a roughing pass (offset +0.25) and finishing pass (offset 0) for their driver gear profile. Run the Simulator for both passes — observe the thin skin in roughing, clean removal in finishing. Compare estimated cut time for one-pass vs two-pass strategy. Justify the extra time cost.
Day 35 (Evaluate — Project Day)	Project 7: Spur Gear Pair Display.

Chapter Summary

- Involute gears produce smooth, constant force transmission during meshing and tolerate small manufacturing errors. The FCGear addon generates mathematically correct involute gear geometry from a small set of parameters.
- The key FCGear parameters are: Module (tooth size — the fundamental compatibility parameter), Teeth (count), Height (gear thickness), Pressure Angle (20° standard — do not change), and Backlash (0 for display, 0.05–0.1 for functional).
- Pitch Diameter = Module × Teeth. This formula determines gear size and centre distance. The Module Matching Rule: two gears only mesh if their Module values are identical.
- For wood CNC: use Module 3 as the minimum. Module 4–5 preferred for functional gears. Smaller Module produces teeth too fragile for timber at CNC cutting speeds.
- Gear Ratio = Driven Teeth ÷ Driver Teeth. For a 2:1 ratio: the driven gear needs exactly twice the teeth of the driver, with the same Module.
- Part Boolean operations work on complete 3D solids: Cut (A−B, subtraction), Fuse (A∪B, union), Common (A∩B, intersection). Use Boolean Cut when the shape to remove is a 3D solid rather than a 2D sketch extrusion.
- Use Part Design Pocket for standard vertical material removal from a 2D sketch. Use Part Boolean Cut for angled cuts, rotated shapes, or any operation where the cutting geometry is a 3D solid rather than a simple sketch profile.
- The roughing + finishing strategy uses two Profile passes: a roughing pass with positive offset (+0.25 mm) at standard feed, then a finishing pass at zero offset and 60–70% feed rate. The roughing pass removes bulk material; the finishing pass cuts to exact dimension with a smooth surface.

Key Vocabulary — Chapter 13

Term	Definition
Involute profile	The mathematically defined tooth curve used in standard spur gears. Ensures smooth force transmission and tolerance of centre distance variation.
FCGear	A community addon workbench for FreeCAD that generates mathematically correct gear geometry from user-specified parameters.

Module (m)	The fundamental gear sizing parameter. Module = Pitch Diameter ÷ Number of Teeth. Two gears must have identical Module values to mesh.
Pitch Diameter	The effective working diameter of a gear, equal to Module × Teeth. The circle along which meshing contact occurs.
Teeth (z)	The number of gear teeth. Combined with Module, determines the pitch diameter and gear ratio.
Pressure Angle	The angle of force transmission at the tooth contact point. Standard value: 20°. Must be identical for both gears in a meshing pair.
Backlash	A small intentional gap built into tooth spaces to prevent jamming during rotation. Set to 0 for display gears, 0.05–0.1 for functional gears.
Gear ratio	The ratio of speeds between driver and driven gears. Gear Ratio = Driven Teeth ÷ Driver Teeth.
Centre distance	The distance between the rotation axes of two meshing gears. Centre Distance = (Pitch Diameter 1 + Pitch Diameter 2) ÷ 2.
Module matching rule	Two gears can only mesh if their Module values are identical — the fundamental gear compatibility requirement.
Part Boolean Cut	A Part workbench operation that subtracts one solid from another: A − B. Works on any two 3D solids, not just sketch extrusions.
Part Boolean Fuse	A Part workbench operation that merges two solids into one: A ∪ B. The boundary between the originals is dissolved.
Part Boolean Common	A Part workbench operation that keeps only the volume shared by both solids: A ∩ B. Used for intersection analysis.
Roughing pass	A CAM Profile pass with positive offset (typically +0.25 mm) that removes bulk material at standard feed rate, leaving a thin skin for the finishing pass.
Finishing pass	A CAM Profile pass with zero offset that cuts to the exact model dimension at reduced feed rate (60–70% of roughing), producing a clean, accurate surface.
Profile offset	An Extra Offset value in the CAM Profile operation that shifts the toolpath away from the nominal model edge. Positive = leaves material (roughing). Zero = cuts to exact dimension (finishing).

Looking Ahead — Chapter 14: 3D Modelling — Datum Planes & the Loft Feature

Chapter 14 enters the advanced modelling territory of 3D organic forms. You will learn to create Datum Planes at arbitrary Z heights and angles, attach sketches to them, and use the Part Loft feature to blend smoothly between multiple cross-section profiles. This is the foundation for Project 8 — the Organic Relief Carving Panel — and for all 3D surface CAM work in Weeks 8–10.

CHAPTER 14 of 24

PART FOUR — INTERMEDIATE SKILLS

3D Modelling — Datum Planes & the Loft Feature

Datum planes, Part Loft for organic surfaces, Chamfer and Fillet on 3D forms, and Project 8 design

Chapter Learning Objectives

By the end of this chapter, you will be able to:

- Create Datum Planes at specified Z-offsets from the XY plane for use as Loft section attachment surfaces
- Create Datum Planes at specified angles to a reference face
- Attach a fully constrained Sketch to a Datum Plane
- Apply Part Loft to blend between two or more 2D section profiles on different datum planes
- Explain the difference between a Ruled Surface Loft and a smooth Loft
- Control Loft profile alignment by manipulating the start vertex of each section
- Build a 3-section and 4-section Loft for complex organic forms
- Combine a Loft solid with Part Design Pocket and Pad features using Boolean Union
- Apply Chamfer and Fillet edge treatments to Loft-derived solids and explain when to use each
- Recognise and resolve Loft twist caused by misaligned section start vertices

Every CNC part in this course until now has been flat — cut from a horizontal sheet of material. Datum planes and Part Loft change that. They give you the tools to design forms that rise, taper, and flow in three dimensions.

Part Loft is the gateway to genuinely 3D form-making in FreeCAD. It generates a solid by blending smoothly between a series of 2D cross-section profiles drawn on different planes at different heights. A circle at ground level, a smaller circle at 20 mm, and a point at 40 mm produces a smooth cone-like form. A wide oval at the base, a narrower oval at mid-height, and a sharp point at the apex produces an organic leaf or petal shape.

Before Part Loft can work, you need somewhere to draw those cross-section profiles. That is where Datum Planes come in. A Datum Plane is a reference surface that can be positioned anywhere in space and used as the attachment surface for a Sketch. Datum Planes are the scaffolding that enables multi-level 3D modelling.

This chapter covers Datum Planes and Part Loft in full, then returns to Chamfer and Fillet with a focus on their application to Loft-derived organic forms — which present different challenges to the simple block geometry of earlier chapters.

Datum Planes — Reference Geometry for 3D Design

In Part Design, every Sketch must be attached to a surface — either one of the three standard origin planes (XY, XZ, YZ), a flat face of an existing model body, or a Datum Plane. For the first seven weeks of this course, the XY plane and model faces were sufficient: all geometry existed in a single horizontal layer. Part Loft requires sketches at multiple different heights, and those heights are not faces of an existing model — they are locations in empty space. Datum Planes fill that role.

> 📖 **Key Term: Datum Plane**
>
> A reference surface created in Part Design that can be positioned at any location and orientation in 3D space, then used as the attachment surface for a Sketch. Datum Planes are not part of the physical model — they are invisible in the final part — but they are essential scaffolding for Loft sections, angled features, and any sketch that cannot be attached to an existing model face.

Z-Offset Datum Planes

A Z-offset datum plane is attached to the XY plane at a specified distance above it. This is the standard type used for Loft sections — you create a stack of datum planes at increasing Z heights, then draw a cross-section profile on each one.

1. Part Design → Create a datum → Create a datum plane. The attachment dialog opens.
2. In the References section, click the XY plane in the Model tree (or the XY grid in the 3D view).
3. Set Map Mode to 'Flat Face'. The datum plane aligns parallel to the XY plane.
4. In the Attachment Offset section, set the Z field to your desired offset — for example, 20 for a datum 20 mm above the XY plane.
5. Click OK. The Datum Plane appears in the Model tree. Rename it immediately: right-click → Rename. Use descriptive names like 'Datum_Z20' or 'Section_Mid'.
6. Repeat for each height you need. For a 3-section Loft at Z=0, 20, and 40, you need datums at Z=20 and Z=40 (the XY plane itself serves as Z=0).

> 💡 **Tip**
>
> Name your datum planes immediately after creating them. A model tree with 'DatumPlane', 'DatumPlane001', 'DatumPlane002' is nearly impossible to navigate later. Use names describing the height and purpose: 'Datum_Base_Z0', 'Datum_Mid_Z20', 'Datum_Apex_Z40'. Ten seconds now saves minutes of confusion later.

Angled Datum Planes

An angled datum plane is tilted at a specified angle relative to a reference face. Angled datums are used for features that must be oriented at an angle — a sketch for a chamfered slot, a section profile for a Loft that fans outward, or a reference for a Boolean Cut at an angle.

1. In the Datum Plane dialog, click the face or plane you want to rotate relative to.

2. Set Map Mode to 'Flat Face'.
3. In the Attachment Offset section, enter the angle in the appropriate rotation field. Rotating about X tilts front-to-back; rotating about Y tilts left-to-right.
4. Preview the tilted datum in the 3D view. Confirm the orientation before clicking OK.
5. Rename immediately.

> **Note**
>
> For the Week 8 project (Organic Relief Carving Panel), you will only use Z-offset datum planes — angled datums are not required. They are introduced here for completeness and will become relevant in more advanced work beyond this course.

Attaching a Sketch to a Datum Plane

Once datum planes exist, attaching a sketch to one follows exactly the same workflow as attaching a sketch to the XY plane. In Part Design, go to Part Design → Create a new Sketch. The attachment dialog opens. Click your desired datum plane in the Model tree. Click OK. FreeCAD enters the Sketcher with the view oriented to look directly at the datum plane — draw and constrain your sketch as normal.

> **Try It Now**
>
> Open a new FreeCAD document. Part Design → Body.
>
> Create three datum planes: attach each to XY plane with Z offsets of 0, 20, and 40. Rename all three immediately.
>
> Create a Sketch on the Z=0 datum (or XY plane directly). Draw a circle, diameter 80 mm, centred on the origin. Fully constrain and close.
>
> Create a Sketch on the Z=20 datum. Draw a circle, diameter 50 mm, centred on the origin. Fully constrain and close.
>
> Create a Sketch on the Z=40 datum. Draw a Point at the origin (Sketch → Sketcher geometries → Create Point). Close.
>
> Save the file — you now have three section sketches at three different heights, exactly what Part Loft needs.

Planning Your Datum Stack — Section Count and Heights

Before creating any datum planes, spend two minutes planning on paper. Decide how many sections you need, what shape each section will have, and what Z height each datum plane will sit at.

Section Count	Recommended For	Design Consideration
2 sections	Simple tapered forms — a rectangle tapering to a smaller rectangle, or a circle tapering to an ellipse	*Very limited shape control. The Loft interpolates directly between two profiles. Often produces a plain, uninteresting result.*
3 sections (recommended for beginners)	Leaf shapes, simple wave profiles, gentle hill forms, pedestal bases	*The middle section controls the 'fattest' or 'thinnest' point. Moving it up or down the Z stack shifts where maximum width occurs.*
4 sections	More complex organic forms with a defined waist or transition point, undulating surfaces	*Two middle sections create an S-curve effect if positioned with different sizes. Alignment becomes more critical.*
5+ sections	Complex bas-relief, multi-lobed organic forms, wave surfaces	*High control but significantly more setup time. Plan each section on paper before starting. Loft twist risk increases with section count.*

Part Loft — Blending Between Profiles

Part Loft is accessed from the Part workbench (not Part Design). This is an important distinction — Part Loft operates on Part-level objects. In practice for this course, you draw your section sketches within a Part Design Body, then apply Part Loft from the Part workbench to generate the organic solid.

> 📖 **Key Term: Part Loft**
>
> A Part workbench operation that generates a 3D solid by smoothly blending between a series of 2D profile sections. The sections can be any closed profiles (circles, ellipses, polygons, B-splines). Each section must lie on a different plane, and sections are added in order from base to apex. The resulting solid follows the shape described by the profile transitions.

Applying a Basic Loft

1. Switch to Part workbench from the Workbench drop-down.
2. Go to Part → Loft. The Loft dialog opens with an empty Profile Sections list.
3. Click 'Add Section'. Select your base section sketch from the Model tree. It appears in the sections list.
4. Click 'Add Section' again. Select the middle sketch. Repeat for all remaining sections in bottom-to-top order.
5. Add the final (top) section sketch — for a pointed apex this is a single point or very small circle.

6. Review the Solid and Ruled Surface checkboxes (explained below). For most CNC organic forms, leave Ruled Surface unticked.
7. Click OK. FreeCAD generates the Loft solid. Inspect it in the 3D view.

> ⚠ **Important**
> Part Loft requires sections to be added in order — base first, apex last. If you add sections out of order, the Loft will fold back on itself and produce a degenerate shape. Use the up/down arrows in the Loft dialog to reorder sections if needed before clicking OK.

Ruled Surface vs Smooth Loft

The Ruled Surface checkbox in the Loft dialog changes the mathematical character of the generated surface significantly.

Ruled Surface: OFF (default)	Ruled Surface: ON
The Loft blends smoothly between sections using B-spline interpolation. The surface curves gently — the cross-section at any intermediate height is a smooth blend of the sections above and below it. **Best for:** Organic relief carving, decorative forms, bas-relief panels, any surface that should look smooth and natural.	The Loft connects corresponding points on adjacent sections with straight lines. Cross-sections at intermediate heights are linear interpolations — the result looks like a faceted, tent-like surface with sharp ridges at each section level. **Best for:** Geometric tapers where the angular transition is intentional — architectural forms, prismatic shapes, faceted decorative elements.

> 💡 **Tip**
> For the Week 8 project, Ruled Surface should be OFF. The organic leaf or wave form you are designing for the relief carving should be smooth and flowing, not faceted. The Closed checkbox connects the last section back to the first — creating a torus-like shape. Leave Closed unticked for all CNC relief carving work.

Profile Alignment — Preventing Loft Twist

This is the most important — and most frequently overlooked — aspect of working with Part Loft. Misaligned start vertices produce a twisted, spiralling Loft surface. One minute of verification before running the Loft prevents a much longer diagnosis-and-fix session after.

When FreeCAD blends between two sections, it decides which points on each section correspond to each other by matching the start vertex of each section profile. If the start vertices of adjacent sections point in inconsistent directions, the Loft will twist between those sections.

> 📖 **Key Term: Start Vertex**
>
> The first point of a closed sketch profile, from which FreeCAD counts around the profile when blending between sections in a Loft. If Section A's start vertex points left and Section B's points right, the Loft will twist 180° between them. All sections in a Loft should have their start vertices pointing in the same compass direction.

Identifying and Setting Consistent Start Vertices

To see the start vertex of a sketch profile, enter the Sketcher (double-click the sketch in the Model tree). The start vertex appears as a green dot on the profile — slightly different from the other white constraint dots. For a circle, the start vertex is at the 3 o'clock position (rightmost point). For a rectangle, it is wherever you first clicked when drawing.

Circles and regular shapes	Start vertices are mathematically predictable — always at the 3 o'clock (rightmost) position. Reliable for Loft work.
Rectangles and polygons	Start vertex is where you first clicked when drawing. For consistent Lofts, always start drawing from the same corner (bottom-left) for every section.
B-splines	The first control point placed becomes the start vertex. Ensure the first control point is in the same relative position across all sections.
Verifying alignment	Open each section sketch in turn. Find the start vertex (green dot). Note which direction it points. All sections should be consistent — all at the right-hand tip, or all at the left, etc.

Fixing a Twisted Loft

If your Loft result looks twisted — the surface spirals rather than flows — re-enter the problematic sketch and move the start vertex to align with the adjacent sections. In the Sketcher, select the edge of the profile, right-click, and look for 'Set starting point of B-spline' or a similar option (version-dependent — check your FreeCAD v1.1.0 Sketcher menu under Sketch → Sketcher tools).

> 🤔 **Stop and Think**
>
> Before applying your Loft, open each section sketch one at a time. Find the start vertex (green dot) and note which direction it points: left, right, top, bottom — relative to the sketch plane. Compare all sections: are they consistent? If Section 1's start vertex points right and Section 2's points left, you will get a 180° twist. Taking this 60-second verification step before running the Loft saves the much longer process of diagnosing a twisted result after the fact.

Multi-Section Loft — Complex Organic Forms

A 3-section Loft (base, mid, apex) is appropriate for most introductory 3D CNC work. As your skills develop, adding more sections gives you progressively more control over the form's silhouette.

Section Configuration	Profile Shapes	Resulting Form
2 sections	Large circle (base) → small circle (apex)	*A smooth truncated cone — tapers uniformly. Predictable but visually simple.*
3 sections — symmetric	Large circle → medium circle → small circle	*A cone with a gentle S-curve if the middle circle is off-centre. Moving it off-centre creates an asymmetric lean.*
3 sections — leaf	Long pointed oval → shorter oval → point	*A leaf or petal — widest at the base, tapering to a point. The mid section controls whether the leaf is fat or graceful.*
4 sections — waisted	Large oval → small oval → large oval → small circle	*A vase silhouette — widens, narrows, widens, tapers. The two different mid sections create the waist and shoulder.*
4 sections — wave	Flat wide rectangle → thin rectangle → wide rectangle → flat rectangle	*A surface that compresses and expands in width as it rises — an undulating wave or dune form. Effective for decorative relief.*
5 sections — complex organic	Various ovals and ellipses at five heights	*A fully organic, free-flowing form. Alignment critical — misaligned start vertices at any level produce visible twist.*

The Leaf Loft — A Step-by-Step Worked Example

The leaf form is the standard practice exercise for Part Loft — visually satisfying, technically achievable in a single lesson, and directly useful as a relief carving subject for the Week 8 project.

1 **New file setup**
File → New. Part Design → Body. Save as LeafRelief.FCStd.

2 **Create datum planes**
Z=0 uses the XY plane directly. Create Datum_Mid at Z=25, and Datum_Apex at Z=40. Rename both in the Model tree.

3 **Base section (Z=0)**
Part Design → Create a new Sketch → XY Plane. Draw a pointed oval: horizontal construction line 100 mm through the origin, two arcs creating a lens shape — 100 mm wide, 40 mm tall at centre. Fully constrain. Close.

4

Mid section (Z=25)
New Sketch on Datum_Mid. Smaller version of the same oval: 70 mm wide, 28 mm tall. Same orientation — start vertex at the same compass point as the base section (rightmost tip). Fully constrain. Close.

5

Apex section (Z=40)
New Sketch on Datum_Apex. Single Point at the origin (Sketch → Sketcher geometries → Create Point). Close.

6

Check start vertices
Enter each of the two oval sketches. Confirm the start vertex (green dot) is at the same relative position in both — both at the right-hand tip of the horizontal axis.

7

Apply Part Loft
Switch to Part workbench. Part → Loft. Add Section: base sketch. Add Section: mid sketch. Add Section: apex point. Ruled Surface: OFF. Solid: ON. Click OK. Inspect — the result should be a smooth leaf rising from a wide base to a point.

8

Fix if twisted
If the surface spirals, re-enter the mid sketch, find the start vertex, and move it to match the base. Re-run the Loft.

9

Save
File → Save. This Loft solid is the base geometry for Week 8 3D CAM operations in Chapter 15.

Try It Now

Using the leaf Loft above, or a new one, try these experiments:

Change the mid section size: double-click the mid sketch and make it narrower (40 mm instead of 70 mm). Re-run the Loft — observe the more graceful, tapering result.

Move the mid section lower: create a new Datum_Mid2 at Z=10. Use it for the mid sketch and re-run — the widest point now occurs much lower down.

Try Ruled Surface ON: re-run the Loft with Ruled Surface ticked — observe the faceted, tent-like result. Untick to return to the smooth version.

Combining Loft Forms with Part Design Features

The solid produced by Part Loft can be used as the starting point for Part Design operations — Pocket, Chamfer, Fillet, and additional Pads can all be applied to a Loft result. For the Week 8 project, the standard approach is to begin with a flat rectangular base pad (providing the machining floor and the overall panel outline), then use Boolean Union to combine the Loft relief form sitting on top of it. The combined solid is then the subject for the 3D CAM operations in Chapter 15.

> **Note**
> In FreeCAD v1.1.0, there are multiple valid workflows for combining Loft results with Part Design geometry. The approach used in this course — building the Loft separately and combining via Boolean Union — is the most reliable for beginners. The Boolean Union approach always works.

Chamfer and Fillet on Loft-Derived Forms

Chamfer and Fillet were introduced in Chapter 6 for simple block geometry. On a Loft-derived organic form, they serve additional purposes. The base edge of a relief carving panel — where the raised organic form meets the flat machined floor — is often a sharp, fragile transition. A small fillet at this base blends the vertical edge into the horizontal floor, producing a cleaner machined result.

Chamfer Applications on Loft Forms

Panel perimeter chamfer	A 0.3–0.5 mm chamfer on all outer edges of the flat panel base body. This lead-in chamfer makes the panel easier to remove from the wasteboard and removes sharp corners that could splinter.
Base-edge chamfer	A small chamfer (0.3–0.5 mm) where the relief form meets the panel floor if the transition is a hard right-angle. Removes the sharp corner without reducing the apparent sharpness of the relief edge.
Failed chamfer	On Loft-derived solids, FreeCAD sometimes cannot chamfer a specific edge because the resulting geometry would be self-intersecting. If a chamfer fails, reduce the size or switch to a fillet.

Fillet Applications on Loft Forms

Base-blend fillet	A 1–3 mm fillet where the Loft form meets the flat panel floor. Creates a smooth blended transition — the organic form 'grows out of' the panel rather than sitting on it with a hard edge. Aesthetically superior to a chamfer for organic forms.
Top-edge softening	A very small fillet (0.5–1 mm) on the sharp apex or top edge. Softens what would otherwise be an impossibly sharp point — the CNC ball-nose bit cannot machine a true zero-radius point anyway, so this makes the model match the physical machining reality.
Interior vertical edges	If the combined Loft + panel body has interior vertical edges (where a side wall meets the flat panel), a small fillet removes the stress concentration at those corners.

Warning

Large fillets on Loft-derived edges can fail if the requested radius exceeds the minimum curvature of the nearby Loft surface. Start small (0.5 mm) and increase gradually until the shape looks right. Never set a fillet radius larger than the smallest distance between any two edges that would be affected.

Feature Order in the Model Tree

Chamfers and Fillets in Part Design are features — they appear in the Model tree and must come after the geometry they are applied to. The correct order for a combined Loft + panel workflow is:

1	Panel Sketch → Pad (the flat base panel)
2	Datum planes → Section sketches → Part Loft (the organic relief form)
3	Part Boolean Union (combining Loft with the flat panel)
4	Chamfer on panel perimeter edges (if required)
5	Fillet on base-blend edges (where Loft meets floor)
6	Any additional Pockets or Pads for functional features

PROJECT 8 — STEP-BY-STEP WALKTHROUGH

Organic Relief Carving Panel

Design an original 3D organic form using Part Loft — the 3D CAM operations are covered in Chapter 15

Project 8 has two halves: the design side (this chapter) and the CAM and cutting side (Chapter 15). Today you build the 3D model — datum planes, section sketches, Part Loft, Boolean Union with a flat base panel, and Chamfer/Fillet edge treatments. Chapter 15 will take this completed model and set up the Waterline roughing and 3D Surface finishing passes.

Design	An original organic relief form using Part Loft with a minimum of 3 datum plane sections
Material	30–40 mm thick pine or hardwood, 150 × 150 mm minimum — extra depth is essential for 3D relief
Bits	6 mm ball-nose end mill (roughing), 3 mm ball-nose end mill (finishing) — covered in Chapter 15
Day	Day 40 of Week 8 — Project Day (design phase today; CAM and cutting in Chapter 15)

Part A — Design the Relief Form

1	**New file and Spreadsheet** File → New. Part Design → Body. Save as ReliefPanel.FCStd. Spreadsheet with aliases: PanelWidth=150, PanelHeight=150, PanelThickness=35, ReliefHeight=28, MidZ=20, ApexZ=35.
2	**Flat panel base** New Sketch on XY plane. Centred rectangle: =Spreadsheet.PanelWidth × =Spreadsheet.PanelHeight. Close. Pad → =Spreadsheet.PanelThickness.
3	**Create datum planes** Datum_Mid at Z = =Spreadsheet.MidZ. Datum_Apex at Z = =Spreadsheet.ApexZ. Rename both in the Model tree.
4	**Base section sketch (Z=0)** New Sketch on XY plane. Draw a pointed oval (leaf shape): horizontal axis = 110 mm, vertical axis = 60 mm, both pointed ends on the horizontal axis, centred on the origin. Fully constrain. Close.
5	**Mid section sketch** New Sketch on Datum_Mid. Smaller pointed oval: horizontal axis = 80 mm, vertical axis = 44 mm. Same orientation — verify start vertex is at the same compass direction as the base section (both at the right-hand tip). Fully constrain. Close.
6	**Apex section sketch** New Sketch on Datum_Apex. Create a Point at the origin. Close.
7	**Apply Part Loft** Switch to Part workbench. Part → Loft. Add Section: base sketch. Add Section: mid sketch. Add Section: apex point. Solid: ON. Ruled Surface: OFF. Click OK. Inspect — should be a smooth leaf. If twisted, fix start vertex alignment and re-run.
8	**Position Loft on panel** Use Part → Move to position the Loft so its base (Z=0) sits on the top face of the flat panel (Z = PanelThickness), centred.
9	**Boolean Union** Part → Boolean → Union. Select the flat panel body and the Loft solid. Click OK. The two solids merge — the flat panel with the leaf relief rising from its top face.
10	**Chamfer and Fillet** Chamfer: select the four outer perimeter edges of the flat panel base. Size = 0.5 mm. Click OK. Optional Fillet: select the base edge of the leaf form where it meets the panel top face. Radius = 2 mm. This creates the smooth blended base transition.

Part B — CAM Preview (full detail in Chapter 15)

11 **CAM Job**
CAM workbench. CAM → Job. Select the combined solid as model. Stock: X = PanelWidth + 4, Y = PanelHeight + 4, Z = PanelThickness + 2. Stock origin: Z=0 at top of stock.

12 **Tool Library**
Add 6 mm ball-nose end mill (roughing). Add 3 mm ball-nose end mill (finishing). Hardwood feeds: roughing 1200 mm/min, plunge 400 mm/min. Finishing 800 mm/min, plunge 250 mm/min.

13 **Operation 1 — Waterline roughing**
CAM → 3D Surface (Waterline type). 6 mm ball-nose. Step-down 2 mm. Stock to leave: 0.3 mm. Run Simulator — should show horizontal slice cuts descending from the top of the stock. Full parameter detail in Chapter 15.

14 **Operation 2 — 3D Surface finishing**
CAM → 3D Surface (Surface type). 3 mm ball-nose. Step-over 0.8 mm. Final Depth: full surface. Run Simulator — should show dense contour-following toolpaths. Full detail in Chapter 15.

15 **Post-process and runtime check**
Post-process → GRBL. Save as ReliefPanel_35mm_v1.nc. Load in UGS. Check the runtime estimate before cutting — 3D CAM jobs run significantly longer than 2D jobs. The full cutting and evaluation workflow is covered in Chapter 15.

> 📋 **End of Design Phase**
>
> The design phase is complete. Save your FreeCAD file — Chapter 15 opens with this model and walks you through the full 3D CAM setup, tool change management, and evaluation of the finished carved panel.

Week 8, Days 36–40 — Lesson Content Overview

Day 36 (Evaluate)	Datum Planes — reference geometry for 3D design. Students create a stack of three datum planes, attach circle sketches at each height, and observe the three floating cross-sections in the 3D view. Discussion: why can't sections for a Loft be drawn on model faces? (The model does not yet have faces at those heights — datum planes provide attachment surfaces in empty space.)
Day 37 (Evaluate)	Part Loft — blending between profiles. Guided Loft using the three circle sections from Day 36. Students experiment with Ruled Surface ON vs OFF and compare results side by side. The twist exercise: deliberately misalign one section's start vertex, produce a twisted Loft, identify the problem vertex, and fix it.

Day 38 (Evaluate)	Chamfer and Fillet on 3D forms. Students apply a 2 mm fillet to the base edge of their leaf Loft and observe how the leaf 'grows out of' the panel. Failure test: try a 5 mm fillet — observe the failure — reduce until it succeeds. Record the maximum feasible radius for that specific geometry.
Day 39 (Evaluate)	3D CAM — Waterline and Surface operations. Students set up the complete 3D CAM job for their leaf relief: Waterline roughing + 3D Surface finishing. Run the Simulator. Justify their step-over and step-down choices for the specific material. Estimate runtime. Full 3D CAM content continues in Chapter 15.
Day 40 (Evaluate — Project Day)	Project 8: Organic Relief Carving Panel — design phase. Full model from Spreadsheet through datum planes, Loft, Boolean Union, and edge treatments. CAM setup and cutting are completed in Chapter 15.

Chapter Summary

- Datum Planes are reference surfaces positioned at any Z height or angle in 3D space. They provide attachment surfaces for Sketches where no physical model face exists — essential scaffolding for all Part Loft work. Name datum planes immediately and descriptively.
- Z-offset datum planes are the standard type for Loft sections. Create a stack at increasing Z heights before drawing any section sketches. For a 3-section Loft at Z=0, 20, and 40: you need datums at Z=20 and Z=40 (the XY plane serves as Z=0).
- Part Loft (accessed from the Part workbench) generates a 3D solid by blending between a series of 2D section profiles in order — base first, apex last. Sections out of order produce a degenerate folded shape.
- Ruled Surface OFF produces smooth, B-spline-blended transitions between sections — the correct setting for all organic CNC relief forms. Ruled Surface ON produces flat-faced, faceted transitions — use only for intentionally geometric tapers.
- Start vertex alignment is the most critical Loft quality control step. All section sketches must have their start vertices pointing in the same compass direction. Misalignment causes visible twist — the surface spirals instead of flowing. Verify all start vertices before running the Loft.
- For the Week 8 project, the recommended workflow is: flat panel body (Pad) → Loft sections (datum planes + sketches) → Part Loft → Boolean Union (combining Loft with flat panel) → Chamfer/Fillet (edge treatments).
- Chamfer is preferred for lead-in edges and geometric panel perimeters. Fillet is preferred for the base-blend where the Loft form meets the panel floor — it creates a smooth organic transition rather than a hard step. Start with small radii (0.5 mm) and increase — large fillets on Loft geometry can fail if the radius exceeds the minimum curvature of the nearby surface.
- The 3D CAM operations (Waterline roughing and 3D Surface finishing) for the completed relief panel model are covered in Chapter 15.

Key Vocabulary — Chapter 14

Term	Definition
Datum Plane	A reference surface created in Part Design that can be positioned at any location and orientation in 3D space. Not part of the physical model — used as an attachment surface for Sketches in locations where no model face exists.
Z-offset datum	A Datum Plane attached to the XY plane with a specified vertical offset. The standard type for Loft section positioning — creates a plane floating above the base at a defined height.
Angled datum	A Datum Plane tilted at a specified angle to a reference face. Used for angled features, chamfered slots, and section profiles that must fan outward.
Map Mode	The attachment mode setting in the Datum Plane dialog that determines how the plane aligns to its reference. 'Flat Face' is the most commonly used mode for both Z-offset and angled datums.
Part Loft	A Part workbench operation that generates a 3D solid by smoothly blending between a series of 2D profile sections. Sections must be on different planes and added in order from base to apex.
Section profile	A 2D sketch attached to a datum plane that defines the cross-sectional shape of the Loft at that height. The Loft blends continuously between all section profiles.
Ruled Surface	A Loft option that connects adjacent sections with straight lines (faceted result) rather than smooth B-spline blending. OFF for organic forms; ON for geometric tapers.
Start vertex	The first point of a closed sketch profile, used by Part Loft as the starting reference when blending between sections. All sections in a Loft must have their start vertices pointing in the same direction to avoid twist.
Loft twist	A defect in a Part Loft result where the surface spirals rather than flowing smoothly between sections. Caused by inconsistent start vertex alignment between adjacent section sketches.
Boolean Union	A Part workbench operation that merges two separate solids into one unified body. Used in the Week 8 workflow to combine the flat base panel (Part Design Pad) with the organic Loft relief form.
Base-blend fillet	A fillet applied at the edge where a Loft relief form meets a flat panel floor, creating a smooth blended transition. Aesthetically superior to a hard step for organic forms in hardwood.
Stock to leave	A 3D CAM roughing setting that deliberately leaves a thin skin of material (typically 0.3 mm) on all surfaces for the finishing pass to remove cleanly. Ensures the finishing pass cuts into undisturbed material.

Looking Ahead — Chapter 15: 3D CAM — Waterline, Surface & Relief Machining

Chapter 15 picks up exactly where this chapter ends. Your completed Organic Relief Carving Panel model is the starting point. Chapter 15 covers ball-nose end mill selection, Waterline roughing configuration, 3D Surface finishing with step-over control, scallop height and surface quality, multi-pass strategy, tool change management for 3D jobs, and the complete cutting and evaluation workflow for Project 8.

CHAPTER
15
of 24

PART FIVE — ADVANCED SKILLS

3D CAM — Waterline, Surface & Relief Machining

How the CNC machine follows curved surfaces, ball-nose bits, scallop height, multi-pass strategy, and Project 8

Chapter Learning Objectives

By the end of this chapter, you will be able to:

- Explain the fundamental difference between 2D and 3D CAM
- Describe why a ball-nose end mill is required for 3D surface machining
- Configure a CAM Job Stock for an organic 3D form
- Create a Waterline roughing operation with correct step-down and boundary settings
- Create a 3D Surface operation in both roughing and finishing modes
- Explain the relationship between step-over, scallop height, and surface quality
- Estimate 3D CAM job runtime and plan accordingly
- Configure the two-pass strategy (Waterline roughing + 3D Surface finishing) for Project 8
- Manage two tool changes in UGS for a multi-pass 3D job
- Post-process a 3D CAM job and verify it correctly in the UGS visualiser

When CNC Stops Being Flat

Everything you have cut so far has had flat surfaces and straight walls. With 3D CAM, the tool chases a curved surface — moving in X, Y, and Z simultaneously to trace every contour of an organic form.

Every project from Chapter 1 through Chapter 14 has involved 2D CAM: the cutting tool moves in X and Y to trace paths, and Z only changes between passes (moving down to a new depth). The surface produced is flat — a series of vertical walls and horizontal floors.

3D CAM is categorically different. The tool follows the actual three-dimensional surface of the model — moving in X, Y, and Z simultaneously to trace every curve, slope, and contour. The result is a machined surface that replicates the original design in three dimensions: a carved leaf, a bas-relief portrait, a flowing organic form. It is the difference between a box and a sculpture.

3D CAM is also significantly more demanding than 2D CAM: longer setup time, much longer run time, more complex tool change management, and a new set of parameters (step-over, scallop height, waterline depth) that do not exist in 2D work. This chapter covers all of it — from the conceptual difference to the practical Project 8 workflow.

	2D CAM	3D CAM
What the tool follows	2D edges, profiles, and enclosed area outlines on a flat plane	The actual 3D contoured surface of the model — every hill, valley, and curve
Cutting bit shape	Flat end mill is standard	Ball-nose end mill required — a flat bit cannot follow curved surfaces cleanly
Z movement	Primarily vertical at the start and end of each pass — Z stays constant during cutting moves	Z changes continuously as the tool traces the 3D surface — it moves in X, Y, and Z simultaneously
Typical operations	Engraving, Profile, Pocket Shape — all 2D	Waterline, 3D Surface, Pencil Trace — all 3D
Job setup complexity	Stock matches material size — straightforward	Stock must be a rectangular block slightly larger than the model in all directions
Typical run time	Minutes to tens of minutes	Tens of minutes to several hours — plan accordingly
Surface result	Flat, clean, machined faces	Smooth curved surface with a slight ripple (scallop) between adjacent passes

The Ball-Nose End Mill — Why It Is Non-Negotiable

A standard flat end mill has a flat cutting face at its tip. When following a curved surface, the corner of the flat tip would dig into the material at any slope — it cannot blend smoothly into curves. The result would be a faceted, chipped surface rather than a smooth carving.

A ball-nose end mill has a hemispherical (half-sphere) cutting tip. The curved tip can follow any surface angle smoothly — flat, sloped, or curved — because it always contacts the material at a tangent point rather than a hard edge. The ball shape is what makes smooth 3D surface machining possible.

> 📖 **Key Term: Ball-nose end mill**
>
> A cutting bit with a hemispherical (half-sphere) tip, used for 3D surface machining. The curved tip traces curved surfaces smoothly without the chipping and gouging that a flat end mill would produce on non-vertical surfaces.

6 mm ball-nose	Standard roughing bit for 3D CAM in this course. Removes material quickly in Waterline roughing passes. At 1.5 mm step-over finishing, produces a surface ready for light sanding.
3 mm ball-nose	Fine finishing bit. Used for Pencil Trace and high-quality finishing passes (step-over 0.8 mm or less). Produces a much smoother surface but takes 3–5× longer per pass than the 6 mm bit.
Shank size	Both balls in this course use 3.175 mm (1/8-inch) shanks to fit the CNC 4040 Pro Max's ER11 collet. Confirm your bit's shank diameter before inserting it.
Cutting speed	Ball-nose bits in wood: 15,000–18,000 RPM for 6 mm. 18,000–21,000 RPM for 3 mm. Higher RPM for smaller bits — they need more cuts per revolution to maintain chip load.

> **Tip**
>
> Always inspect ball-nose bits before a 3D job. A chipped or worn ball tip will leave a track mark on the surface — a straight groove where the damaged part of the ball passes. Hold the bit up to a strong light and look at the hemisphere from below. If the edge is not perfectly smooth and consistent, replace the bit before starting a long job.

3D CAM Job Setup — Stock for Organic Forms

Setting up the CAM Job for a 3D model requires more careful stock configuration than a flat 2D part. The stock must enclose the entire model — not just its base footprint, but its full three-dimensional height — and must be positioned correctly so that the Z=0 origin corresponds to the top surface of the physical material.

Configuring Stock for a 3D Model

1. Switch to CAM workbench. Go to CAM → Job. Select your 3D body (the Loft form from Chapter 14 exercises).
2. Double-click Stock in the Job tree to open its settings.
3. Set X and Y dimensions: the stock should be at least 10–15 mm larger than the model's footprint in each direction. This provides safe clearance for the tool on entry and exit.
4. Set Z height: the stock height must equal the maximum height of your 3D model plus 2–5 mm. If your Loft peaks at 25 mm above the base, your stock should be at least 28–30 mm tall.
5. Set the Stock origin: the most common configuration is Z=0 at the TOP of the stock (the model's highest point). FreeCAD positions the stock so the model sits within it with the top of the model flush with the top of the stock.
6. Check the Simulator preview — the model should be completely enclosed within the stock block.

> **Important**
>
> If the Stock Z height is too small (shorter than the model's maximum height), the Waterline and 3D Surface operations will not generate passes deep enough to fully machine the model.
>
> Always verify the stock height against your actual material thickness before cutting. A model that is 25 mm tall requires material at least 28 mm thick.

Waterline Operation — Horizontal Slice Roughing

The Waterline operation is named after the technique it uses: it makes horizontal slices through the model at regular Z intervals — like the waterline marks on a boat hull. Each slice produces a cutting pass at that Z height. The result is a 'staircase' of horizontal passes that approximates the 3D form, removing most of the material while leaving the staircase pattern on the surface for the finishing pass to smooth away.

Waterline is the preferred roughing strategy for 3D work in this course. It is computationally efficient, predictable, and works well on any organic form regardless of surface direction. The Waterline is to 3D roughing what ZigZag is to 2D pocket clearing — fast, simple, and reliable.

Configuring a Waterline Operation

1. Go to CAM → Waterline (this may also appear as 3D Surface with Pattern set to Waterline in some FreeCAD v1.1.0 installations).
2. In Base Geometry, click Add and select the 3D body you are machining.
3. In the Tool section, select your 6 mm ball-nose end mill.
4. Set Step Down: the vertical distance between adjacent waterline slices. For 3D roughing in timber: 2–3 mm. A finer step-down produces more slices and a smoother staircase, but takes longer.
5. Set the Boundary: choose Stock to restrict the tool to within the stock boundaries. This prevents the tool from making unnecessary passes outside the material.
6. Set Feed Rate: 1200–1400 mm/min for roughing in timber with a 6 mm ball-nose.
7. Set Plunge Rate: 400 mm/min.
8. Click OK. The Waterline toolpath appears — a set of concentric horizontal passes descending from the top of the model.

Step Down 1 mm	Very fine staircase — many passes, smooth roughing result. Rarely needed for initial roughing.
Step Down 2 mm	Good roughing balance — recommended starting value for most 3D relief work.
Step Down 3 mm	Coarser staircase — fewer passes, faster. Acceptable for simple organic forms. May be aggressive for sharp features.
Step Down 5+ mm	Very coarse — only appropriate for very simple curved surfaces where the staircase is not a concern.

> 💡 **Tip**
>
> The Waterline roughing pass intentionally leaves a rough staircase surface. This is correct and expected. Do not be alarmed by the jagged appearance in the Simulator — that is exactly what roughing should look like. The finishing pass removes the staircase entirely.

3D Surface Operation — Finishing the Contours

The 3D Surface operation makes the tool follow the exact three-dimensional contours of the model, moving in X, Y, and Z simultaneously to trace every curve. Unlike Waterline (which makes horizontal slices), the 3D Surface tool path follows the surface topography directly — climbing hills, descending valleys, curving around transitions.

3D Surface has two modes: Roughing (controlled by step-down, like Waterline but following the surface) and Finishing (controlled by step-over — the horizontal spacing between adjacent surface-following passes). For this course's standard workflow, Waterline handles roughing and 3D Surface in Finishing mode handles the final quality pass.

Configuring a 3D Surface Finishing Operation

1. Go to CAM → 3D Surface.
2. In Base Geometry, select the 3D body.
3. Select Tool: 6 mm ball-nose (for the finishing pass at 1.5 mm step-over) or 3 mm ball-nose (for a fine-detail finish at 0.8 mm step-over).
4. Set Pattern/Direction. Options include: X (parallel passes in X direction), Y (parallel passes in Y), Spiral (outward from centre), Offset (concentric following model outline). For most organic forms: X or Spiral produces the most natural-looking result.
5. Set Step Over: the horizontal distance between adjacent passes. This is the most important quality parameter (see the Scallop Height section below). For standard finishing: 1.5 mm.
6. Set Start Depth: 0 (top of stock).
7. Set Final Depth: the full model depth (e.g., -25 mm for a model 25 mm tall).
8. Set Feed Rate: 700–800 mm/min for finishing with a 6 mm ball-nose.
9. Click OK. The finishing toolpath appears — dense parallel lines following the surface contours.

Scallop Height — The Key to Surface Quality

Every 3D surface machined with a ball-nose bit has a pattern of tiny ridges between adjacent passes. Understanding scallop height lets you control surface quality deliberately rather than hoping for a good result.

When a ball-nose bit makes two adjacent passes with a step-over between them, the space between the passes is not fully removed — a tiny ridge of material, called a scallop, remains between each pair of passes. The height of this scallop determines how smooth the surface feels and looks. Lower scallop height = smoother surface. Higher scallop height = visible ripple between passes.

Scallop Height — What It Is and Why It Matters

```
Ball-nose bit profile (side view):
      Pass 1          Pass 2          Pass 3
         ⌒               ⌒               ⌒       ← Ball-nose paths
──────^──────────^──────────^─────  ← Material surface
      ↑  scallop  ↑  scallop  ↑      ← Ridges left between passes
```

Each scallop is a small ridge of material left between adjacent ball-nose passes. Smaller step-over = closer passes = lower scallops = smoother surface. But more passes = much longer run time.

The scallop height is determined by the step-over and the ball-nose radius. The relationship is: smaller step-over → closer passes → lower scallops → smoother surface. But the trade-off is dramatic: halving the step-over roughly quadruples the number of passes — and quadruples the runtime.

The table below gives practical guidance for a 6 mm ball-nose bit (radius = 3 mm). Use it to choose a step-over that gives you the surface quality you need within an acceptable run time.

Step-Over	Scallop Height	Surface Quality	When to Use
0.4 mm	**<0.003 mm**	Mirror-smooth. Identical to sanding with 400 grit. 3–5× longer than 1.5 mm setting.	*Trophy plaques, fine art, items where no sanding is desired*
0.8 mm	**~0.013 mm**	Very smooth. Light sanding to 220 grit produces a perfect finish.	*High-quality display pieces, clock faces, detailed relief panels*
1.5 mm	**~0.047 mm**	Smooth. Visible under raking light, disappears with light sanding (180 grit).	*Standard finishing pass — the default for most 3D work in this course*
3.0 mm	**~0.188 mm**	Noticeable ripple. Requires significant sanding to remove.	*Roughing pass finish quality — acceptable as roughing only*
5.0 mm	**~0.521 mm**	Coarse staircase pattern. Requires aggressive sanding or a further finishing pass.	*Rough bulk removal only — never use as a final finish*

> **Tip**
>
> For Project 8 (Organic Relief Carving Panel), start with a 1.5 mm step-over. After cutting, sand lightly with 180-grit paper — the surface will become very smooth with minimal sanding. If you want a no-sand result for a display piece, reduce step-over to 0.8 mm and accept the longer run time.

Runtime Estimation — Planning for Long Jobs

3D CAM jobs are fundamentally different from 2D jobs in one critical way: they can take hours rather than minutes. A flat name badge takes 15 minutes. A 3D relief carving of similar physical size can take 90 minutes for roughing and another 60–120 minutes for finishing. You must know the estimated runtime before you start.

⏱ **Runtime Reality**

UGS shows an estimated runtime after you load a .nc file. Look at this number BEFORE starting any 3D job.

A typical Project 8 job (150 × 150 mm, 25 mm deep): roughing ≈ 25–40 min, finishing at 1.5 mm step-over ≈ 45–75 min. Total with tool change: 70–120 minutes.

You cannot leave the machine unattended for any of this time. Plan your session accordingly.

If the estimated runtime exceeds your available supervised time, reduce the model size, increase step-over, or split the job across two sessions (save the finishing pass G-code separately).

How to Check Runtime in UGS

1. Post-process your job and save the .nc file.
2. Open UGS. Load the .nc file.
3. In the UGS interface, look for a runtime estimate in the file info area (typically shown near the file name or in the machine status panel). It appears after the file is loaded and the visualiser has processed the toolpath.
4. If UGS does not show a runtime, count the number of toolpath lines in the file — divide by your expected lines-per-minute (approx. 800–1200 for typical 3D finishing) to estimate.
5. Always add 20% to any estimate to account for tool changes, setup time, and the occasional pause.

Strategies for Managing Long Run Times

- Run roughing and finishing as separate G-code files — if your session is interrupted after roughing, you can return and run only the finishing pass.
- Increase step-over for an initial 'proof' cut (e.g., 3 mm step-over for a quick test) before committing to the fine finishing pass.
- Reduce the model size. A 100 × 100 mm carving at 1.5 mm step-over takes roughly 60% of the time of a 130 × 130 mm carving at the same setting.
- If the job must run in one session, ensure you have uninterrupted access to the machine for the full estimated time plus buffer.

The Multi-Pass Strategy — Roughing, Finishing, and Pencil Trace

Professional 3D CNC relief machining uses a sequence of passes rather than a single operation. Each pass has a different priority — speed, quality, or detail recovery — and together they produce a result that no single pass can achieve alone. The three-pass strategy is the professional standard, though for many course projects the two-pass version (roughing + finishing) is sufficient.

Pass	Operation	Bit	Setting	Surface	Effect on Material
PASS 1	**Roughing — Waterline**	*6 mm ball-nose*	Step-down: 2–3 mm	Fast, staircase surface	Removes ~80–90% of material
PASS 2	**Scallop Finishing — 3D Surface**	*6 mm ball-nose*	Step-over: 1.5 mm	Smooth, final-quality surface	Removes staircase, achieves final form
PASS 3	**Pencil Trace (optional)**	*3 mm ball-nose*	Traces ridges and valleys	Crisp detail in fine geometry	Recovers detail in narrow areas the larger bit missed

When Two Passes Are Enough

For most Project 8 outcomes — a decorative leaf, a wave, an abstract organic form — the two-pass strategy (Waterline roughing + 3D Surface finishing) produces an excellent result that requires only light sanding. The Pencil Trace third pass is justified when:

- The model has narrow ridges, fine lettering, or deep narrow valleys that the 6 mm finishing bit physically cannot reach.
- The piece will be displayed without any sanding — the Pencil Trace recovers the last fine details.
- The piece is a commission or a high-value item where maximum quality is required regardless of run time.

> 📋 **Note**
>
> Chapter 16 (Advanced CAM Strategies) covers the Pencil Trace and Rest Machining in detail. For Project 8, the two-pass strategy is the requirement. The Pencil Trace is introduced as an optional extension for students who want to explore further.

Tool Change Management for Multi-Pass 3D Jobs

A two-pass 3D job with different bits (6 mm roughing, 3 mm finishing) requires a tool change between passes. In a 2D job, a tool change pauses the job briefly and resumes quickly. In a 3D job, the tool change requires particular care because re-zeroing Z with a different bit is critical — the finishing pass must start exactly at the same Z=0 as the roughing pass. Any error in Z zero will produce a surface that is either too high (leaving a skin of uncut material) or too low (cutting into previously finished surfaces).

The Tool Change Procedure for 3D Jobs

1. After the roughing pass completes, the spindle stops (M5 command in G-code). UGS will pause the job if M6 (tool change) is in the file, or the job will complete and you load the finishing pass file separately.
2. Jog Z up to give clear working height (at least 50 mm above the workpiece — the 3D form has height, so you need more clearance than a flat job).
3. Remove the roughing bit. Insert the finishing bit (3 mm ball-nose).
4. Critical: re-zero Z precisely. Lower the Z slowly until the 3 mm ball-nose tip just touches the original material surface level. Use the paper method. The material surface is now partially carved — find the highest original surface level (usually at the very edge of the stock, where no material has been removed) and zero there.
5. Zero Z. Raise Z 5 mm.
6. If the finishing pass is a separate file: load the finishing .nc file in UGS. Verify the visualiser shows the finishing toolpath. Do not re-zero X or Y — they have not changed.
7. Resume the finishing pass.

> ⚠ **Important**
>
> The highest risk in multi-pass 3D tool changes is Z zero error. Even 0.2 mm of Z error will leave a visible step in the finished surface where the finishing pass starts slightly too high or too low.
>
> Always zero Z from the original stock surface level — the undisturbed flat area at the edge of the material, not from the carved surface itself.

Post-Processing and UGS Verification for 3D Jobs

Post-processing a 3D CAM job follows the same procedure as any other job — select the GRBL post-processor and save the .nc file. However, 3D jobs require additional verification steps in UGS because the toolpath is far more complex than a 2D job and errors are harder to spot visually.

Verifying a 3D Job in UGS

- Load the .nc file in UGS. Wait for the 3D visualiser to fully render the toolpath — this may take 10–30 seconds for a complex 3D job.
- Rotate the visualiser to view the toolpath from multiple angles. Does it clearly show the shape of the model being carved?
- Check that the toolpath lines are densely packed across the entire model surface — gaps or missing areas suggest a Base Geometry selection error.
- Check that rapid moves (yellow lines) stay above the stock surface. Any yellow line that dips into the material represents a crash waiting to happen.

- Check the depth: the deepest toolpath lines should reach the bottom of your model (the base of the carved form, not the base of the stock — the carving does not go all the way through).
- Note the estimated runtime. Confirm you have sufficient supervised time available before pressing Send.

> **Tip**
>
> Save the roughing pass and finishing pass as separate .nc files: CarvedLeaf_Rough_v1.nc and CarvedLeaf_Finish_v1.nc. This gives you flexibility — you can run the roughing pass in one session, verify the result, and run the finishing pass in the next session without starting the entire job over.

PROJECT 8 — STEP-BY-STEP WALKTHROUGH

Organic Relief Carving Panel

A 3D Loft-based organic form machined using Waterline roughing and 3D Surface finishing — the first true 3D CNC project

Project 8 is the first fully three-dimensional CNC project in the course. The model is the Part Loft form designed in Chapter 14 (a leaf, wave, petal, or similar organic shape). The CAM strategy is Waterline roughing followed by 3D Surface finishing. The result is a sculptural relief carving — a piece that looks and feels fundamentally different from everything you have made so far.

Design	Organic relief form from Chapter 14 — Loft with 3+ datum plane sections
Size	150 × 150 mm footprint, 20–30 mm peak height (above the flat base)
Material	30–40 mm thick pine, cedar, or hardwood. MDF is acceptable but produces less attractive grain contrast.
Bits	6 mm ball-nose (roughing Waterline + finishing 3D Surface), optional 3 mm ball-nose (fine finishing)
CAM passes	Pass 1: Waterline roughing (6mm, step-down 2.5mm). Pass 2: 3D Surface finishing (6mm, step-over 1.5mm)
Estimated time	Roughing: 25–40 min. Finishing: 50–80 min. Total: 75–120 min.
Day	Day 40 of Week 8 — Project Day

Part A — Verify Your Chapter 14 Model

1 **Open your Loft model from Chapter 14**
Open the LeafRelief.FCStd (or equivalent organic Loft model) from Chapter 14. Verify the model looks correct: a smooth 3D organic form sitting on a flat base, peak height 20–30 mm. If you have not completed Chapter 14, create a simple 3-section Loft now: oval at Z=0 (100×60 mm), smaller oval at Z=20 (40×25 mm), and a point at Z=25.

2 **Check model dimensions**
In the Properties panel → Data → Placement, verify the model origin is at Z=0 (base level). Measure the peak height — the model must fit within your 30–40 mm thick material.

Part B — CAM Setup

3 **Switch to CAM, create Job**
CAM → Job. Select your Loft body. Configure Stock: X = model_width + 20, Y = model_depth + 20, Z = material_thickness (e.g., 35 mm). The model should be centred within the stock horizontally, and the model base should sit at the stock bottom.

4 **Add both ball-nose tools**
Tool Library: T1 = BallNose6mm (Type: Ball End Mill, Diameter: 6 mm). T2 = BallNose3mm (Diameter: 3 mm, optional).

5 **Pass 1 — Waterline Roughing**
CAM → Waterline (or 3D Surface → Waterline pattern). Tool: T1 BallNose6mm. Step Down: 2.5 mm. Boundary: Stock. Feed: 1300 mm/min. Plunge: 400 mm/min. Click OK. Estimated time in Simulator: note it.

6 **Pass 2 — 3D Surface Finishing**
CAM → 3D Surface. Tool: T1 BallNose6mm. Pattern: X (parallel passes). Step Over: 1.5 mm. Start Depth: 0. Final Depth: full model depth. Feed: 750 mm/min. Plunge: 250 mm/min. Click OK. Estimated time: note it. Total job time = Pass 1 + Pass 2 + tool change overhead.

7 **Optional Pass 3 — Fine Finishing**
If you have a 3 mm ball-nose and want maximum quality: CAM → 3D Surface. Tool: T2 BallNose3mm. Step Over: 0.8 mm. Feed: 550 mm/min. Click OK. Note the estimated time — this pass is often 45–90 min on its own.

8 **Run full Simulator**
CAM → CAM Simulator. Run all passes in sequence. After roughing simulation: does the staircase surface approximate your organic form? After finishing simulation: does the smooth contoured surface match your design? If the finishing pass has visible gaps or missed areas, check the Base Geometry selection.

9 **Post-process — save two separate files**
Select Job. CAM → Post Process. grbl. Save roughing as LeafRelief_Rough_v1.nc and finishing as LeafRelief_Finish_v1.nc. Two separate files gives you flexibility.

Part C — Machine

10 **Pre-cut checklist — all sections, extra care on Z**
For 3D work: also check that material is thick enough (≥ peak model height + 5 mm). Check that clamps do not obstruct the Z-axis travel range needed for the model height.

11 **Secure material — screws preferred**
For long 3D jobs, screw the material down from the sides rather than relying on clamps alone — a 90-minute job stresses fixturing more than a 15-minute job. Add double-sided tape under the material as supplementary adhesion.

12 **6mm ball-nose, set X/Y/Z zero**
Insert 6mm ball-nose. Jog to the centre of the material (centred design = centred zero). Set X and Y zero. Lower Z using paper method to the flat top surface of the material. Set Z zero. Raise Z 5 mm.

13 **Load roughing file, verify, run Pass 1**
Load LeafRelief_Rough_v1.nc in UGS. Verify 3D visualiser — Waterline passes should be visible as stacked horizontal rings. Note estimated runtime. PPE on. Send. Supervise entire roughing pass.

14 **Inspect roughing result**
After roughing: brush away chips. Look at the carved surface. You should see a clear staircase approximation of your leaf or organic form. If the form is recognisable — roughing succeeded. If the surface looks flat or wrong — stop and investigate the Z zero or model height before continuing.

15 **Tool change: re-zero Z from original surface level**
Jog Z up 60 mm. Same 6mm ball-nose (no physical tool change unless you are using the 3mm for finishing). The bit has not changed — but RE-ZERO Z is still required if there was any spindle removal. Use the flat edge of the stock (undisturbed surface) for Z zero — not the carved area.

16 **Load finishing file, verify, run Pass 2**
Load LeafRelief_Finish_v1.nc. Visualiser should show dense parallel passes following the surface contours. Send. The finishing pass is slower — the sound of the machine will be softer and more consistent than roughing.

17 **Optional: tool change to 3mm, run Pass 3**
If using the 3mm ball-nose for fine finishing: jog up, swap bits, re-zero Z from flat stock edge (critical), load fine finishing file, send.

18 **Release and finish**
Spindle off. Z raised. Remove screws/clamps. Carefully lift the carved piece. Sand with 180-grit very lightly — mostly to remove any fuzz. Progress to 240-grit. Apply Danish oil or clear varnish. Multiple thin coats bring out the grain depth through the carved contours.

🏆 **Project 8 Complete — Your First Relief Carving**

Project 8 produces a piece that could not be made without CNC. The smooth organic surface, the precise depth gradation, the clean base — these are characteristics of 3D machining that no hand tool can replicate at this quality and consistency.

If you are not happy with the surface quality: note the step-over value you used and reduce it by 30% for the next attempt. The difference between 1.5 mm and 1.0 mm step-over is dramatic — and now you understand exactly why.

Week 8, Days 36–40 — Lesson Content Overview

Day 36 (Evaluate)	Datum Planes and Part Loft (Chapter 14 content). Students create their Loft model for Project 8. Instructor connects Chapter 14's 3D modelling to Chapter 15's 3D CAM — establishing the model-to-machine pipeline for 3D work.
Day 37 (Evaluate)	3D CAM fundamentals — Waterline vs 3D Surface, ball-nose requirements, scallop height. Students configure a Waterline roughing pass on a simple dome shape. Run the Simulator. Measure estimated runtime. Discussion: how does changing step-down from 2 mm to 1 mm affect the runtime and result?
Day 38 (Evaluate)	3D Surface finishing operation. Students add a finishing pass to their Day 37 job. Compare 1.5 mm vs 0.8 mm step-over in the Simulator — observe the visual difference and the runtime difference. Which would they choose for Project 8, and why?
Day 39 (Evaluate → Create)	Runtime management and tool change procedure for 3D jobs. UGS verification of a 3D toolpath. Students post-process their Project 8 CAM job into two separate files (roughing and finishing). Verify both in UGS. Complete the pre-cut checklist for a 3D job.
Day 40 (Create — Project Day)	Project 8: Organic Relief Carving Panel.

Chapter Summary

- 3D CAM moves the tool in X, Y, and Z simultaneously to follow curved surfaces. This is fundamentally different from 2D CAM, which only changes Z between passes. 3D CAM requires a ball-nose end mill, longer setup, and significantly longer run times.
- A ball-nose end mill has a hemispherical tip that can follow any surface angle smoothly. It is non-negotiable for 3D surface work. Inspect ball-nose bits before any long job — a chipped tip leaves a visible track mark on the surface.
- 3D CAM Stock must enclose the full 3D height of the model. Set Stock Z height = model peak height + 2–5 mm. The model base sits at the stock base; Z=0 is at the top of the stock (top of the material surface).
- Waterline operation makes horizontal slice passes at regular Z intervals — a staircase approximation of the 3D surface. It is the preferred roughing strategy. Step-down of 2–3 mm is the practical range for timber.
- 3D Surface operation follows the actual contoured surface of the model. In Finishing mode, it is controlled by step-over — the horizontal spacing between adjacent passes. Standard finishing step-over: 1.5 mm.
- Scallop height is the tiny ridge left between adjacent ball-nose passes. It determines surface quality. The scallop height decreases as step-over decreases, but run time increases steeply. 1.5 mm step-over is the practical sweet spot for this course.
- Always check the estimated runtime in UGS before starting a 3D job. Long jobs (60+ minutes) require full supervised time planning. Save roughing and finishing as separate .nc files for session flexibility.
- Multi-pass tool changes in 3D jobs require Z re-zeroing from the flat stock edge — not from the carved surface. Any Z zero error produces a visible step in the finished carving.

Key Vocabulary — Chapter 15

Term	Definition
3D CAM	Computer-Aided Manufacturing operations where the cutting tool moves in X, Y, and Z simultaneously to follow the contours of a three-dimensional surface.
Ball-nose end mill	A cutting bit with a hemispherical tip, required for 3D surface machining. The curved tip traces curved surfaces without chipping or gouging.
Waterline	A 3D CAM roughing operation that makes horizontal slice passes at regular Z intervals — named after waterline marks on a boat hull. Produces a staircase approximation of the 3D form.
3D Surface	A 3D CAM operation where the tool follows the exact contoured surface of the model. Used in Finishing mode (step-over based) to produce the final smooth surface.
Scallop	A tiny ridge of material left between adjacent ball-nose passes. Its height determines surface quality — lower scallop = smoother surface.
Scallop height	The height of the ridge left between adjacent ball-nose passes. Related to step-over: smaller step-over = lower scallop = smoother surface, but more passes and longer run time.
Step-over	In 3D Surface finishing: the horizontal distance between adjacent passes. Controls surface quality — 1.5 mm is standard for this course.

Step-down	In Waterline roughing: the vertical distance between adjacent horizontal slice passes. 2–3 mm is standard for 3D roughing in timber.
Multi-pass strategy	A 3D machining approach using multiple operations in sequence: Waterline roughing (speed), 3D Surface finishing (quality), optional Pencil Trace (detail recovery).
Pencil Trace	An optional third CAM pass that traces ridges, valleys, and sharp feature edges to recover fine detail not reached by the finishing bit. Covered in Chapter 16.
Runtime estimate	The projected time to complete a G-code job, shown in UGS after a file is loaded. Must be checked before starting any 3D job.
Boundary mode	A Waterline setting that restricts the toolpath to within the Stock boundaries, preventing unnecessary passes outside the material.

Looking Ahead — Chapter 16: Advanced CAM Strategies

Chapter 16 builds on the 3D CAM foundation with three advanced strategies: Adaptive pocket clearing (constant chip load for deeper/harder pockets), Pencil Trace (ridge-following detail pass), and Rest Machining (targeting only areas the previous larger bit could not reach). Together these form the three-pass professional standard for maximum-quality 3D relief machining.

CHAPTER
16
of 24

PART FIVE — ADVANCED SKILLS

Advanced CAM Strategies

Adaptive clearing, scallop finishing, pencil trace, rest machining, and the 3-pass professional workflow

Chapter Learning Objectives

By the end of this chapter, you will be able to:

- Explain the mechanical difference between ZigZag and Adaptive (trochoidal) pocket clearing at the tool-material interface
- Describe chip load and tool engagement angle and explain why Adaptive controls them where ZigZag does not
- Configure a CAM Pocket Shape operation with Adaptive strategy and set appropriate step-over and engagement values
- Apply the decision framework to choose between ZigZag and Adaptive for any given scenario
- Describe the scallop finishing strategy and explain what 'constant cusp height' means
- Create a CAM Pencil Trace operation and explain which surfaces benefit from it
- Configure rest machining on a 3D finishing pass to target only areas the roughing bit could not reach
- Build a complete 3-pass CAM strategy (roughing, scallop, pencil) and estimate its total runtime
- Evaluate quality versus time tradeoffs and justify strategy choices for different project types

The CNC machine does exactly what the G-code tells it to do. The quality of the result, the life of the cutting tool, and the time it takes are all determined before a single line of code is sent — in the CAM strategy decisions you make here.

The CAM operations you have used up to this point — Profile, Pocket Shape with ZigZag, Engraving, and the basic 3D Waterline and Surface passes from Chapter 15 — are the standard tools for the majority of CNC routing work. They are correct, capable, and sufficient for most tasks.

This chapter introduces three operations that move beyond the standard toolkit: Adaptive clearing (also called trochoidal milling), scallop finishing, and pencil trace. Together, these three strategies form a professional-grade 3-pass CAM workflow used in production CNC routing, mould-making, and precision woodworking worldwide. Understanding when and why to apply them — not just how — is the Evaluate-level skill that distinguishes a capable CNC operator from someone who has merely memorised a button sequence.

The chapter closes with the decision framework you will use for the rest of this course and beyond: given a specific pocket, material, and quality target, which strategy is the correct choice? There is always a correct answer — and the reasoning is always rooted in the physics of what happens between a spinning tool and a piece of timber.

ZigZag Clearing — Understanding the Baseline

Before explaining Adaptive, it is worth understanding precisely what ZigZag clearing does and why it works well for many CNC tasks — but not all of them. ZigZag is not a bad strategy. It is the correct strategy for specific situations. Choosing Adaptive when ZigZag would be sufficient wastes simulation time and introduces complexity without benefit.

> 📖 **Key Term: ZigZag Clearing**
>
> A pocket clearing strategy where the tool moves in a series of parallel passes across the pocket floor, alternating direction on each pass like a lawn mower. Each pass removes a strip of material equal to the step-over width. Also called raster or parallel clearing. Standard in FreeCAD's Pocket Shape operation.

In ZigZag mode, the tool traverses the full pocket width, lifts to the opposite side, steps over by the step-over distance, and traverses back. For shallow pockets in soft material — an MDF recess 3 mm deep — it works perfectly well. The cutting forces are manageable and the result is clean.

ZigZag's Limitation — Corners and Depth

At the end of each ZigZag pass, the tool must turn and begin the next pass. At this turning point, particularly at pocket corners, the full side of the tool is briefly in contact with the material — a condition called full-width engagement. This means the chip load (the amount of material each cutting flute removes per rotation) spikes sharply, deflecting the tool sideways, causing chatter, and producing a visible tool mark at the corner absent from the straight portions of the pass.

In deeper pockets or harder materials, this corner spike becomes severe enough to break small end mills, strip flutes from a bit, or pull the workpiece from the wasteboard. ZigZag's constant engagement problem is manageable in shallow MDF work. In 15 mm deep hardwood pockets — precisely the kind encountered in the Week 9 clock project — it becomes a genuine machining risk.

Adaptive Clearing — Constant Chip Load Through Arc Motion

> **Key Term: Adaptive Clearing (Trochoidal Milling)**
>
> A pocket clearing strategy where the tool moves in small arc (trochoidal) motions rather than straight passes. The arc motion is designed so that the tool never engages more than a specified percentage of its diameter at any point, maintaining a constant chip load regardless of pocket geometry — including at corners. Also called trochoidal milling, HSM clearing, or in FreeCAD, the 'Adaptive' option in CAM Pocket Shape operations.

The Physics — Chip Load and Tool Engagement

> **Chip Load vs Engagement Angle**
>
> **Chip Load**
>
> The thickness of material each cutting flute removes per revolution, in millimetres. Determined by feed rate and spindle speed. Too low = rubbing (heat, poor surface, rapid tool wear). Too high = tool overload (chatter, deflection, breakage). Correct chip load = efficient, clean cutting.
>
> **Tool Engagement Angle**
>
> The arc of contact between the rotating tool and the material, measured in degrees. A full-slot cut has 180° engagement. A 50% step-over pass has approximately 120°. Adaptive's trochoidal arc limits this to a set maximum — typically 20–40°.
>
> ZigZag's problem: engagement angle varies constantly — minimal mid-pass, maximal at corners and full-slot entry. Chip load therefore varies constantly, producing inconsistent cutting forces. Adaptive's solution: the arc motion is calculated so the engagement angle never exceeds the set maximum. Chip load is therefore constant throughout — including at corners.

Practical Benefits of Adaptive

Extended tool life	Consistent, manageable cutting forces mean the tool wears evenly and slowly. In production scenarios, Adaptive can extend tool life 3–5× compared to ZigZag in the same material.
Higher feed rates	Because peak load never exceeds the design limit, the feed rate can be increased above what ZigZag would tolerate at the same step-over. The tool moves faster but cuts more gently.
Less heat	Full-width engagement in ZigZag traps chips in the cut zone, re-cutting them and generating heat. Adaptive's open arc motion ejects chips cleanly.
Better wall quality	Consistent engagement produces a consistent surface finish on pocket walls. ZigZag's variable engagement produces micro-variations, particularly at corners.

Configuring Adaptive Strategy in FreeCAD

1. Create a new Pocket Shape operation (CAM → Pocket Shape) or double-click an existing one to edit it.
2. In the operation properties, find the Operation Type field. Set it to 'Adaptive' (not the default 'Pocket').
3. Set Step Over as a percentage of tool diameter: 50–65% for hardwood, 60–75% for MDF.
4. Set Step Down — in Adaptive mode a more aggressive step-down is safe: 3–5 mm per pass for a 6 mm end mill in hardwood (vs 1.5 mm with ZigZag).
5. Set Lift Distance (how far the tool rises between arcs): 0.1–0.3 mm reduces air-cutting time.
6. Optionally set a Finish Depth pass at reduced feed rate for a cleaner pocket floor.
7. Run the CAM Simulator. Adaptive toolpaths look like overlapping arcs and spirals — this is correct. Verify the entire pocket floor is covered.

> **Tip**
> The Adaptive simulation in FreeCAD's CAM Simulator takes noticeably longer to compute than ZigZag — the trochoidal arc calculations are mathematically more complex. For large pockets, allow 30–90 seconds. Do not interrupt the computation.

Adaptive Step-Over — Practical Values by Material

Material	Recommended Step-Over	Notes
MDF, soft pine, cedar	**60–70% of tool diameter**	Soft material tolerates higher engagement. 70% is typical for efficient MDF clearing.
Hardwood (oak, jarrah, spotted gum)	**50–60% of tool diameter**	Harder material requires lower engagement to prevent heat and tool stress. 55% is a good starting point.
Plywood, LVL	**55–65% of tool diameter**	Variable grain direction means some passes cut with the grain, some against. Middle range is safest.
Dense hardwood (ironbark, merbau)	**40–50% of tool diameter**	Very dense material — test with 45% and adjust based on sound and surface quality.

When to Use Adaptive vs ZigZag — The Decision Framework

Scenario	Strategy	Why
Shallow pocket (under 5 mm), soft material	**ZigZag**	*Fast, simple, low tool wear at shallow depths. Adaptive's trochoidal arc motion is unnecessary overhead.*
Deep pocket (8 mm+), any material	**Adaptive**	*Full-width ZigZag cuts at depth generate force spikes that deflect the tool. Adaptive's constant engagement prevents this.*
Hard material (hardwood), any depth	**Adaptive**	*Hard material resists cutting force. Constant chip load reduces heat, prevents chip re-cutting, extends tool life significantly.*
Production run — 50+ identical parts	**Adaptive**	*Tool life cost over a long run dwarfs marginal time savings of ZigZag. Adaptive pays for itself by reducing tool changes.*

Prototype — one-off piece, soft material	ZigZag	*Speed to result matters more than tool life for a single prototype. ZigZag is faster to set up and simulate.*
Narrow channel (width < 2× tool diameter)	Adaptive	*ZigZag can't fit a full-width pass in a narrow channel. Adaptive's arc motion adapts to narrow geometry cleanly.*
Wide open area, soft material	Either	*Wide open areas in soft material don't stress the tool regardless of strategy. ZigZag is marginally simpler to configure.*

Stop and Think

Before setting up any pocket operation, ask yourself three questions:

1. How deep is the pocket relative to the tool diameter? If depth exceeds 1.5× tool diameter, Adaptive should be your default choice.

2. What material am I cutting? If it is harder than pine, Adaptive is strongly preferred.

3. How many of these parts will I cut? If more than 10–15, the tool life savings from Adaptive almost certainly justify the slightly longer simulation setup time.

If all three answers point to Adaptive, the choice is clear. If all three point to ZigZag, use ZigZag. The interesting cases are in the middle — and the judgement you develop here is what the Evaluate level of Bloom's Taxonomy demands.

Scallop Finishing — Constant Cusp Height

In Chapter 15 you used a 3D Surface finishing pass with a fixed step-over. This works well on surfaces of roughly uniform steepness. But on a surface with mixed angles — steep in one area, nearly flat in another — a fixed step-over produces inconsistent results: flat areas show near-invisible scallops while steep areas show much taller, more visible ridges for the same step-over distance.

Scallop finishing solves this by targeting a constant cusp height rather than a constant step-over. The step-over automatically adjusts based on the local surface angle — smaller on steep sections, larger on flat sections — to maintain consistent finish quality across the entire surface.

Key Term: Cusp Height

The height of the ridge left between two adjacent finishing passes by the ball-nose end mill. On a flat surface, cusp height depends on step-over and ball radius. On a steep surface, the same step-over produces a taller cusp. Scallop finishing controls the target cusp height, varying step-over automatically to maintain it.

Cusp Height and Step-Over — Practical Reference

Ball-Nose Diameter	Step-Over	Cusp Height	Visible Result
6 mm (r=3)	**0.5 mm**	`~0.010 mm`	Mirror-smooth. Runtime very high.
6 mm (r=3)	**0.8 mm**	`~0.027 mm`	Near-invisible ridges. Good for display pieces.
6 mm (r=3)	**1.2 mm**	`~0.060 mm`	Fine ridges visible in raking light. Typical finishing quality.
6 mm (r=3)	**2.0 mm**	`~0.170 mm`	Visible corrugation. Acceptable for paint-finish pieces.
3 mm (r=1.5)	**0.4 mm**	`~0.027 mm`	Excellent finish quality. Half the runtime of 6 mm at same cusp height.
3 mm (r=1.5)	**0.8 mm**	`~0.110 mm`	Moderate quality. Fast. Appropriate for rough finishing passes.

Configuring Scallop Finishing in FreeCAD

1. CAM → 3D Surface. Select the model body. Set a 3 mm ball-nose end mill.
2. In the operation properties, find the Pattern parameter. Change to 'Scallop'.
3. Set the step-over value — in Scallop mode this is interpreted as the target cusp height. Quality display piece: 0.05 mm. Standard finish: 0.1 mm.
4. Run the CAM Simulator. The toolpath will show variable spacing — passes closer together on steep areas, further apart on flat areas. This is correct Scallop behaviour.
5. Check the estimated runtime. Scallop may be faster or slower than fixed step-over depending on the surface geometry.

> 📋 **Note**
>
> Not all versions of FreeCAD CAM expose Scallop as a named pattern — in some builds it appears as 'Offset' with an option to use constant scallop height. If you cannot find a dedicated Scallop mode in your FreeCAD v1.1.0 build, use a fixed step-over of 0.8 mm with the 3D Surface operation. This approximates scallop quality for surfaces of moderate steepness.

Pencil Trace — Following Ridges for Maximum Detail

Even the finest scallop finishing pass leaves one category of surface detail unresolved: ridges, inside corners, and sharp convex features where the ball-nose tool's radius prevents it from reaching fully. A 3 mm ball-nose cannot machine the base of a narrow V-groove, the sharp crease between two surfaces, or the tight inner corner of a recessed pocket. These areas are left with a radius equal to half the tool diameter — 1.5 mm for a 3 mm bit.

Pencil trace automatically identifies these ridge lines and sharp feature lines on the model surface and generates a toolpath that follows them precisely with a smaller tool. The result is crisp, resolved detail where scallop finishing left a rounded, unresolved transition.

📖 **Key Term: Pencil Trace**

A 3D CAM finishing operation that detects ridge lines, inside corners, and sharp concave features on the model surface and generates a toolpath following those lines with the cutting tool. Used as a final detail pass after scallop or surface finishing. The tool follows the exact 3D path of the ridge, machining only the residual material left by the previous larger tool. Also called pencil milling, ridge following, or deburr operation.

Where Pencil Trace Makes a Visible Difference

✅ Pencil trace adds visible value	❌ Pencil trace adds little value
Sharp creases between two surfaces — the ridge between a leaf's centre vein and its outer surface	Smoothly flowing organic forms with no sharp internal features — a round hill or dome has no ridges to trace
Inside corners where two faces meet at a sharp angle — typical in architectural bas-relief	Flat floors and walls with square corners — better addressed by a flat finishing pass
Fine decorative engraving details the finishing ball-nose could not fully resolve	Very soft material where the ball-nose already achieves near-sharp results
Trophy plaques, name plates, and presentation pieces examined closely at arm's length	Parts that will be painted or filled — fine surface detail will be obscured

Configuring Pencil Trace in FreeCAD

1. Select a tool smaller than the previous finishing pass. If finishing used a 3 mm ball-nose, use a 1.5 mm or 2 mm ball-nose for pencil trace.
2. CAM → Deburr (or CAM → 3D Surface with Pencil pattern mode, version-dependent). Select the model body.
3. Set the Depth parameter: 0.1–0.3 mm below the previously finished surface removes the residual radius left by the finishing tool.
4. Set feed rate to 50–70% of the finishing pass feed rate — the tool is following complex 3D curves.
5. Run the Simulator. Verify the thin toolpath lines follow the ridge features you intended to sharpen. Unexpected pencil trace lines on smooth areas indicate model geometry issues.

⚠️ **Important**

Pencil trace with a very small tool (1.5 mm or smaller) in hard material carries a real tool breakage risk. The tool follows complex 3D curves at variable depths — any unexpected geometry spike or chip in the wood can snap it. Always verify the 3D Surface finishing pass was completed correctly before running pencil trace. A roughed but unfinished surface with deep steps will break a 1.5 mm tool almost immediately.

Rest Machining — Targeting Unreached Areas

📖 **Key Term: Rest Machining**

A CAM strategy that generates a toolpath targeting only the material a previous, larger tool could not reach — the 'rest' of the material remaining after that operation. By telling the CAM system the previous tool's diameter, the software calculates exactly where uncut material remains and generates a toolpath only for those areas, avoiding re-cutting already-finished surfaces and reducing runtime significantly.

Without rest machining, a finishing pass must cover the entire model surface — even areas already machined to final quality by the previous pass. On a complex surface with both open areas and tight detail, the open areas finish quickly but dense toolpath coverage is needed for detail areas. A finishing pass covering everything wastes significant time re-traversing already-clean open areas.

Rest machining tells the CAM software: 'The previous operation used an 8 mm ball-nose. Here is a 3 mm ball-nose. Only generate toolpaths for areas where the 8 mm tool could not fit.' The result targets detail areas where residual material remains and skips open areas already finished by the larger tool.

Configuring Rest Machining in FreeCAD

1. Note the tool used in the previous roughing or first finishing pass — its diameter and type.
2. Create a new 3D Surface finishing operation with a smaller tool (half or less of the previous tool's diameter).
3. In the operation properties, find the Rest Machining or 'Previous Operation' reference field. Select the previous operation from the list.
4. Run the Simulator. The rest machining pass should cover only the tight areas the previous tool could not reach. Open areas showing no toolpath is correct and expected.
5. Compare the runtime: rest machining targeting only tight areas is typically 40–70% faster than a full-surface finishing pass.

📋 **Note**

Rest machining availability varies between FreeCAD versions. If it is not available as a named feature in your v1.1.0 build, simulate it manually: run the larger-tool finishing pass, then add a smaller-tool pass with the machining boundary restricted to the known tight areas of the model.

The 3-Pass Professional Strategy

Roughing, scallop finishing, and pencil trace combine into a complete professional-grade 3D CAM workflow — the standard approach for high-quality relief carving, trophy plaques, and any CNC work where the finished surface will be examined closely.

Pass	Type	Operation	Tool	Key Settings	Purpose	Leaves
PASS 1	Roughing	Waterline or Adaptive	**6–8 mm ball-nose**	Step-down 2 mm, Step-over 50%, Stock-to-leave 0.3 mm	Remove bulk material rapidly, leaving a staircase approximation of the final form	*0.3 mm stock remaining on all surfaces*
PASS 2	Scallop Finishing	3D Surface — Scallop mode	**3–4 mm ball-nose**	Step-over 0.8–1.2 mm, Scallop mode or constant cusp height	Follow the exact surface contours — removes the roughing staircase to a near-finished surface	*0.05–0.1 mm scallop height on all surfaces*
PASS 3	Pencil Trace	Deburr / Pencil operation (optional)	**2–3 mm ball-nose**	Pencil pattern, automatic ridge detection, depth 0.1–0.3 mm	Follow ridges, corners, and sharp detail lines that scallop finishing could not fully resolve	*Clean ridges and crisp corner transitions*

3-Pass Workflow — Runtime and Quality Comparison

Strategy Configuration	Typical Runtime	Surface Quality	Best For
Roughing only	**8–15 min**	Coarse — not a finished surface	*Internal test of form before final design decisions*
Roughing + Fixed step-over finishing (1.5 mm)	**30–45 min**	Good — visible scallop in raking light	*Functional prototypes, painted pieces, practice projects*
Roughing + Scallop finishing (0.8 mm cusp)	**50–80 min**	Very good — near-invisible on most surfaces	*Standard display-quality pieces, most client work*
Roughing + Scallop + Pencil Trace	**70–100 min**	Professional — sharp features fully resolved	*Commissions, trophies, high-value display pieces*
Roughing + Scallop (0.05 mm) + Pencil + Rest	**120–180 min**	Exhibition-quality — mirror finish hardwood	*Exceptional pieces where runtime is no constraint*

When Is the Third Pass Justified?

The pencil trace pass adds setup complexity and runtime. Apply the following criteria before adding it to a project:

The piece will be examined closely	Display pieces, trophies, plaques, and gifts held and inspected up close benefit from pencil trace. Parts that will be installed, painted, or used functionally do not.
The model has genuine sharp features	A smooth organic form with no creases, ridges, or inside corners will show no visible improvement from pencil trace. Analyse the model — count its sharp features before committing.

The runtime is acceptable	For a one-off commission piece, 20 extra minutes is trivial. For a production run of 20 identical pieces, 20 × 20 = 400 minutes of extra machine time requires a cost-benefit evaluation.
The tool can reach the features	If the tightest features require a 1 mm tool and you only have 3 mm tools, pencil trace will not fully resolve them. Know your tool inventory before designing features that require it.

> **Tip**
>
> For the Week 9 clock face project, roughing + scallop finishing at 0.8–1.0 mm cusp height is the appropriate standard. The hour marker recesses are cut with Adaptive pocket. The 3-pass strategies in this chapter become relevant if you add a 3D relief element to the clock face — which is optional in Week 9 but required in the Week 11 Certification Project.

Week 9, Days 41–42 — Lesson Content Overview

Chapter 16 covers Days 41–42 of Week 9. Days 43–45 continue in Chapter 17 (TechDraw Workbench and Project 9 — Decorative Wall Clock Face).

Day 41 (Evaluate)	Adaptive pocket clearing. Students set up a 15 mm deep hardwood pocket using ZigZag first, observe the toolpath pattern and estimated runtime, then switch to Adaptive strategy with the same tool and step-down. Compare toolpath patterns and runtimes. Scenario analysis: given 5 pocket scenarios with varied depth, material, and quantity, each student recommends and justifies ZigZag or Adaptive for each. Production calculation: for 50 identical parts, does Adaptive save or cost time overall?
Day 42 (Evaluate)	Multi-pass 3D CAM — pencil trace and rest machining. Students add a pencil trace operation to their Week 8 leaf relief, predict which surfaces it will target, then run the Simulator and compare prediction to result. Configure rest machining: create a 3D Surface finishing pass using the previous Waterline operation as the reference. Compare runtime with and without rest machining. Full 3-pass setup: roughing + scallop + pencil — run all three in the Simulator and record estimated runtime for each pass.

Chapter Summary

- ZigZag clearing moves in parallel passes. Its limitation is variable tool engagement — at corners and full-slot conditions, chip load spikes sharply. This is acceptable in shallow MDF work but problematic in deep hardwood pockets.
- Adaptive (trochoidal) clearing uses arc motions to maintain a constant maximum tool engagement angle throughout the entire operation. The result is consistent chip load, reduced heat, longer tool life, and the ability to use higher feed rates than ZigZag at the same step-over.
- The Adaptive decision framework centres on three questions: How deep? What material? How many parts? Deep pockets, hard materials, and production quantities all favour Adaptive. Shallow, soft, one-off pieces are fine with ZigZag.
- Scallop finishing targets a constant cusp height rather than a constant step-over distance. On surfaces with mixed steepness angles, it automatically adjusts the step-over to maintain consistent finish quality across the entire surface. Configure it using the Scallop pattern mode in the 3D Surface operation.
- Pencil trace detects and follows ridge lines, inside corners, and sharp concave features on the model surface. It is the third pass in the professional 3-pass strategy and is only justified when the piece has genuine sharp features that scallop finishing has left visibly unresolved.
- Rest machining generates a toolpath targeting only the areas a previous, larger tool could not reach. It is typically 40–70% faster than a full-surface finishing pass by skipping already-machined open areas.
- The 3-pass professional strategy: Pass 1 = Waterline/Adaptive roughing (large ball-nose, fast material removal) → Pass 2 = Scallop finishing (medium ball-nose, consistent cusp height) → Pass 3 = Pencil trace (small ball-nose, sharp detail). Not all projects need all three passes — evaluate each project independently before committing to the full workflow.

Key Vocabulary — Chapter 16

Term	Definition
Adaptive clearing	A pocket clearing strategy using trochoidal (arc) motions to maintain a constant maximum tool engagement angle throughout. Prevents chip load spikes at corners, extends tool life, and enables higher feed rates than ZigZag in hard materials or deep pockets.
Trochoidal milling	The mathematical motion principle behind Adaptive clearing — the tool moves in a series of small overlapping arcs. Also called HSM clearing (high-speed machining clearing).
Chip load	The thickness of material each cutting flute removes per revolution, in mm. Correct chip load produces efficient, clean cutting. Too low causes heat and tool wear; too high causes deflection and breakage.
Tool engagement angle	The arc of contact between the rotating tool and material, measured in degrees. Adaptive clearing controls this to a set maximum (typically 20–40°), maintaining consistent chip load throughout.
Scallop finishing	A 3D CAM finishing strategy that targets a constant cusp height rather than a constant step-over. The step-over varies automatically based on local surface angle, producing consistent finish quality across surfaces of mixed steepness.
Cusp height	The height of the ridge left between adjacent ball-nose finishing passes. Lower cusp height = smoother surface = more passes = longer runtime. For display quality: aim for 0.05–0.1 mm cusp height.

Pencil trace	A 3D CAM finishing operation that detects and follows ridge lines, inside corners, and sharp concave features on the model surface. Used as a final detail pass with a small ball-nose tool after scallop finishing.
Rest machining	A CAM strategy that generates a toolpath targeting only the material a previous, larger tool could not reach. Avoids re-cutting already-finished surfaces and typically reduces runtime by 40–70% compared to a full-surface pass.
3-pass strategy	The professional-grade 3D CAM workflow: Roughing (fast bulk removal, large tool) → Scallop finishing (consistent quality, medium tool) → Pencil trace (sharp detail, small tool). Not all projects require all three passes.
Full-width engagement	A cutting condition where the full side of the tool is in contact with material — occurs at corners in ZigZag clearing and causes chip load spikes, chatter, and tool stress. Adaptive clearing prevents full-width engagement by design.

Looking Ahead — Chapter 17: TechDraw Workbench — Engineering Documentation

Chapter 17 covers the TechDraw workbench — creating dimensioned 2D engineering drawings directly from 3D models, inserting projection groups, applying first-angle projection (Australian standard AS 1100), adding dimensions and annotations, completing the title block, and exporting to PDF. It also introduces the CNC Setup Sheet — a TechDraw-based document that puts all machining parameters on a single printable page — culminating in Project 9, the Decorative Wall Clock Face.

CHAPTER 17 of 24

PART FIVE — ADVANCED SKILLS

TechDraw Workbench — Engineering Documentation

Technical drawing views, dimensioning, annotations, title blocks, PDF export, and the CNC Setup Sheet

Chapter Learning Objectives

By the end of this chapter, you will be able to:

- Explain why engineering drawings matter in a CNC workflow
- Create a TechDraw drawing page from an A4 template
- Insert a Projection Group to place Front, Top, Right, and Isometric views in one operation
- Identify first-angle and third-angle projection and state the Australian standard
- Apply at least six dimension types: horizontal, vertical, radius, diameter, angle, and smart
- Add text annotations and leaders to document machining notes on a drawing
- Complete all standard title block fields accurately
- Export a finished drawing to PDF
- Create a CNC Setup Sheet using TechDraw annotations for a complete job
- Design and cut the Decorative Wall Clock Face using Polar Pattern and TechDraw documentation

Why Engineering Drawings in a CNC Course?

Every machined part is described twice: once in the G-code that tells the machine what to cut, and once in the engineering drawing that tells everyone else what the part is, what it is made from, and how it is dimensioned.

FreeCAD's TechDraw workbench creates 2D technical drawings directly from your 3D models. These drawings serve several practical purposes in a CNC workflow: they document the finished design so it can be reproduced months later without re-opening the FreeCAD file; they communicate the design to others — teachers, clients, collaborators — in a universally readable format; and when annotated with machining notes, they become a CNC Setup Sheet that contains everything an operator needs to run a job without consulting the CAM file.

An engineering drawing is also a verification tool. Placing a finished model into TechDraw and dimensioning it reveals any inconsistency between your intended design and what was actually modelled. If a dimension looks wrong on the drawing, it was wrong in the model — and catching it on paper is cheaper than catching it on the machine.

This chapter covers the complete TechDraw workflow: creating a page, inserting views, dimensioning, annotating, filling the title block, and exporting to PDF. The project applies all of it to the Decorative Wall Clock Face — a design that combines Polar Pattern geometry from Chapter 9 with a full TechDraw documentation page that also serves as the machining setup sheet.

The TechDraw Interface

Switch to the TechDraw workbench from the Workbench drop-down. The interface consists of three areas: the Model tree (same as always — your 3D model parts are listed here), the 2D page view (a white canvas representing the drawing sheet), and the Properties panel (shows properties of whatever is selected on the drawing).

Important: TechDraw works on a flat 2D canvas. You are not editing the 3D model here — you are creating a 2D representation of it. Changes to the 3D model update the drawing automatically if Link Dimensions are used. Plain dimensions added manually do not update automatically — which is why the Link Dimension tool is preferred for parametric designs.

Page view	The white 2D canvas showing the drawing sheet. Zoom and pan here just like the 3D view. The drawing border and title block are visible at the page edges.
View objects	Projected 2D outlines of the 3D model, placed on the page. Each view is a separate object in the Model tree and can be selected, moved, and scaled independently.
Dimension objects	Measurement annotations placed on the page. Linked dimensions track the model; unlinked dimensions are static.
Annotation objects	Text notes, leaders, surface finish symbols, and other markup placed on the page for documentation purposes.
Templates	Pre-designed page borders with title blocks, logo areas, and revision tables. FreeCAD ships with standard A4, A3, A2, A1, and A0 templates. Custom templates can be made using Inkscape SVG.

Creating a Drawing Page

Step 1 — Insert a New Page Using a Template

1. Switch to TechDraw workbench.
2. Go to TechDraw → Pages → Insert Page using Template.
3. A file browser opens. Navigate to FreeCAD's built-in template folder (this is inside the FreeCAD installation directory — the browser opens there automatically).
4. Select A4_LandscapeUS.svg for landscape A4 (preferred for multi-view drawings), or A4_Portrait.svg for portrait orientation.
5. Click Open. A new Page object appears in the Model tree and the blank drawing canvas opens.

> **Tip**
>
> A4 Landscape suits most CNC part drawings — it provides enough horizontal space for a Projection Group with Front, Top, and Side views side by side. Use A3 Landscape for large parts that need 1:1 scale on paper.

Inserting Views — Projection Groups

The fastest and most organised way to place views is the Projection Group — a single operation that creates a Front view, plus aligned Top, Right, Left, and Bottom views, all locked in their correct relative positions. It also optionally adds an isometric view.

Inserting a Projection Group

1. In the 3D view (or Model tree), select the Body or Part you want to document.
2. Go to TechDraw → Views → Insert Projection Group.
3. A dialog opens. Set the Primary View Direction — this determines which face of the 3D model appears as the Front view. For a flat panel: select 'Front' (looking along the Y axis). For a gear: select 'Top' (looking down).
4. Tick the views you want: Front (required), Top, Right. Tick Isometric for a 3D overview view.
5. Set Scale — 1:1 for small parts, 1:2 for parts larger than about 150mm.
6. Click OK. FreeCAD places all selected views on the page in their correct positions.
7. Click and drag the entire group to position it well on the page, leaving room for dimensions and the title block.

View Positions — First Angle Projection

The position of each view on the page follows a projection system. Australia uses first-angle projection, also known as the European or ISO system. The diagram below shows where each view appears relative to the Front view in first-angle projection.

First Angle Projection — View Positions on the Drawing Page		
	TOP VIEW *(looking down on the part)*	
RIGHT VIEW *(looking from the right)*	**FRONT VIEW** *(the primary view)*	**LEFT VIEW** *(looking from the left)*
	BOTTOM VIEW *(looking up at the part)*	**ISOMETRIC** *(3D view, no scale)*

> 🇦🇺 **Australian Standard**
>
> Australia uses AS 1100 Technical Drawing Standards, which specifies first-angle projection (the ISO system). In first-angle projection, the Top view appears BELOW the Front view, and the Right view appears to the LEFT of the Front view — the opposite of American third-angle projection. FreeCAD's Projection Group defaults to first-angle when configured correctly. If your views appear in the wrong positions, check: TechDraw → Preferences → General → Projection Angle = First.

Moving and Scaling Individual Views

Move a view	Click the view on the page, then drag it. Note: views in a Projection Group are locked relative to each other — dragging the Front view moves the whole group. To move one view independently, right-click → Toggle Movable.
Rescale a view	Click the view. In the Properties panel → Data → Scale, change the value. 0.5 = half size (1:2). 2.0 = double size (2:1). The view redraws at the new scale.
Toggle edge visibility	Right-click a view → Toggle Frames to show or hide edge lines. Some internal hidden edges can be toggled on with the Show Hidden Lines option.
Refresh a view	If the view does not update after a model change: right-click the view → Update View, or right-click the Page → Redraw Page.

Dimensioning — Measuring the Drawing

A drawing without dimensions tells someone the shape of a part. A drawing with dimensions tells them how to make it. Dimensioning is where the drawing becomes a manufacturing specification.

TechDraw dimensions are placed on the 2D page — they measure the projected view, not the 3D model directly. For this reason it is important to select the correct view when dimensioning:

measure hole diameters from the view that shows them as circles, measure heights from the view that shows them as vertical edges.

Tool Name	What It Does & How to Use It
New Horizontal Dimension	Select two points or a horizontal edge. Places a horizontal measurement with extension lines and a dimension value above the dimension line.
New Vertical Dimension	Select two points or a vertical edge. Places a vertical measurement.
New Radius Dimension	Select a circular edge. Places a radius leader from the edge to a text callout showing the radius value (R xx.xx).
New Diameter Dimension	Select a circular edge. Places a diameter dimension line through the circle centre with Ø prefix. Preferred over radius for holes.
New Angle Dimension	Select two edges that meet at an angle. Places an arc dimension showing the angle between them in degrees.
New Smart Dimension	Select any edge or two points — FreeCAD detects whether to apply horizontal, vertical, radius, or diameter automatically. Fastest for rapid dimensioning.
New Arc Length Dimension	Select an arc edge. Measures the length along the arc (not the chord). Useful for curved path lengths.
New Chamfer Dimension	Select a chamfered edge. Displays the chamfer as size×45° or size×angle notation.
New Extent Horizontal	Selects multiple features and spans an overall horizontal extent dimension across all of them in one operation — useful for part families or arrays.
New Extent Vertical	As above, but for vertical extents.
New Link Dimension	Links a TechDraw dimension directly to the 3D model measurement — if the model changes, the drawing dimension updates automatically (requires constraints).

Dimensioning Workflow — The Right Order

1. Click the view containing the feature you want to dimension (this makes it active).
2. Select the geometry: one edge for length, two points for point-to-point, a circular edge for diameter or radius.
3. Choose the dimension tool from TechDraw → Dimensions menu.
4. The dimension appears. Click and drag the dimension text to position it clearly — avoid overlapping other dimensions or edges.
5. Repeat for all required dimensions. Apply dimensions from the most important features first (overall size, critical holes, locating features) before adding detail dimensions.

> **Tip**
> Dimension from the most critical feature outward — not from every edge to every adjacent edge. A drawing that shows every single measurement is harder to read than one that shows only the dimensions needed to reproduce the part. For CNC parts: dimension overall size, all hole positions and diameters, and any feature that has a tolerance.

Annotations — Notes and Leaders

Annotations add text information to the drawing that cannot be conveyed by dimensions alone. In a CNC context, annotations are particularly useful for noting material specifications, surface finish requirements, machining notes, and tool recommendations directly on the drawing.

Text Annotations

1. TechDraw → Annotations → Insert Annotation.
2. A text box appears on the page. Double-click it to edit the text.
3. Type your note. Press Enter for a new line within the annotation.
4. Click outside the text box to finish. Drag to reposition.

Leaders — Text with a Pointer

A leader is a text note connected to a feature by an arrow line. Use leaders to annotate specific features: 'Bore Ø10 — clock mechanism shaft', 'This face — no finish sanding', 'All holes — drill after CNC'.

1. TechDraw → Annotations → Insert Leader Line.
2. Click the feature you want to point to on the drawing.
3. Click to set the elbow point of the leader line.
4. Double-click to finish the line and enter the annotation text.
5. The leader arrow points to the feature; the text sits at the end of the line.

Surface Finish and Weld Symbols

TechDraw includes Surface Finish (roughness) and Welding symbol tools under TechDraw → Annotations. These are primarily used in metal fabrication documentation. For CNC timber work, plain text annotations are usually sufficient — but surface finish symbols can be useful if you are creating documentation for a client who expects formal engineering drawings.

The Title Block — Drawing Metadata

Every engineering drawing has a title block — a bordered area, usually in the bottom-right corner of the page, containing standard metadata about the drawing and the part it depicts. The title block is the first thing a reader checks to understand what the drawing is, who made it, when, and at what scale.

In FreeCAD TechDraw, the title block is part of the page template. You edit it by double-clicking the text fields directly on the drawing page. Each template has different fields — the A4 standard templates include the fields listed below.

Field	Example Value	Purpose / Notes
Title / Part Name	`Clock Face — 250mm / Project 9`	The name of the part or assembly. Used when filing or searching for the drawing later.
Drawing Number	`CNC-P09-001`	A unique identifier for this drawing. Increment for each revision (CNC-P09-002 for revision 2).
Designer / Author	`Your name`	The person who created the design and drawing.
Date	`Auto-filled or manual`	The date the drawing was created or last revised. Essential for version tracking.
Scale	`1:1 or 1:2 for large parts`	The ratio of drawing size to actual part size. 1:1 = drawn at true size. 1:2 = drawn at half size.
Material	`12mm MDF / Pine`	The material to be used for fabrication — critical information for the machinist (or your own reference).
Sheet	`1 of 1`	Sheet number when a design requires multiple pages. Helps keep multi-page sets organised.
Revision	`A, B, C...`	Updated each time the drawing is revised. Rev A is the first issue; Rev B is the first revision.

Editing the Title Block

1. In the TechDraw page view, double-click a title block text field.
2. A small text editor opens. Type the new content.
3. Press Enter or click outside to confirm.
4. Repeat for each field. The title block updates immediately.

Exporting — PDF and SVG

Export to PDF

PDF is the standard format for sharing engineering drawings. A PDF drawing can be opened on any device, printed at exact scale, and archived indefinitely without risk of software incompatibility.

1. With the TechDraw page open and active, go to File → Export PDF.
2. In some FreeCAD versions: TechDraw → Pages → Export Page as PDF.

3. Choose a save location and filename. Use a descriptive name that includes the part name, scale, and revision: ClockFace_A4_1to1_RevA.pdf.
4. Click Save. The PDF is created at the exact page dimensions of the template.

Export as SVG

SVG export produces a vector file that can be opened in Inkscape for further editing — adding colour, logos, or decorative elements. Use SVG when you need to customise the drawing appearance beyond what TechDraw's tools support.

1. TechDraw → Pages → Export Page as SVG.
2. The SVG file includes all views, dimensions, and annotations as editable vector objects.
3. Open in Inkscape to add colour fills, adjust line weights, or overlay additional artwork.

The CNC Setup Sheet — TechDraw's Most Practical CNC Use

A CNC Setup Sheet is an annotated engineering drawing that contains everything an operator needs to run a job: dimensions, material, tools, operations sequence, zero point, and G-code filename — all on one printed page that lives next to the machine.

This is where TechDraw's value in a CNC course becomes most concrete. A setup sheet is not a formal engineering drawing — it is a working document. It combines the dimensioned drawing of the part with practical machining notes added as annotations and leaders. When the setup sheet is printed and pinned next to the CNC 4040 Pro Max, the operator never needs to open FreeCAD, UGS, or the G-code file to understand the job.

The table below shows what a typical CNC setup sheet annotation block looks like, and what information each entry conveys to the operator.

TechDraw Annotation — CNC Setup Notes	Purpose
`Material: 12mm MDF, 260×260mm`	Operator knows what to clamp
`Bit: 6mm end mill, 3mm end mill`	Tools required for the job
`Zero: centre of part, Z from top surface`	Unambiguous machine zero location
`Operations: 1=Engrave, 2=Pocket, 3=Profile`	Correct sequence, no guessing
`Feeds: 900mm/min, Plunge: 250mm/min`	No need to check CAM file separately
`Tabs: 4 per profile, chisel before sanding`	Post-cut reminder
`G-code file: ClockFace_12mm_v2.nc`	Exact file name to load in UGS

> 📋 **Note**
> For this course, creating a setup sheet for every project is optional until Chapter 21 (Workshop Tool Rack), where a complete setup sheet is a required deliverable. Starting from Project 9, practise creating setup sheets as part of your project documentation — it is a habit that pays dividends when you return to a job weeks or months later.

PROJECT 9 — STEP-BY-STEP WALKTHROUGH

Decorative Wall Clock Face

Design, document, and cut a 250mm polar-patterned clock face with full TechDraw setup sheet

Project 9 has two equally important deliverables: a cut clock face that is accurate enough to accept a clock mechanism, and a complete TechDraw documentation page that doubles as the CNC setup sheet. Both are assessed. The clock face itself combines Polar Pattern from Chapter 9 with new TechDraw documentation skills from this chapter.

Design	250mm diameter clock face, 12 hour marker recesses (Polar Pattern, 30° spacing), centre bore, decorative chamfered edge
Material	12–18mm MDF or pine — 12mm MDF for a lightweight display piece, 18mm pine for a more substantial wall-mounted item
Bit	6mm end mill for engraving and profiles; 3mm end mill optional for fine detail hour markers
Operations	1 — Engrave/Pocket (hour markers), 2 — Profile (outer circle), 3 — Drill/Pocket (centre bore)
Day	Day 45 of Week 9 — Project Day
Assessment	CNC cut clock face + printed TechDraw setup sheet (all fields complete, min 6 dimensions, setup notes annotated)

Part A — Clock Face Design

1 — **New file — ClockFace.FCStd**
File → New. Save. Switch to Part Design. New Body, rename 'ClockFace'.

2 — **Sketch the clock face base**
New sketch on XY plane. Draw a circle at the origin, diameter 250mm. Close sketch. Part Design → Pad, 15mm (or your material thickness + 3mm for tool clearance in CAM).

3	**Sketch and pocket the centre bore** New sketch on the top face. Circle at origin, diameter 10mm (standard clock shaft size — confirm with your clock mechanism before cutting). Part Design → Pocket → Through All.

4 **Sketch the hour marker recesses — one position first**
New sketch on top face. Draw a small rectangle: 8mm wide × 3mm tall, positioned at the 12 o'clock position (Y = 115mm from centre — about 10mm inside the clock edge). This will be the 12 o'clock hour marker slot. Constrain it carefully: centred on the Y axis, positioned at Y=115.

5 **Apply Polar Pattern — 12 positions**
With the Pocket feature (for the hour marker) selected in the Model tree: Part Design → Polar Pattern. Axis: Z axis. Occurrences: 12. Angle: 360°. FreeCAD places 12 evenly-spaced identical hour marker pockets, one every 30°.

> **Tip**
>
> For the quarter-hour markers (3, 6, 9 o'clock positions), you can add a second, slightly wider and deeper marker pocket patterned at 4 occurrences × 90° before running the 12-occurrence pattern. Apply the 4-occurrence quarter markers first, then the 12-occurrence standard markers — the quarter positions will be slightly more prominent, matching traditional clock face design.

6 **Optional — chamfer the clock face edge**
Select the top circular edge of the pad. Part Design → Chamfer, 3mm × 45°. Softens the edge and gives a finished appearance.

7 **Optional — add a mounting recess or hook hole**
On the back face (bottom of the pad), sketch a keyhole or two 4mm holes for wall mounting screws, positioned symmetrically. Pocket 8mm deep (not through).

Part B — CAM Setup

8 **Create CAM Job**
CAM → Job. Model: ClockFace body. Stock: 260×260mm, thickness = material. Zero: centre of part, Z from top surface.

9 **Tool Library**
T1 = EndMill6mm (primary). T2 = EndMill3mm (optional, for finer hour markers).

10 **Operation 1 — Pocket: hour markers**
CAM → Pocket Shape. Select all 12 hour marker recesses. T1. Step Down: 1.5mm. Final Depth: −3mm. Feed: 900. Zigzag pattern.

11 **Operation 2 — Profile: outer circle**
CAM → Profile. Select outer circular edge. Side: Outside. T1. Step Down: 1.5mm. Final Depth: −18mm (through material). Tabs: 4, 3mm wide, 3mm high. Feed: 900. Plunge: 250.

12 **Operation 3 — Pocket: centre bore**
CAM → Pocket Shape. Select centre bore circle. T1. Step Down: 1.5mm. Through All. Feed: 600. Plunge: 150.

13 **Simulator check, post-process**
Run Simulator. Verify hour marker positions (they should be evenly spaced — if any are missing, check the Polar Pattern was applied correctly). Post-process → grbl → ClockFace_15mm_v1.nc.

Part C — TechDraw Documentation

14 **Switch to TechDraw — create A4 Landscape page**
TechDraw → Pages → Insert Page using Template → A4_LandscapeUS.svg.

15 **Insert Projection Group**
Select the ClockFace body. TechDraw → Views → Insert Projection Group. Views: Front + Top + Isometric. Scale: 1:2 (the 250mm clock face is too large for 1:1 on A4). Tick Isometric. Place on left two-thirds of the page.

16 **Add dimensions — minimum 6**
Add: (1) Overall diameter — Smart Dimension on the outer circle in the Front view, (2) Centre bore diameter — Diameter Dimension on the bore in the Top view, (3) Hour marker width — Horizontal Dimension on one marker, (4) Hour marker depth — Vertical Dimension on the marker recess, (5) Marker radial position from centre — Dimension from centre point to marker centre, (6) Chamfer — Chamfer Dimension on the front view edge. More dimensions are better — cover all critical features.

17 **Add setup sheet annotations**
Using Insert Annotation and Leader Line tools, add: Material spec (12mm MDF), Bit information (T1=6mm, T2=3mm), Zero point (centre, Z from top), G-code filename (ClockFace_15mm_v1.nc), Operations sequence (1=Hour markers, 2=Profile, 3=Bore), Post-cut note (Chisel tabs, sand 120→220→320 grit, Danish oil).

18 **Complete the title block**
Fill in: Title = Clock Face 250mm, Drawing Number = CNC-P09-001, Designer = your name, Date = today, Scale = 1:2, Material = 12mm MDF, Sheet = 1 of 1, Revision = A.

19 **Export PDF — print and pin**
File → Export PDF. Save as ClockFace_A4_1to2_RevA.pdf. Print it. This is your setup sheet — it goes next to the machine.

Part D — Cutting and Assembly

20 **Pre-cut checklist, clamp, set zero**
Clamp 260×260mm material. Set zero at the centre of the stock — use a V-bit to find the exact centre point, then swap to the 6mm end mill and re-zero Z.

21 **Run the job — 3 operations**
Op 1 (hour marker pockets) → Op 2 (outer circle profile) → Op 3 (centre bore). Supervise each. The circular profile cut on Op 2 requires a clean, continuous path around the full 250mm diameter — listen for any chatter.

22	**Release, sand, finish, assemble clock mechanism** Chisel tabs. Sand: 120 → 220 → 320 grit. Apply Danish oil or clear lacquer. Once dry, press-fit the clock mechanism shaft through the centre bore. Install battery and set the hands. Hang it up.

🏆 **Project 9 — Precision plus Documentation**

Project 9 produces two deliverables that represent two different kinds of value: the physical clock face demonstrates your CNC precision; the TechDraw setup sheet demonstrates your ability to document and communicate a design. Both will be useful beyond this course — the clock face on a wall, the setup sheet habit in every future job.

Week 9, Days 41–45 — Lesson Content Overview

Day 41 (Evaluate)	TechDraw introduction. Students create their first drawing page from an A4 template, insert a Projection Group from any previously completed model (from Projects 1–8), and correctly place Front, Top, and Isometric views. Discussion: first-angle vs third-angle projection, the Australian Standard AS 1100, and the projection symbol.
Day 42 (Evaluate)	Dimensioning. Students apply all six dimension types (horizontal, vertical, radius, diameter, angle, smart) to their drawing from Day 41. Class challenge: add dimensions to a provided drawing without any numbers — see who produces the most complete, readable, non-overlapping dimension set in 20 minutes. Discuss: what is the minimum set of dimensions to fully describe a CNC part?
Day 43 (Evaluate)	Annotations, leaders, and title blocks. Students create a full setup sheet for any project from earlier in the course. Required: at least 5 annotations covering material, tools, zero point, G-code filename, and one post-cut note. All title block fields completed. Export to PDF and submit digitally.
Day 44 (Create)	Clock face design. Students build the ClockFace model: base pad, centre bore, hour marker sketch, Polar Pattern (revisiting Chapter 9). Independent challenge: design a variant with Roman numeral-style bar markers (different recess dimensions for hours vs quarter-hours). CAM setup completed.
Day 45 (Create — Project Day)	Project 9: Decorative Wall Clock Face. Full project including TechDraw documentation. Assessment: cut quality, Polar Pattern accuracy (12 evenly-spaced markers), TechDraw drawing completeness (minimum 6 dimensions + all title block fields + setup annotations), and PDF export submitted.

Chapter Summary

- TechDraw creates 2D engineering drawings from 3D FreeCAD models. Its role in CNC work extends beyond formal documentation to include the CNC Setup Sheet — a single annotated drawing that tells an operator everything needed to run a job.
- Create a drawing page using TechDraw → Pages → Insert Page using Template. A4 Landscape is the standard page size for most CNC part documentation in this course.
- Use Projection Group to place multiple views in a single operation: Front, Top, Right, and optional Isometric. Australia uses first-angle projection (AS 1100 / ISO standard): Top view sits below Front view; Right view sits to the left of Front view.
- TechDraw provides 11 dimension tools including Horizontal, Vertical, Radius, Diameter, Angle, Smart, and Link Dimension. Smart Dimension auto-detects the appropriate type. Link Dimension tracks the 3D model — if the model updates, the drawing dimension updates too.
- Annotations (text notes) and Leaders (annotated pointer lines) add machining notes, material specifications, and setup instructions directly onto the drawing. A complete CNC setup sheet includes: material, tools, zero point, operations sequence, feeds, G-code filename, and post-cut notes.
- The title block records drawing metadata: part name, drawing number, designer, date, scale, material, sheet number, and revision. Fill in every field — the drawing number and revision fields are essential for version tracking.
- Export to PDF via File → Export PDF. Use a filename that includes the part name, scale, and revision. Print the PDF and pin it at the machine — this is the setup sheet.
- Project 9 (Decorative Wall Clock Face) combines Polar Pattern (12 × 30° hour markers) with full TechDraw documentation. Both the physical cut piece and the PDF setup sheet are assessed deliverables.

Key Vocabulary — Chapter 17

Term	Definition
TechDraw workbench	The FreeCAD workbench for creating 2D engineering drawings from 3D models. Produces dimensioned, annotated drawings suitable for manufacturing, documentation, and archiving.
Drawing page	The 2D canvas in TechDraw representing a physical sheet of paper at a defined size (A4, A3, etc.). Contains views, dimensions, annotations, and the title block.
Template	A pre-designed SVG file used as the drawing border. Contains a structured title block, page border, and optional logo/revision areas.
Projection Group	A TechDraw operation that inserts multiple aligned views (Front, Top, Right, Isometric) in a single step, positioned correctly relative to each other.
First-angle projection	The ISO/Australian/European standard for view placement on engineering drawings. The Top view sits below the Front view; the Right view sits to the left of the Front view. Specified in AS 1100.
Third-angle projection	The American standard for view placement. The Top view sits above the Front view; the Right view sits to the right of the Front view. Do not use for Australian drawings.
View	A 2D projected outline of the 3D model placed on the drawing page, representing the part as seen from a specific direction.

Smart Dimension	A TechDraw dimensioning tool that automatically applies the appropriate dimension type (horizontal, vertical, radius, or diameter) based on the geometry selected.
Link Dimension	A TechDraw dimension that is directly linked to the 3D model constraint. If the model's parameter changes, the drawing dimension updates automatically.
Annotation	A text note placed directly on a drawing page, independent of any specific geometric feature. Used for general notes, machining specifications, and material callouts.
Leader	An annotation connected to a specific feature on the drawing by a line with an arrowhead. Used to label specific features: bore diameters, surface specifications, machining notes.
Title block	The bordered area, typically in the bottom-right corner, containing standard drawing metadata: part name, drawing number, designer, date, scale, material, sheet, and revision.
CNC Setup Sheet	An annotated engineering drawing that includes all information an operator needs to run a CNC job without consulting the CAM file: material, tools, zero point, operation sequence, G-code filename, and post-cut notes.
Revision	A letter (A, B, C...) that increments each time a drawing is formally updated. Rev A is the initial release. Revision tracking prevents operators from running a job using an outdated drawing.

Looking Ahead — Chapter 18: The Expressions Editor — Advanced Parametric

Chapter 18 takes parametric design beyond the Spreadsheet workbench into FreeCAD's Expressions Editor — a formula system that can drive any property in the model using mathematical expressions, conditional logic, and references to external data. This is the final parametric tool, completing the journey from basic constraints (Chapter 5) through Spreadsheet aliases (Chapter 8) to fully expression-driven models.

PART FIVE

Advanced Skills

Weeks 8–10 • Chapters 15 through 21 • Bloom's Levels: Evaluate → Create

📖 **About Part Five**

Part Five covers Weeks 8 through 10 — the advanced skills stage of the course. These chapters are concerned with synthesis: taking the discrete skills developed across the first seven weeks and combining them into complete, professional, independently-executed workflows. Part Five operates at the Evaluate and Create levels of Bloom's Taxonomy. You are no longer following instructions — you are making consequential design and strategy decisions and producing work that demonstrates mastery.

CHAPTER
18
of 24

PART FIVE — ADVANCED SKILLS

The Expressions Editor

Advanced parametric formulas, trigonometric positioning, inter-body expressions, and fully expression-driven design

Chapter Learning Objectives

By the end of this chapter, you will be able to:

- Access the Expressions Editor from any FreeCAD property or constraint field using the blue = button
- Write expressions using mathematical operators: addition, subtraction, multiplication, division, power, and common functions
- Convert degrees to radians for use in trigonometric expressions
- Apply sin() and cos() to calculate X and Y positions of features on a circle of known radius and angle
- Use the polar coordinate conversion formula to position any feature at a specified angular position
- Create an inter-body expression that links one Part Design Body's dimension to another Body's parameter
- Write a scale-factor expression that multiplies all model dimensions proportionally from a single cell
- Detect and resolve circular reference errors in an expression chain
- Design a model where every dimension is calculated from two or fewer input values — zero hard-coded numbers

In Chapter 8, you replaced typed numbers with spreadsheet references. In Chapter 18, you replace spreadsheet references with mathematical expressions. The model stops being a collection of dimensions and becomes a system of equations — a design that calculates itself.

The Spreadsheet workbench gave you parametric design: the ability to change one value and have the entire model update. But the spreadsheet approach has a limitation — it requires you to manually calculate any derived values and enter them as separate cells. The calculation is external to the model; you do it, then you type the result.

The Expressions Editor takes this further. It embeds mathematical calculations directly inside any FreeCAD field — constraint dialog, Pad length, placement coordinate, ShapeString size — without requiring a separate spreadsheet cell for every derived value. More importantly, it provides access to mathematical functions beyond basic arithmetic: trigonometric functions, square roots, and the constant pi. This makes it possible to calculate positions on circles, gear pitch distances, proportional scaling, and any geometry that requires more than addition and subtraction.

This chapter works through the Expressions Editor from first principles — what it is, how to access it, and what it can do — before building to the two most practically useful capabilities for CNC design: trigonometric positioning and inter-body expressions. Both directly support the Week 10 Synthesis project and the Week 11 Certification Project.

The Expressions Editor — FreeCAD's Formula Engine

Key Term: Expressions Editor

FreeCAD's built-in formula interpreter, accessible from any numeric property field by clicking the blue = button. Accepts mathematical expressions, references to spreadsheet aliases, references to other model properties, and mathematical functions including trigonometric functions. Results are evaluated in real time and update automatically when any referenced value changes.

Accessing the Expressions Editor

The blue = button appears wherever FreeCAD accepts a parametric value: Sketcher constraint dialogs, Pad/Pocket length fields, Placement position and rotation fields, ShapeString height, and Spreadsheet cells. Click the = button next to any numeric field to open the expression editor for that field.

1. Open any dialog containing a numeric field — for example, double-click a dimensional constraint in the Sketcher.
2. Look for the small blue = button to the right of the numeric value field.
3. Click =. The field changes to expression entry mode and the Expressions Editor dialog opens.
4. Type your expression. FreeCAD evaluates it in real time and shows the computed result below the input field.
5. Verify the computed value. If it shows an error, read the error message and correct the syntax.
6. Click OK. The expression is stored. The field shows the computed value and a small = icon indicating it is formula-driven.

Tip

Once a field contains an expression, hover over the value in the Properties panel to see the full expression as a tooltip. To edit it, click = again. To remove the expression and return to a plain number, clear the expression field and type the value directly.

Mathematical Operators and Functions

FreeCAD's expression engine supports standard mathematical notation. The tables below list every operator and function you will need in this course.

Arithmetic Operators

Operator	Example	Result / Meaning
+	`Spreadsheet.Width + 10`	Width plus 10 mm
−	`Spreadsheet.Height − 2 * Spreadsheet.BorderInset`	Height minus border on both sides
*	`Spreadsheet.Module * Spreadsheet.Teeth`	Gear pitch diameter (Module × Teeth)
/	`Spreadsheet.PanelWidth / 2`	Half the panel width — for centring calculations
^	`Spreadsheet.SideLength ^ 2`	Side length squared — for area or diagonal calculations
( )	`(Spreadsheet.Width + Spreadsheet.Height) / 2`	Average of width and height
−	`−Spreadsheet.EngravingDepth`	Negative depth — for Final Depth fields in CAM

Mathematical Functions

Syntax	What It Computes	CNC Use Case
`sqrt(x)`	Square root of x	*Hypotenuse: sqrt(Width^2 + Height^2)*
`abs(x)`	Positive magnitude, regardless of sign	*Ensuring a depth value is always positive*
`sin(x)`	Sine of x in radians	*Y position of a feature on a circle at angle x*
`cos(x)`	Cosine of x in radians	*X position of a feature on a circle at angle x*
`tan(x)`	Tangent of x in radians	*Calculating opposite side length from angle and adjacent*
`asin(x)`	Angle whose sine is x, in radians	*Finding the angle that produces a given vertical position*
`acos(x)`	Angle whose cosine is x, in radians	*Finding the angle for a given horizontal position*
`atan(x)`	Angle whose tangent is x, in radians	*Calculating angle from known opposite and adjacent sides*
`atan2(y,x)`	Full-circle angle from x,y coordinates	*Converting Cartesian coordinates back to an angle*
`pi`	3.14159265... (mathematical constant π)	*All radian conversions and circular calculations*

> **Note**
>
> All trigonometric functions in FreeCAD's Expressions Editor work in radians, not degrees. The conversion is: radians = degrees × pi / 180. This appears in every trigonometric expression you write and quickly becomes automatic.

The Radians–Degrees Conversion

Converting degrees to radians in FreeCAD expressions

```
Radians = degrees * pi / 180

30°  →  30 * pi / 180  =  0.5236 radians
45°  →  45 * pi / 180  =  0.7854 radians
90°  →  90 * pi / 180  =  1.5708 radians
180° →  180 * pi / 180 =  pi     radians
360° →  360 * pi / 180 =  2*pi   radians

Always use  * pi / 180  after any degree value in a trig expression.
```

Trigonometric Positioning — Features on a Circle

The single most useful application of trigonometry in FreeCAD CNC design: calculating the X and Y positions of features arranged on a circle — hour markers, bolt holes, decorative elements arranged at specific angular positions.

A position on a circle is most naturally described in polar coordinates: a radius (how far from the centre) and an angle (which direction from the positive X axis). FreeCAD's placement system uses Cartesian coordinates (X, Y) — so to place a feature at a polar position, you must convert from polar to Cartesian using the fundamental relationship:

Polar Coordinate Conversion — Feature at angle θ on radius R from origin

```
X position = R × cos(θ × pi / 180)
Y position = R × sin(θ × pi / 180)

FreeCAD expression syntax (θ stored in spreadsheet alias 'Angle'):
  X:  Spreadsheet.Radius * cos(Spreadsheet.Angle * pi / 180)
  Y:  Spreadsheet.Radius * sin(Spreadsheet.Angle * pi / 180)

Example: Angle = 30°, Radius = 80 mm
  X = 80 × cos(30 × pi/180) = 80 × 0.866 = 69.3 mm
  Y = 80 × sin(30 × pi/180) = 80 × 0.500 = 40.0 mm
```

Clock Face Example — 12 Hour Marker Positions

On a standard clock face, the 12 hour positions are at 30° intervals (360° ÷ 12 = 30°). In standard mathematical convention, 12 o'clock is at 90°, 3 o'clock is at 0°, 6 o'clock is at 270°, and 9 o'clock is at 180°.

Hour	Angle (°)	X = 80×cos(θ×π/180)	Y = 80×sin(θ×π/180)	Note
12	90°	0.0 mm	80.0 mm	*12 o'clock — top of dial*
1	60°	40.0 mm	69.3 mm	
2	30°	69.3 mm	40.0 mm	
3	0°	80.0 mm	0.0 mm	*3 o'clock — right of dial*
4	330°	69.3 mm	−40.0 mm	
6	270°	0.0 mm	−80.0 mm	*6 o'clock — bottom of dial*
9	180°	−80.0 mm	0.0 mm	*9 o'clock — left of dial*

Making the Angle Parametric

The real power emerges when the angle itself is a spreadsheet value. Instead of hard-coding 30 in the expression, reference a spreadsheet alias:

Parametric angular positioning

```
Spreadsheet setup:
  MarkerRadius = 80   (mm – distance from centre to marker)
  HourAngle    = 30   (degrees – change per marker)

X expression:  Spreadsheet.MarkerRadius * cos(Spreadsheet.HourAngle * pi /
180)
Y expression:  Spreadsheet.MarkerRadius * sin(Spreadsheet.HourAngle * pi /
180)

HourAngle = 0   → 3 o'clock position
HourAngle = 30  → 2 o'clock position
HourAngle = 60  → 1 o'clock position
HourAngle = 90  → 12 o'clock position
```

> **Tip**
>
> For a clock face or any fully polar design, create one cell called AngleStep = 30 and another called CurrentAngle = n * AngleStep where n is the position number. Changing AngleStep from 30 to 36 converts the clock to a 10-position dial without editing any positioning expression.

Gear Centre Distance — A Practical Trig Application

With expressions, the correct centre distance between two meshing gears can be calculated automatically from the gear parameters — no manual measurement or separate calculation needed.

Worked Example — Automatic Gear Centre Distance

Problem: Two spur gears must mesh. Gear 1 has 15 teeth, Module 3. Gear 2 has 30 teeth, Module 3. What is the correct centre-to-centre distance?

Formula: Centre Distance = (PitchDiameter1 + PitchDiameter2) / 2, where PitchDiameter = Module × Teeth

Spreadsheet setup

```
GearModule      = 3
Teeth1          = 15
Teeth2          = 30
PD1             = GearModule * Teeth1  →  45 mm
PD2             = GearModule * Teeth2  →  90 mm
CentreDist      = (PD1 + PD2) / 2     →  67.5 mm

Gear 2 Body Placement → X field:  =Spreadsheet.CentreDist
→ Gear 2 is placed at exactly 67.5 mm from Gear 1 automatically.
→ When Module or Teeth changes, the centre distance updates instantly.
```

Try It Now

Open the gear pair model from Week 7 (or create two simple cylinder bodies to represent gears).

In the Spreadsheet, create aliases: GearModule = 3, Teeth1 = 15, Teeth2 = 30, PD1 = GearModule * Teeth1, PD2 = GearModule * Teeth2, CentreDist = (PD1 + PD2) / 2.

Select Gear 2. In Properties → Data → Placement → Position → x, click the = button. Type: Spreadsheet.CentreDist. Press Enter. Gear 2 jumps to exactly 67.5 mm.

Test the parametric update: change Teeth2 from 30 to 40. Refresh. Gear 2 moves to a new position (CentreDist = (45 + 120)/2 = 82.5 mm). The gear spacing updates from the tooth count — no manual repositioning.

Inter-Body Expressions — Linking Dimensions Across Bodies

In Week 6 you used a shared Spreadsheet to link multiple Part Design Bodies to the same parameter values. Inter-body expressions provide a complementary capability: directly referencing one body's property from another body's expression, without needing an intermediate spreadsheet cell.

Inter-Body Expression Syntax

```
Syntax:  BodyName.FeatureName.ConstraintName

Examples:
  FrontPanel.Sketch.Width            → Width constraint from FrontPanel's
Sketch
  SidePanel.Pad.Length               → Pad length from SidePanel body
  GearDriver.InvoluteGear.teeth      → Tooth count of the driver gear

In a constraint expression:
  FrontPanel.Pad.Length * 0.75       → SidePanel height = 75% of FrontPanel
  GearDriver.InvoluteGear.teeth * 2 → DrivenGear teeth = 2× driver (2:1
ratio)
```

When to Use Inter-Body Expressions vs Shared Spreadsheet

Situation	Better Approach	Why
Multiple bodies sharing several common dimensions (a flat-pack box where all panels share thickness and clearance)	**Shared Spreadsheet**	*The spreadsheet provides one place to manage all shared values. Changing one cell updates all bodies simultaneously.*
Body B's dimension should always be a fixed ratio of Body A's (e.g., a shelf divider always 75% of shelf height)	**Inter-body expression**	*The relationship is inherent to the design — BodyA.Sketch.Height * 0.75 makes the relationship self-documenting.*
Gear B's tooth count should always be exactly double Gear A's to maintain a 2:1 ratio	**Inter-body expression**	*GearDriver.InvoluteGear.teeth * 2 is clearer and more reliable than a separate cell that might drift out of sync.*
A complex assembly where many bodies have independent dimensions with occasional cross-references	**Combination**	*Use shared spreadsheet for common values, inter-body expressions for specific ratios.*

Naming Your Model for Inter-Body Expressions

Inter-body expressions depend on the names of bodies, features, and constraints in the Model tree. FreeCAD's default names — Body, Body001, Sketch, Sketch001 — are unusable in practice. Before writing inter-body expressions, rename everything descriptively:

Bodies	Right-click → Rename. Use: FrontPanel, SidePanel, Lid, GearDriver, GearDriven.
Sketches	Rename to reflect what they define: OutlineSketch, HoleSketch, ProfileSketch.
Constraints	In the Sketcher, double-click a dimensional constraint and enter a name in the Name field: Width, Height, Radius, Depth. Named constraints become referenceable in inter-body expressions.

> ⚠ **Important**
>
> If you rename a body, sketch, or constraint after an inter-body expression has already been written, the expression will break — it references the old name. Always establish your naming convention before writing cross-references. If you must rename, update every expression that referenced the old name.

Scale Factor Expressions — Proportional Resizing

A scale factor expression is a single spreadsheet value that multiplies all dimensions in a model proportionally. Instead of changing 20 individual dimension values to resize a model by 20%, you change one ScaleFactor cell from 1.0 to 1.2 and every dimension updates automatically.

Scale Factor Pattern — Scalable Sign Model

```
Spreadsheet setup:
  BaseWidth     = 150   (mm — design width at scale 1.0)
  BaseHeight    = 80    (mm — design height at scale 1.0)
  BaseThickness = 12    (mm — material thickness at scale 1.0)
  ScaleFactor   = 1.0   (dimensionless — change this to resize everything)

Derived cells (calculated, not manually entered):
  ActualWidth     = BaseWidth     * ScaleFactor
  ActualHeight    = BaseHeight    * ScaleFactor
  ActualThickness = BaseThickness * ScaleFactor

Sketch constraint expressions:
  Width:  =Spreadsheet.ActualWidth
  Height: =Spreadsheet.ActualHeight
  Pad:    =Spreadsheet.ActualThickness

To produce 120% scale: set ScaleFactor = 1.2
  → Width = 180 mm, Height = 96 mm, Thickness = 14.4 mm
```

> 💡 **Tip**
>
> For CNC work, a scale factor approach is most useful when a client requests a piece in a different size, or when testing a design at 50% scale before committing to full-size material. Set ScaleFactor = 0.5, verify the form, then set it back to 1.0 for the production cut. CAM operations that reference model geometry update automatically — only the Stock dimensions need manual adjustment.

Circular References — Detecting and Avoiding Dependency Loops

A circular reference occurs when an expression depends (directly or indirectly) on the value it is trying to compute. FreeCAD detects circular references and displays an error — they cannot be resolved by any amount of computation.

> 📖 **Key Term: Circular Reference**
>
> An expression dependency loop where value A depends on value B, which depends (directly or through intermediate steps) on value A. Since the computation has no starting point, it cannot be resolved. The fix is always to redesign the expression chain so that values flow in one direction only: inputs → derived values → outputs.

Common Circular Reference Patterns and Fixes

Pattern	Example	Fix
Direct self-reference	Width cell contains: =Width + 10	Remove the self-reference. Separate the input (Width = 150) from derived value (WiderWidth = Width + 10).
Two-cell loop	CellA = CellB * 2, CellB = CellA / 2	One cell must be the input (plain value). The other references it. Decide which is the independent variable.
Inter-body loop	BodyA.Height references BodyB.Height, BodyB.Height references BodyA.Height	One body must have an independent height. The other body's height references the first — one direction only.
Indirect loop through spreadsheet	Spreadsheet.A = Body.Sketch.Width, Sketch.Width = Spreadsheet.A	The constraint and spreadsheet cell are referencing each other. The spreadsheet is always the source; the sketch is always the consumer.

> 📐 **Preventing Circular References**
>
> **The data flow rule:**
>
> Inputs (plain values or spreadsheet cells without expressions) → Derived values (computed from inputs) → Constraint expressions (read from derived values or inputs) → Placement expressions (read from constraints or derived values). Nothing flows backwards up this chain. Maintaining this strict one-directional flow prevents all circular references.

Building a Fully Expression-Driven Design

A fully expression-driven design has zero hard-coded numbers in any constraint, feature, or placement field. Every value is either a direct spreadsheet alias reference or a calculated expression derived from those references. The spreadsheet itself has only the minimum number of independent input values — ideally just a handful of fundamental dimensions and a ScaleFactor — and all other values are calculated.

This represents the top of the parametric design hierarchy:

Level	Name	Description
Level 0	Fixed-dimension model	All values typed directly into constraints. No parametric capability. Changing any dimension requires editing individual sketches.
Level 1	Spreadsheet-driven	Some or all sketch constraints reference spreadsheet aliases. Changing a spreadsheet value updates those constraints. (Chapters 8–9)
Level 2	Fully parametric with derived values	All constraints reference spreadsheet aliases, including compound formulas. No hard-coded numbers in the model. (Chapter 8 mastery level)
Level 3	Expression-driven with trig and inter-body	Circular positions calculated automatically, multi-body relationships expressed as formulas. (This chapter)
Level 4	Fully expression-driven	Minimum number of independent inputs — all other values calculated. Changing 2 inputs changes the entire model.

For the Week 10 Synthesis project (Workshop Tool Rack) and the Week 11 Certification Project (Keepsake Memory Box), Level 3 or Level 4 is the expectation. The assessment criteria explicitly require all critical dimensions to be spreadsheet-driven — a model with even one hard-coded number in a key constraint receives feedback to correct it before the project is accepted.

Quick Reference — All Expression Syntax

Spreadsheet References

Expression	Result / Meaning
Spreadsheet.Width	Direct alias reference — the value of the 'Width' cell
Spreadsheet.Width * 0.75	75% of Width
Spreadsheet.Width - 2 * Spreadsheet.Margin	Width minus margin on both sides
Spreadsheet.PanelWidth / 2	Half-width — for centring calculations
-Spreadsheet.EngravingDepth	Negative depth — for Final Depth fields in CAM

Trigonometric Positioning

Expression	Result / Meaning
Spreadsheet.Radius * cos(Spreadsheet.Angle * pi / 180)	X coordinate at given angle on circle
Spreadsheet.Radius * sin(Spreadsheet.Angle * pi / 180)	Y coordinate at given angle on circle
(Spreadsheet.PD1 + Spreadsheet.PD2) / 2	Gear centre distance from pitch diameters
sqrt(Spreadsheet.A ^ 2 + Spreadsheet.B ^ 2)	Hypotenuse of right triangle with sides A and B

Inter-Body References

Expression	Result / Meaning
FrontPanel.Sketch.Width	Width constraint named 'Width' from FrontPanel's sketch
FrontPanel.Pad.Length	Pad length from FrontPanel
FrontPanel.Sketch.Width * 0.75	SidePanel height = 75% of FrontPanel width
GearDriver.InvoluteGear.teeth * 2	Driven gear tooth count = 2× driver (2:1 ratio)

Week 10, Day 46 — Lesson Content Overview

Chapter 18 covers Day 46 — the first teaching day of Week 10. Days 47–50 are covered in Chapters 19–21.

Day 46 (Create)	Expressions Editor — advanced parametric formulas. Students use cos() and sin() with radian conversion to position a feature on a circle by changing a Spreadsheet.Angle alias — then watch it move to the mathematically correct position. Gear centre distance exercise: set up spreadsheet aliases for Module and Teeth, link Gear 2's X placement to the CentreDist expression, change Teeth2 and observe automatic repositioning. Inter-body link: write FrontPanel.OutlineSketch.Width * 0.75 in a side panel sketch and verify it tracks the front panel width. Independent challenge: design a radially symmetric piece where all element positions are calculated from exactly two input values — Radius and Count — with no manual placement and no hard-coded angles.

Chapter Summary

- The Expressions Editor is accessible from any numeric FreeCAD field via the blue = button. It evaluates mathematical expressions in real time and stores the formula rather than just the result — the value updates automatically whenever any referenced value changes.
- All trigonometric functions in FreeCAD work in radians. Always append * pi / 180 after any degree value: sin(30 * pi / 180) = 0.5. Forgetting this conversion produces completely wrong results with no error message.
- Polar coordinate conversion is the fundamental circular positioning tool: X = Radius × cos(angle × pi/180), Y = Radius × sin(angle × pi/180). This formula places any feature at any angular position on a circle from two input values — replacing all manual circular measurement.
- Inter-body expressions reference another body's properties by name: BodyName.FeatureName.ConstraintName. This requires descriptive names for all bodies, sketches, and constraints — FreeCAD's default names (Body001, Sketch002) cannot be used in expressions.
- Scale factor expressions multiply all model dimensions from a single cell. Setting ScaleFactor = 0.5 produces a half-scale test cut; ScaleFactor = 1.2 produces a 120% variant — all dimensions update automatically without editing individual sketches.
- Circular references occur when expression A depends on expression B which depends on expression A. Prevent them with the data flow rule: inputs → derived calculations → constraint expressions → placement expressions. Nothing flows backwards.
- A Level 4 fully expression-driven model has zero hard-coded numbers in any constraint or feature. All values are either direct spreadsheet alias references or mathematical expressions derived from those aliases. The Certification Project requires Level 3 or Level 4.

Key Vocabulary — Chapter 18

Term	Definition
Expressions Editor	FreeCAD's built-in formula interpreter, accessible from any numeric property field via the blue = button. Evaluates mathematical expressions in real time and stores the formula, updating automatically when referenced values change.

Radian	The standard mathematical unit for angles used by FreeCAD's trigonometric functions. One full circle = 2π radians = 360°. Conversion: degrees × π / 180 = radians.
pi	The mathematical constant π (3.14159...). Available in FreeCAD expressions by typing 'pi'. Used in all radian conversions and circular calculations.
Polar coordinates	A way of describing a 2D position using a radius (distance from centre) and an angle (direction from positive X axis), rather than Cartesian X and Y values.
Polar conversion	The formula that converts polar coordinates (R, θ) to Cartesian (X, Y): X = R × cos(θ × π/180), Y = R × sin(θ × π/180). The core tool for circular feature positioning.
Inter-body expression	A FreeCAD expression that directly references a property from a different Part Design Body using the syntax: BodyName.FeatureName.ConstraintName. Requires all bodies, sketches, and constraints to have descriptive names.
Scale factor	A dimensionless spreadsheet alias (typically 1.0) that multiplies all dimension aliases proportionally. Setting it to 0.5 halves all model dimensions; setting it to 1.2 increases them by 20%.
Circular reference	An expression dependency loop where value A depends on value B which depends on value A. FreeCAD cannot resolve circular references and displays an error. Prevented by maintaining strict one-directional data flow.
Data flow rule	The design principle that prevents circular references: inputs (plain values) → derived calculations → constraint expressions → placement expressions. Nothing flows backwards in this chain.
Level 4 parametric	A fully expression-driven model with zero hard-coded numbers in any constraint or feature. All values are either direct spreadsheet references or calculated expressions. The target parametric level for the Certification Project.

Looking Ahead — Chapter 19: Assembly Concepts & Bill of Materials

Chapter 19 covers the FreeCAD v1.1.0 Assembly workbench — creating constrained assemblies of multiple part files, applying Fixed and Revolute joints, solving the assembly to verify spatial fit, and generating a CNC Bill of Materials. It also introduces the Tool Rack component preview that feeds into Chapter 21's synthesis project.

CHAPTER
19
of 24

PART FIVE — ADVANCED SKILLS

Assembly Concepts & Bill of Materials

FreeCAD v1.1.0 Assembly workbench, joints, component links, BOM creation, and the Tool Rack component preview

Chapter Learning Objectives

By the end of this chapter, you will be able to:

- Explain why assembly modelling is useful in a CNC workflow
- Create an Assembly container and insert components as Links
- Apply a Ground constraint to the anchor component
- Add at least three joint types: Fixed, Revolute, and Slider
- Solve the assembly and interpret any constraint errors
- Check for visual clashes between components
- Explain the difference between Links and embedded copies in an Assembly
- Create a CNC Bill of Materials listing part name, quantity, material, stock size, bit, and G-code filename
- Embed a BOM spreadsheet in a TechDraw drawing page
- Build an assembly preview for the Workshop Tool Rack (Project 10)

Why Assembly in a CNC Workflow?

An assembly tells you whether the parts you designed actually fit together — before you cut a single piece of timber. It is the cheapest possible prototype.

Every CNC project in this course has involved designing individual parts and then cutting them. For simple single-piece projects (a name badge, a gear), there is no assembly to worry about. For multi-piece projects — a finger joint box, a tool rack, a cabinet — the parts must fit together as a system, and that system needs to be verified before committing any timber to the machine.

FreeCAD's Assembly workbench (introduced in v1.0 and refined in v1.1.0) allows you to bring individual part files into a single Assembly document, constrain them with joints, and verify the assembled result. If two panels clash (overlap in space), or a shelf is 5mm too wide to fit between the side panels, the assembly reveals it. The fix is a spreadsheet alias change — seconds of work. Without an assembly, you discover the same problem after cutting, which costs material, time, and machine use.

The second major topic in this chapter — the Bill of Materials — is the practical output of an assembly. Once all components are listed and counted in the assembly, generating a cut list (what pieces to cut, from what material, at what size) is a straightforward exercise. For the Workshop Tool Rack in Chapter 21, a complete BOM is a required deliverable alongside the physical rack.

The Assembly Workbench — Overview

The Assembly workbench is a native FreeCAD workbench — it does not need to be installed from the Addon Manager. Switch to it from the Workbench drop-down. The menu bar gains an Assembly menu containing all tools for creating assemblies, inserting components, and managing joints.

FreeCAD's Assembly workbench uses a constraint-based solver: you define the relationships between parts (this face is fixed, this axis is shared, this edge is parallel to that edge) and the solver calculates the position and orientation of each component that satisfies all constraints. This is the same approach used by professional CAD tools like SolidWorks and CATIA.

Assembly container	A new document object that holds all components and joint definitions. It appears in the Model tree as 'Assembly'. Double-click to make it active.
Component (Link)	A reference to an external .FCStd file inserted into the assembly. Changing the original part file updates the component in the assembly automatically.
Ground	A constraint that fixes one component completely in space — it becomes the immovable reference. All other components are positioned relative to it.
Joint	A constraint that defines the allowable relative motion between two components. Different joint types allow different combinations of motion.
Solver	The mathematical engine that calculates component positions satisfying all joint constraints. Run it with Assembly → Solve.
Clash detection	A visual check for components that overlap in space — indicating they physically cannot occupy the same volume. Run via Assembly → Detect Clashes (if available in your FreeCAD version) or inspect visually in the 3D view.

Creating an Assembly — The Five-Step Workflow

Assembly Workflow — Five Steps		
1	**Create Assembly**	Assembly → Create Assembly. A new Assembly object appears in the Model tree. This is the container for all components.
2	**Insert Components**	Assembly → Insert Component. Browse to the .FCStd file containing a part. Repeat for each part. Components appear as Links — they reference the original file.
3	**Ground One Part**	Right-click the first inserted component → Ground. This fixes it in space as the reference anchor. All other parts are positioned relative to the grounded part.

4	Add Joints	Assembly → Add Joint. Select the joint type, then select the features (faces, axes, points) on each component that define the joint. The solver positions the parts accordingly.
5	Solve & Verify	Assembly → Solve. FreeCAD calculates all joint constraints and positions components. Check for clashes — parts overlapping in space.

Step 1 — Creating the Assembly Container

1. With any FreeCAD document open (or a blank new document), switch to the Assembly workbench.
2. Go to Assembly → Create Assembly.
3. A new 'Assembly' object appears in the Model tree. This is the container.
4. Save the file immediately with a descriptive name: ToolRack_Assembly.FCStd.

Step 2 — Inserting Components as Links

Components in an Assembly are inserted as Links — references to external part files — rather than as embedded copies. This is the key design decision in FreeCAD's Assembly system, and it has significant practical advantages.

	✅ Links (App::Link) — the correct approach	❌ Copies — avoid in Assembly
Storage	References the original part file — no geometry duplication	Embeds a full copy of the geometry in the assembly file
File size	Assembly file stays small	Assembly file grows with every part added
Updates	Edit the original part → all assembly instances update instantly	Edit the original → the copy in the assembly does NOT update
BOM	One entry per Link reference — can be counted automatically	Copies blur the part count — harder to automate
CNC use	Change a shelf width in the part file → all shelf Links in the assembly update immediately	Change shelf width → must manually update every embedded copy

1. Go to Assembly → Insert Component.
2. A file browser opens. Navigate to the .FCStd file containing your first part.
3. Select the file and click Open. The component appears in the assembly at the origin.
4. Repeat for every part that belongs in the assembly. Each part file is inserted as a separate Link.
5. In the Model tree, each Link shows the original filename. If you have two identical shelf panels, insert the shelf file twice — you will get two Links, both referencing the same part.

> 💡 **Tip**
>
> For repeated components (identical shelf panels, matching side panels), insert the same part file multiple times. Each insertion creates a new Link pointing to the same source file. This means changing the source file updates all instances simultaneously — exactly the same benefit as Polar Pattern, but for entire parts.

Step 3 — Grounding the First Component

Before adding joints, one component must be grounded — fixed absolutely in space. The grounded component is the assembly's anchor. In a CNC furniture assembly, ground the largest structural piece: the back panel, the base, or the main frame.

1. Right-click the first component in the Model tree.
2. Select Ground from the context menu.
3. A small lock icon appears next to the component in the Model tree.
4. The component is now fixed. All subsequent joint constraints position other components relative to it.

Step 4 — Adding Joints

Joints define how two components are connected and what motion (if any) is allowed between them. For CNC furniture assemblies, Fixed and Slider joints cover most situations — panels are either fixed to each other or must slide into position.

Joint Type	Colour	Degrees of Freedom — What Motion it Allows & CNC Use Cases
Fixed	●	Locks two components together completely. No relative motion possible. Use to permanently attach parts that do not move relative to each other: panels to frames, hardware to bodies.
Revolute	●	Allows rotation around a single axis. One component turns relative to another — used for hinges, gear shafts, rotating mechanisms, clock hands.
Cylindrical	●	Allows both rotation AND translation along an axis. A component can spin and slide. Use for bolts in slots, shafts that can move axially.
Slider	●	Allows translation along one axis only — no rotation. Use for drawers, sliding panels, linear guides.
Ball	●	Allows rotation in any direction (3 degrees of rotational freedom). No translation. Use for ball-socket joints, casters.
Angle	●	Constrains the angle between two features (faces, axes) to a specific value. Use when two parts must be held at a defined angle to each other.
Gear	●	Links two revolute joints with a gear ratio. When one shaft turns, the other turns at the calculated ratio. Perfect for documenting the gear pair from Project 7.
Parallel	●	Forces two axes or faces to remain parallel. No position constraint — the component can move, but its orientation is locked. Use for panels that must remain flat.

Adding a Joint — The General Procedure

1. Go to Assembly → Add Joint. A dialog opens.
2. Select the Joint Type from the drop-down: Fixed, Revolute, Slider, etc.
3. The dialog asks you to select two features — one from each component. Features can be: faces (flat surfaces), axes (cylindrical surfaces or sketch axes), points (vertices), or edges.

4. Click the first feature on Component A, then the first feature on Component B. The dialog shows the selected features.
5. For Fixed and Slider joints: select a flat face on each component that should be in contact. For Revolute joints: select the cylindrical face of a bore on one component and the cylindrical face of a matching shaft on the other.
6. Click OK. The solver positions the components according to the joint constraint.

Step 5 — Solve and Verify

1. Go to Assembly → Solve. FreeCAD's constraint solver runs.
2. If all constraints are satisfiable, the components snap into their correct positions.
3. If a constraint error appears, one of the joints is over-constrained (too many restrictions) or under-constrained (not enough). Read the error message — it identifies which joint is the problem.
4. Inspect the assembly visually. Look for: (a) parts that appear in the wrong position or orientation, (b) parts that visually overlap (clash), (c) gaps where there should be contact.

> **Assembly Rule**
> The purpose of an assembly in a CNC workflow is verification, not simulation. You are not checking that the joints move correctly under load — you are checking that all parts fit together in space without overlapping, and that the assembled dimensions match your design intent. A simple visual check after Solve is usually sufficient for furniture-scale CNC projects.

Interpreting and Fixing Assembly Problems

Common Constraint Errors

Over-constrained joint	Too many constraints applied — two joints are contradicting each other. Fix: remove one of the conflicting joints. Check: did you accidentally add a Fixed joint where a Revolute or Slider was intended?
Under-constrained component	Not enough joints — the component still has free degrees of motion. Fix: add another joint to restrict the remaining motion. Symptom: the component drifts or floats when the assembly is solved.
No solution found	The solver could not position the parts to satisfy all constraints simultaneously. Usually means a dimensional mismatch: the part dimensions don't allow the joints to be satisfied. Check the part files for the relevant dimensions.
Clash (visual overlap)	Two parts occupy the same physical space — they cannot both exist there. Causes: incorrect shelf depth, wrong clearance, mis-sized housing. Fix the dimension in the source part file, refresh the assembly.

Bill of Materials — The CNC Cut List

The Bill of Materials (BOM) is the bridge between the digital assembly and the physical workshop. It answers the fundamental question: what do I need to cut, and in what quantities?

A Bill of Materials is a structured list of every component in an assembly, with the information needed to produce each one. In a traditional engineering context this includes part numbers, revision levels, and supplier codes. In a CNC woodworking context the relevant fields are simpler and more practical: what to call the part, how many you need, what material it is, how large a piece of stock you need, how thick it is, what bit to use, and what G-code file to load.

The BOM is usually the last document you create before going to the machine. With a BOM in hand, you can prepare all your stock pieces in advance — marking them out, cutting them to rough size on the table saw if needed — so the CNC session runs efficiently without stopping to measure up more material mid-job.

CNC Bill of Materials — Required Fields

Part Name	Qty	Material	Stock Size (mm)	Thick (mm)	Primary Bit	G-code File
Back Panel	1	12mm MDF	600 × 450	12	6mm EM	`ToolRack_Back_v1.nc`
Top Shelf	1	12mm MDF	600 × 120	12	6mm EM	`ToolRack_Top_v1.nc`
Mid Shelf	1	12mm MDF	600 × 100	12	6mm EM	`ToolRack_Mid_v1.nc`
Cleat Rail × 2	2	18mm Pine	600 × 60	18	6mm EM	`ToolRack_Cleat_v1.nc`
Side Panel × 2	2	12mm MDF	450 × 80	12	6mm EM	`ToolRack_Side_v1.nc`
Chisel Rack Strip	1	12mm MDF	400 × 80	12	3mm EM	`ToolRack_Chisel_v1.nc`

Example BOM for the Workshop Tool Rack (Project 10 — Chapter 21). All G-code filenames follow the project naming convention.

> 📖 **Key Term: Bill of Materials (BOM)**
> A structured list of every component in an assembly with the information needed to manufacture each one. For CNC work: part name, quantity, material, stock dimensions, material thickness, primary bit, and G-code filename.

Creating the BOM in a FreeCAD Spreadsheet

The most practical approach in FreeCAD is to maintain the BOM as a Spreadsheet object within the Assembly file. The spreadsheet can be manually populated (type in the values) or partially automated by linking cells to component properties using expressions.

1. With the Assembly file open, switch to the Spreadsheet workbench.

2. Create a new Spreadsheet: Spreadsheet → Create Spreadsheet. Rename it 'BOM'.
3. In Row 1, type column headers: Part Name, Qty, Material, Stock Size (mm), Thickness, Bit, G-code File.
4. Fill in one row per component. For repeated components (e.g., two side panels), use Qty=2 and one row — do not create duplicate rows.
5. Save the spreadsheet. It is now part of the assembly file.

Embedding the BOM in a TechDraw Drawing

A BOM spreadsheet can be embedded directly into a TechDraw drawing page as a formatted table. This creates a formal documentation package: the dimensioned drawing on one side of the page, and the cut list on the other.

1. Switch to the TechDraw workbench. Open or create a drawing page (A4 Landscape recommended for a BOM alongside a view).
2. Go to TechDraw → Views → Insert Spreadsheet View.
3. Select the BOM spreadsheet from the dialog.
4. A table representation of the spreadsheet appears on the drawing page. Drag it to position — typically below or beside the main views.
5. Resize the table by adjusting cell widths in the Spreadsheet workbench and refreshing the TechDraw view.
6. Add a title annotation above the table: 'Bill of Materials — Workshop Tool Rack'.
7. Export to PDF. The resulting document contains both the engineering drawing and the cut list on one page.

> **Tip**
>
> Keep the BOM spreadsheet and the part files in the same project folder. If you move files, the Links in the assembly will break and the BOM spreadsheet references will also fail. Folder structure discipline (established in Chapter 4) is essential for multi-file assembly projects.

Workshop Tool Rack — Assembly Preview (Project 10 Preparation)

The Workshop Tool Rack (Chapter 21, Project 10) is the course's synthesis project — it requires every major skill developed in the course. Before Chapter 21's project day, building an assembly preview of the Tool Rack in this chapter serves two purposes: it verifies that all planned components fit together, and it generates the BOM that the Chapter 21 project requires as a deliverable.

The Tool Rack design used in this preview is a guide — you may vary the dimensions, number of shelves, and configuration to suit your own workshop storage needs. The assembly and BOM process applies to any configuration.

Tool Rack Component Overview

Back Panel	The primary vertical surface, 600mm wide × 450mm tall × 12mm MDF. Grounded as the anchor component. All other parts attach to or are positioned relative to this panel.
Top Shelf	Horizontal shelf at the top, 600mm × 120mm × 12mm. Attaches to the back panel with a Fixed joint on the rear face and the top edge.
Mid Shelf	Horizontal shelf at mid-height, 600mm × 100mm × 12mm. Same attachment logic as Top Shelf.
Side Panels × 2	Vertical end pieces, 450mm × 80mm × 12mm. Fixed to the left and right edges of the Back Panel.
French Cleat Rails × 2	Angled-front strips, 600mm × 60mm × 18mm pine. Attach along the back panel at specific vertical positions — the angled edge faces forward.
Chisel Rack Strip	A row of chisel holes, 400mm × 80mm × 12mm, with Polar Pattern holes along its length. Attaches to the mid shelf top face.

Assembly Walkthrough — Tool Rack Preview

1 — Create separate part files for each component
In previous sessions you may have already designed these components as part of Chapter 21 preparation. If not, create simplified placeholder bodies (basic pads at the correct dimensions) for each component listed above. Save each as its own .FCStd file: BackPanel.FCStd, TopShelf.FCStd, etc.

2 — Create the Assembly file
New file → Assembly → Create Assembly. Save as ToolRack_Assembly.FCStd in the same project folder as all component files.

3 — Insert all components as Links
Assembly → Insert Component, repeat for each part file. Insert SidePanel.FCStd twice (gives two Link instances for both side panels).

4 — Ground the Back Panel
Right-click BackPanel in the Model tree → Ground. The back panel is now the fixed anchor.

5 — Fix the Top Shelf to the Back Panel
Assembly → Add Joint → Fixed. Select the rear face of the Top Shelf and the top region of the front face of the Back Panel. Click OK. Solve. The shelf should snap into position across the top of the back panel.

6 — Fix the Mid Shelf at the correct height
Same process as Step 5. Before adding the joint, set the Offset value in the joint dialog to position the shelf at the desired height (e.g., 200mm from the bottom of the back panel).

7 **Fix both Side Panels**
Add a Fixed joint for each side panel. Select the inner face of the side panel and the corresponding vertical edge face of the back panel. After solving, both side panels should stand vertically at the left and right ends of the back panel.

8 **Fix French Cleat Rails**
Position each cleat rail at the correct vertical position on the back panel front face. Fixed joints, with offset values matching your design.

9 **Solve and inspect**
Assembly → Solve. Inspect the result: Do all components appear at their correct positions? Is there any visual clash (overlap)? Rotate the view and inspect all corners and joints closely.

10 **Create the BOM spreadsheet**
Switch to Spreadsheet workbench. Add Spreadsheet to the assembly file. Fill in one row per component as per the BOM table shown earlier in this chapter. Include all 6 columns: Part Name, Qty, Material, Stock Size, Thickness, Bit, G-code File.

11 **Embed BOM in TechDraw — create documentation page**
Create an A4 Landscape TechDraw page. Insert a simple isometric view of the assembly (TechDraw → Insert View → select the Assembly object). Insert the BOM spreadsheet view below. Complete the title block: ToolRack Assembly — BOM, your name, today's date. Export to PDF.

> 🔧 **The Value of the Assembly Preview**
>
> This assembly preview is not just a preparation exercise — it is the engineering validation step that professional CNC operators complete before committing any material. If the assembly reveals a dimensional problem now, fixing it takes minutes. If the same problem is discovered on the machine, it costs timber, time, and tool wear.
>
> Keep this assembly file. In Chapter 21 (the synthesis project), you will update it with the finished, fully-detailed component files rather than the placeholder bodies — and the BOM will update automatically via the Links.

Week 10, Days 46–48 — Lesson Content Overview

Chapter 19 covers the first three days of Week 10. Days 49–50 are covered in Chapter 20 (Advanced UGS Workflow & Machine Management).

Day 46 (Create)	Assembly workbench introduction. Students create an assembly of any two parts from earlier in the course — the gear pair from Project 7 is ideal (ground the driver gear frame, add a Revolute joint for the driven gear, add a Gear joint linking the two shafts). Solve the assembly. Discussion: when does adding an assembly add real value vs adding complexity? What is the minimum assembly that usefully validates a design?
Day 47 (Create)	Assembly for the Tool Rack preview. Students insert all Tool Rack component files, ground the back panel, and apply Fixed joints for shelves and side panels. Solve and inspect. If any component does not fit (dimensional error), identify the dimension to change, update the source file, and refresh the assembly. This replicates the real pre-production validation workflow.
Day 48 (Create)	Bill of Materials. Students create a BOM Spreadsheet for the Tool Rack assembly (all 6 fields, all components). Embed the BOM in a TechDraw page alongside an isometric assembly view. Export to PDF. Independent challenge: add a 'Total stock area required' row at the bottom of the BOM that uses a formula to sum (Stock Length × Stock Width × Qty) for all rows — giving the total MDF/timber area in mm^2.

Chapter Summary

- FreeCAD v1.1.0's native Assembly workbench inserts components as Links — references to external part files. When the source part file changes, all linked instances in the assembly update automatically. This is the correct approach; embedding copies is less efficient and breaks the update link.
- The five-step assembly workflow: Create Assembly → Insert Components (as Links) → Ground one component → Add Joints → Solve and verify.
- Ground fixes one component as the immovable reference for the whole assembly. All other parts are positioned relative to it. In a furniture assembly, ground the largest structural piece.
- Eight joint types are available: Fixed, Revolute, Cylindrical, Slider, Ball, Angle, Gear, and Parallel. For CNC furniture: Fixed (permanent attachment) and Slider (sliding panel insertion) cover most needs. Revolute is essential for rotating mechanical assemblies like the gear pair.
- The assembly solver produces three outcomes: correct positioning (all constraints satisfied), constraint errors (over or under-constrained — fix the joint definitions), or no solution (dimensional mismatch — fix the part dimensions).
- A Bill of Materials is a structured list of all components needed to manufacture an assembly. For CNC work, the essential fields are: Part Name, Quantity, Material, Stock Size, Thickness, Primary Bit, and G-code Filename.
- A BOM Spreadsheet can be embedded directly into a TechDraw page using TechDraw → Insert Spreadsheet View. This creates a single PDF document containing both the dimensioned drawing and the cut list.
- The Tool Rack assembly preview in this chapter is a pre-production validation exercise. Discovering dimensional problems in the assembly takes minutes to fix; discovering them after cutting takes timber and time.

Key Vocabulary — Chapter 19

Term	Definition
Assembly workbench	The native FreeCAD workbench for creating constrained assemblies of multiple part files. Introduced in FreeCAD v1.0; available without addon installation in v1.1.0.
Assembly container	The document object created by Assembly → Create Assembly. It holds all component Links and joint definitions for the assembly.
Component (Link)	A reference to an external .FCStd part file inserted into an assembly. Changes to the source file automatically update the linked component in the assembly.
Ground	An assembly constraint that fixes one component completely in space — no translation or rotation is possible. All other components are positioned relative to the grounded part.
Fixed joint	An assembly joint that locks two components together with no relative motion. Used for permanently attached parts: shelves to panels, hardware to bodies.
Revolute joint	An assembly joint that allows rotation around a single shared axis. Used for hinges, gear shafts, rotating mechanisms.
Slider joint	An assembly joint that allows translation along a single axis with no rotation. Used for drawers, sliding panels, and linear guides.
Gear joint	An assembly joint that links two revolute joints with a defined ratio. When one shaft rotates, the other rotates at the calculated gear ratio.
Constraint solver	The mathematical engine within the Assembly workbench that calculates component positions satisfying all defined joint constraints.
Clash	A condition where two components occupy the same physical volume — they are overlapping in space. Clashes indicate dimensional errors in the part designs.
Bill of Materials	A structured list of all components in an assembly with information required to manufacture them. CNC BOM fields: Part Name, Quantity, Material, Stock Size, Thickness, Bit, G-code Filename.
Spreadsheet View	A TechDraw tool (TechDraw → Insert Spreadsheet View) that embeds a FreeCAD Spreadsheet as a formatted table directly on a drawing page — used to display a BOM alongside an engineering drawing.
Cut list	A practical synonym for Bill of Materials in a CNC or woodworking context. Emphasises the manufacturing intent: a list of pieces to cut, from specified materials, at specified sizes.
Over-constrained	A joint configuration where too many constraints are applied, creating contradictions the solver cannot resolve. Symptom: solver error message. Fix: remove one conflicting joint.

Looking Ahead — Chapter 20: Advanced UGS Workflow & Machine Management

Chapter 20 covers the UGS features and machine management techniques that allow experienced operators to work more efficiently and handle complex jobs with confidence: coordinate system management, saved zero positions, feed rate override strategies, mid-job recovery, and machine calibration checking. This knowledge underpins the Workshop Tool Rack synthesis project in Chapter 21.

CHAPTER
20
of 24

PART FIVE — ADVANCED SKILLS

Advanced UGS Workflow & Machine Management

UGS configuration, GRBL limits, real-time overrides, job recovery, the Job Setup Sheet, and post-cut evaluation

Chapter Learning Objectives

By the end of this chapter, you will be able to:

- Verify and configure UGS connection settings for the CNC 4040 Pro Max: port, baud rate, and status polling interval
- Explain the difference between GRBL soft limits and hard limits and verify that both are correctly configured
- Use feed rate override and spindle speed override in UGS during a live cutting operation
- Apply the 10-step job recovery procedure correctly after an e-stop or mid-job interruption
- Complete a Job Setup Sheet for every CNC cutting session using the standard template
- Carry out a structured post-cut evaluation: dimensional measurement, photography, and result documentation
- Manage a CNC project archive using the standard file naming and version control conventions
- Complete the pre-cut checklist independently before any cutting operation

The CNC machine does not care whether you are experienced or inexperienced. It executes the G-code. Your job is to set up the conditions for that execution correctly, manage it while it runs, recover cleanly when something unexpected happens, and document what occurred so you can reproduce or improve the result.

Chapter 20 takes you beyond the standard UGS workflow into the machinery of UGS itself: the settings that configure how it communicates with the machine, the real-time controls that let you adjust a job while it runs, and the recovery procedures that allow you to resume a job after something goes wrong.

The second half of the chapter covers the professional workflow that surrounds every cutting session: the Job Setup Sheet, the pre-cut checklist, the post-cut evaluation, and the file management system that keeps your project history organised. These are not optional extras — they are the practices that distinguish a professional CNC operator from someone who gets lucky occasionally.

UGS Configuration for the CNC 4040 Pro Max

UGS connects to the CNC 4040 Pro Max via a USB serial connection. The machine's GRBL controller appears to your computer as a serial (COM) port on Windows or a /dev/tty device on macOS and Linux. Getting these settings correct is a one-time setup task, but verifying them whenever you change computers or encounter a connection problem is an important diagnostic skill.

Connection Settings — Step by Step

1. Launch UGS. Look for the connection toolbar at the top — port selector, baud rate field, and Connect button.
2. Power on the CNC 4040 Pro Max. Connect the USB cable. Wait 5–10 seconds for GRBL to initialise.
3. Click the port drop-down. On Windows: look for COM3–COM6. On macOS: /dev/tty.usbmodem or /dev/tty.usbserial. If multiple ports appear, unplug the USB cable and note which port disappears — that is your machine.
4. Set baud rate to 115200. This is the standard baud rate for GRBL 1.1 on the CNC 4040 Pro Max. A different baud rate produces garbled communication or no response.
5. Click Connect. The UGS console should show a GRBL version string: 'Grbl 1.1h ["$" for help]'. If you see this, the connection is established.
6. Note the GRBL version. GRBL 1.1h is standard for the 4040 Pro Max. Earlier versions have different $-parameter numbering — consult your instructor if you see an older version.

Port	COM3–COM6 (Windows) or /dev/tty.usbmodem (macOS). Identify by unplugging and re-plugging.
Baud Rate	115200 — standard for GRBL 1.1. Do not change.
Status poll interval	200 ms — how often UGS requests machine position. Faster causes GRBL buffer overflow; slower makes position display laggy.

Accessing and Verifying GRBL Settings

GRBL's firmware settings — the $-parameters that control machine behaviour — can be read from UGS's console at any time. Type $$ and press Enter. GRBL responds with all current settings. Compare the values to the table below to verify the CNC 4040 Pro Max is correctly configured.

$-Param	Setting	CNC 4040 Pro Max Value	What It Controls
`$0`	**Step pulse time**	**10 µs**	Duration of motor driver step pulse — do not change
`$1`	**Step idle delay**	**25 ms**	How long motors stay energised after a move — do not change
`$10`	**Status report mask**	**1**	What data UGS requests from GRBL — leave at default
`$20`	**Soft limits**	**1 (enabled)**	Whether GRBL checks position against soft limit values before moving
`$21`	**Hard limits**	**0 or 1**	Whether limit switch triggers halt the machine immediately

$22	**Homing cycle**	**1 (enabled)**	Whether the machine must home before operations are allowed
$110	**X max rate**	**3000 mm/min**	Maximum X axis movement speed
$111	**Y max rate**	**3000 mm/min**	Maximum Y axis movement speed
$112	**Z max rate**	**500 mm/min**	Maximum Z axis movement speed — lower for safety
$130	**X max travel**	**350 mm**	X axis soft limit boundary — should match machine bed
$131	**Y max travel**	**350 mm**	Y axis soft limit boundary — should match machine bed
$132	**Z max travel**	**80 mm**	Z axis soft limit boundary — should match spindle travel

> **Tip**
>
> Screenshot the $$ output at the start of every new course module and save it in your project folder. If someone accidentally changes a GRBL setting, comparing the current $$ output to a known-good screenshot is the fastest way to identify what changed.

GRBL Soft Limits and Hard Limits

The CNC 4040 Pro Max has a defined working volume — the envelope within which the spindle can safely move without crashing into the machine frame. GRBL's limit system enforces this boundary in software and optionally in hardware.

Soft Limits ($20=1)	Software-based position checking. GRBL compares the planned move to the machine's known position. If the move would exceed the configured travel limits ($130/$131/$132), GRBL refuses the move and raises an alarm before it happens. Requires homing active ($22=1). Best practice: always enabled. Response: alarm raised, machine stops before moving.
Hard Limits ($21=1)	Hardware limit switches mounted at the ends of each axis. When triggered, GRBL immediately halts all motion. Acts as a last-resort backup after soft limits. Can cause missed steps and position loss if triggered at high speed. Response: immediate all-axes halt.
Soft vs Hard	Soft limits protect against G-code errors and incorrect zeros — they stop a bad move before it starts. Hard limits are the physical emergency stop of last resort. Both should be enabled simultaneously. NEVER disable soft limits to work around errors — investigate the root cause instead.

Safety — Read Before Operating

NEVER disable soft limits ($20=0) or homing ($22=0) as a 'workaround' for limit errors. The correct fix is to investigate why the G-code is requesting moves outside the machine envelope — not to remove the protection that is catching the problem.

A machine with soft limits disabled will drive the spindle into the frame, snap the bit, strip stepper motor couplings, and potentially destroy the machine. The repair cost far exceeds the inconvenience of fixing the root cause of a soft limit error.

Clearing a Limit Alarm

1. Read the alarm — UGS console shows ALARM:1 (hard limit), ALARM:2 (soft limit violation), or ALARM:8 (homing failure).
2. Type $X in the UGS console and press Enter. This unlocks GRBL from the alarm state — status changes from 'Alarm' to 'Idle'.
3. Investigate the cause: wrong zero position? G-code coordinates outside machine travel? Fix the root cause before resending.
4. After any alarm — particularly a hard limit alarm — re-home the machine ($H or the Home button) before resuming.

Real-Time Overrides — Adjusting a Job While It Runs

GRBL 1.1 supports real-time override commands — adjustments to feed rate and spindle speed that take effect immediately during a running job without stopping or restarting the G-code. UGS exposes these as sliders in the Machine Control panel.

Key Term: Feed Rate Override (FRO)

A real-time GRBL command that scales the programmed feed rate by a percentage without modifying the G-code. A 70% override on a 1000 mm/min programmed feed produces actual motion at 700 mm/min. The G-code feed rate value is unchanged — the override is a temporary scaling factor applied at the firmware level.

Override Control	How to Adjust	When to Use It
Feed Rate Override (FRO)	Feed Rate slider in UGS Machine Control panel. Range: 10–200% of programmed feed rate.	*Reduce to 70–80% when entering a difficult cut area — tight corner, hard grain, knot. Increase if tool sounds like it is rubbing. Restore to 100% after the difficult section.*
Spindle Speed Override (SSO)	Spindle Speed slider in UGS. Range: 10–200% of programmed RPM. Only active if your machine has spindle speed control via UGS.	*Reduce when cutting softwood to prevent burning. Most operators leave SSO at 100% for routine work.*

Feed Hold (Pause)	Press the Feed Hold button (or F key). The machine decelerates smoothly to a stop without losing position. Spindle continues running.	*When you need to pause to check progress, clear chips, or investigate an unexpected sound. Not a substitute for e-stop.*
Resume after Feed Hold	Press the Resume/Play button. Machine re-accelerates smoothly from the held position.	*After a Feed Hold — verify the workpiece is still correctly positioned and the spindle is running before resuming.*
Reset / Unlock	Press the Reset button after an e-stop or alarm. Clears alarm state but does NOT re-home the machine.	*Required before any new operation after an alarm state. Always re-verify position after a reset — the machine may have lost steps.*

When to Use Feed Rate Override — Practical Scenarios

Scenario	Action	Why
Tool sounds like it is chattering in a corner	Reduce FRO to 70–80% before the corner	Reduces instantaneous chip load at the direction change — the most common cause of corner chatter
Entering a knot or significantly harder grain	Reduce FRO to 60–70% and listen	Hard grain areas require lower chip load. The tool sound will tell you — smooth vs labouring
Cutting sound is clean and consistent throughout	Leave FRO at 100%	Do not adjust what is working. Over-reducing causes rubbing in soft material
Tool seems to be rubbing — high heat, glazed surface	Increase FRO to 110–120%	Too-slow feed causes rubbing rather than chip formation. More common in hard material than MDF
A long straight finishing pass on a flat floor	Leave FRO at 100%	Finishing passes are designed for 100%. Reducing unnecessarily increases runtime without improving quality
Testing a new material or bit for the first time	Start at 70%, increase if all sounds well	Caution is appropriate when you don't have data for the specific combination

> 🤔 **Stop and Think**
>
> The feed rate override is a diagnostic tool as much as an adjustment tool. When you change the FRO and listen to the result, you are gathering real-time data about the cutting conditions.
>
> Developing an ear for the cutting sound is one of the most valuable skills you can build. A healthy cut sounds like a clean, consistent hiss or whir. Too-fast cutting sounds harsh or crackles. Too-slow cutting sounds like a dull rubbing or humming. Chatter sounds like a rapid rhythmic vibration — you can feel it through the machine table. Once you have heard all of these, you will never need to guess again.

Job Recovery — Resuming After an Interruption

Jobs stop mid-cut more often than beginners expect. Power fluctuations, accidental e-stop presses, USB cable disconnections, computer sleep modes, and tool breakages all cause mid-job interruptions. The ability to resume a job cleanly is a professional skill that takes practice.

The challenge is that GRBL operates statelessly — it processes G-code line by line and does not remember what it has already done. If you simply send the job again from the beginning, it will re-machine already-cut areas, which at best wastes time and at worst crashes the tool into raised geometry or previously cut tabs.

10-Step Job Recovery Procedure		
1	Stop the spindle	Send M5 in the UGS console, or press the physical e-stop. Do not move the machine while the spindle is spinning.
2	Note the line number	In UGS, look at the progress counter. Note the current G-code line number. Write it on the Job Setup Sheet immediately — you will need it to resume.
3	Raise Z safely	Use the Z+ jog button to raise the spindle well clear of the workpiece — at least 20 mm above the highest point of the material.
4	Diagnose the cause	Identify why the job stopped. E-stop triggered accidentally? Power flicker? UGS connection drop? Fix the root cause before resuming.
5	Check the workpiece	Is the partially machined piece still correctly positioned? Have any clamps moved? Is the zero reference still valid?
6	Re-verify zeros	If there is any doubt about XY position, re-zero XY using the original reference corner. If the bit has not changed, Z zero should still be valid.
7	Re-home if needed	If the machine lost position due to a power interruption or missed steps, re-home (press the Home button) before relying on any position coordinates.
8	Find restart line	In UGS, navigate to the noted line number. Back up 2–3 lines from where it stopped to ensure no material is missed.
9	Start from line	Use File → Run From Line and enter the line number. UGS will skip to that point and resume from there.
10	Monitor closely	Watch the first few moves carefully after resuming. The tool should engage the material at the correct position and depth. If anything looks wrong, pause immediately.

Preventing Avoidable Job Interruptions

Disable computer sleep	Prevent the OS from sleeping during a job. Windows: Power Options → Never sleep. macOS: System Preferences → Energy Saver → prevent sleep. A computer waking mid-cut resets the USB connection.
Use a dedicated USB cable	A short, high-quality cable with ferrite cores reduces interference. Long, thin, or damaged cables cause intermittent disconnections that look like random job stops.
Secure the cable	Tape or cable-tie the USB cable so it cannot be snagged or pulled accidentally. The most common non-technical cause of job interruptions is someone bumping the cable.

Disable notifications	Put the computer in 'Do Not Disturb' mode and defer Windows Update before starting a long job. System notifications can compete for CPU resources during USB communication.
Check the e-stop position	Verify the physical e-stop is accessible but not where it can be accidentally triggered. A tool resting on it, or a cable draped over it, has caused more than a few interrupted jobs.

> ● **Warning**
> Never try to recover a job by guessing the line number. If you are not certain which line the job stopped on, restart from the beginning of the current operation (the last tool change or operation start) rather than from an uncertain mid-operation point. A re-cut of a completed pocket is a minor time loss. A crash into unexpected geometry caused by an incorrect restart line can break the tool, damage the workpiece, and potentially damage the machine.

The Job Setup Sheet

The Job Setup Sheet is a completed record of every significant parameter for a cutting session. It is filled in before cutting starts, updated during cutting if any settings change, and completed after cutting with the results. A completed Sheet takes 5–10 minutes to fill in. Its value becomes apparent the first time you need to reproduce a cut, investigate a problem, or explain to a client exactly how a piece was made.

> 📖 **Key Term: Job Setup Sheet**
> A standardised documentation form completed for every CNC cutting session. Records the design file name and version, material, tools, zero point, CAM settings, estimated and actual runtime, issues encountered, dimensional results, and overall assessment. The CNC operator's equivalent of a laboratory notebook — a contemporaneous record that makes every job reproducible.

JOB SETUP SHEET — CNC 4040 Pro Max	
JOB IDENTIFICATION	
Job Name	*Descriptive name: e.g. ClockFace_v2, GearPair_Module3*
Date	*Date of cutting session*
Operator	*Student / operator name*
FreeCAD file	*Full filename with version: ProjectName_v3.FCStd*
G-code file	*Full filename: ProjectName_15mm_v3.nc*
MATERIAL	
Species / grade	*e.g. 18 mm MDF, 12 mm birch ply, 30 mm Victorian ash*
Prepared dimensions	*Actual measured W × H × thickness before cutting*
Moisture content	*Especially relevant for hardwood — use a moisture meter*

TOOLS	
Tool 1	*Bit type, diameter, flutes, OAL: e.g. 6 mm 2-flute flat end mill, 50 mm OAL*
Tool 2	*As above — leave blank if single-tool job*
Tool 3	*As above — leave blank if 2-tool job*
MACHINE SETUP	
Zero point location	*e.g. Top-left corner of workpiece, top face*
Z zero method	*e.g. Paper method, touch plate, fixed reference block*
Spoilboard state	*Flat and clean? Any previous cut marks in cut path?*
Workholding	*Clamps, vacuum, double-sided tape, screws — describe*
CAM SETTINGS	
Feed rates	*List all operations: Engraving 600/150, Pocket 1200/400, Profile 1000/350*
Spindle speed	*RPM for each tool*
Step-down per pass	*Per operation if different*
Estimated runtime	*From UGS visualiser estimate*
RESULTS	
Actual runtime	*Record actual elapsed time*
Issues encountered	*Any tool changes unplanned, e-stops, pauses, retries*
Dimensional check	*Measure 3 critical dimensions — record design value vs actual*
Surface quality	*Describe finish quality per operation*
Overall assessment	*Pass / Minor issues / Fail — and brief reason*
Action for next run	*Any settings to change for next version*

> **Tip**
>
> Keep a physical folder of completed Job Setup Sheets for every piece you have cut. After a dozen projects, you will have a personal reference library of working settings for different materials and tools — far more useful than any generic feeds-and-speeds chart because it reflects your specific machine, your specific tools, and your specific conditions.

Post-Cut Evaluation — Measuring, Photographing, and Documenting

The job does not end when the last G-code line executes. Post-cut evaluation closes the loop between design intent and physical result — it is how you verify what you designed is what was cut, identify discrepancies, and capture the data needed to improve the next run.

Step 1 — Dimensional Measurement

Before removing the workpiece from the spoilboard, measure the three most critical dimensions and compare them to the design values. Use a digital calliper for features under 150 mm and a steel rule for larger dimensions.

What to Measure	Why	Accept / Investigate Threshold
Overall width and height	Verifies the profile cut followed the correct outer boundary. A systematic error indicates bit deflection or incorrect tool diameter in the CAM setup.	*Within 0.3 mm of design value*
Pocket depth (at least one pocket)	Verifies Z zero was set correctly and step-down calculation was correct. Too shallow indicates a Z-zero error or positive error in the Final Depth field.	*Within 0.2 mm of design value*
Joinery fit (if applicable)	Test-fit the joint before the piece leaves the machine. A too-tight fit can be re-cut; a too-loose fit cannot be un-cut.	*Assembles by hand without tools (snug fit)*
Feature spacing (e.g. hole centres)	For patterns — verify spacing between identical features matches the design. A cumulative error can indicate step/mm calibration is off.	*Within 0.5 mm across the full pattern width*
Engraving depth	Measure with a depth gauge or calliper. Too shallow may be illegible; too deep increases risk of burning.	*Within 0.15 mm of target depth*

Step 2 — Photography

In machine (before removal)	Photograph the full workpiece in position showing the surrounding machine context. Documents the condition at the moment of completion.
After removal	Close-up of the machined surface in good lighting. Raking light — a light source at a low angle across the surface — reveals scallop marks, tool marks, and finish quality that are invisible under direct overhead light.
Scale reference	Include a calliper, rule, or coin in at least one photograph to provide a physical scale reference.
Defect documentation	If there are visible defects — tearout, chatter marks, a broken tab, a missed area — photograph them in detail and label the file with what the defect is and where on the piece it occurred.

Step 3 — File Management and Version Control

After cutting, organise all files from the session into the project's version-controlled folder structure. The goal is that any file from any session can be found and reproduced by you or anyone else without ambiguity — even months later.

File Type	Naming Convention	Example / Purpose
FreeCAD model	ProjectName_vN.FCStd	ClockFace_v3.FCStd — design iterations, increment N for every significant change

G-code	ProjectName_MaterialThickness_vN.nc	ClockFace_18mm_v3.nc — ties G-code to specific thickness and model version
G-code variant	ProjectName_MaterialThickness_SizeName_vN.nc	ClockFace_18mm_Large_v1.nc — for different sizes from same parametric model
Job Setup Sheet	ProjectName_YYYYMMDD_vN.pdf	ClockFace_20241115_v3.pdf — date stamp makes history chronologically sortable
Project folder	ProjectName_YYYYMMDD/	ClockFace_20241115/ — one folder per cutting session
Photograph	ProjectName_YYYYMMDD_desc.jpg	ClockFace_20241115_finished.jpg — descriptive suffix distinguishes setup vs result shots

Project Folder Structure

Example folder structure for a complete project session:

```
Projects/ClockFace/
  ClockFace_v1.FCStd        ← First version (archived)
  ClockFace_v3.FCStd        ← Current production design
  ClockFace_18mm_v1.nc      ← First cut (too shallow — archived)
  ClockFace_18mm_v3.nc      ← Production G-code
  20241115_CuttingSession/
    JobSetupSheet_v3.pdf
    ClockFace_20241115_setup.jpg
    ClockFace_20241115_finished.jpg
    ClockFace_20241115_defect.jpg  ← Chatter on marker 3
```

Important

Never delete earlier versions of .FCStd or .nc files, even if they were cut incorrectly. Old files are your audit trail. If a piece cut three weeks ago needs to be reproduced and something has changed, having version 1 through 3 allows you to identify which version was cut. Disc space is cheap; remaking a piece because you deleted the file is expensive.

The Complete Pre-Cut Checklist

This checklist integrates all machine setup, safety, and workflow verification steps covered in this chapter and throughout the course. Print it, laminate it, and keep it at the machine. Complete it for every job before sending a single line of G-code. No exceptions.

Complete Pre-Cut Checklist — Every Job, No Exceptions		
✓	Category	Check
☐	Machine	Machine is homed and position is verified
☐	Machine	Soft limits are enabled ($20=1 confirmed in GRBL settings)
☐	Machine	Spoilboard is flat, clean, and free of debris in the cut path
☐	Material	Workpiece is correctly positioned and securely clamped
☐	Material	Material dimensions measured and match the CAM stock settings
☐	Material	No knots, voids, or defects in the planned cut path
☐	Tool	Correct tool is installed and fully seated in the collet
☐	Tool	Collet nut is tight (use spanners — not finger-tight)
☐	Tool	No visible damage or wear on the cutting edges
☐	Zeros	X/Y zero is set at the correct reference point on the workpiece
☐	Zeros	Z zero is set correctly — paper method or touch plate confirmed
☐	G-code	Correct .nc file is loaded in UGS — check filename and version
☐	G-code	3D Visualiser in UGS shows expected toolpaths — all operations visible
☐	G-code	Estimated runtime has been noted on the Job Setup Sheet
☐	Safety	PPE on: safety glasses, hearing protection, dust mask or extraction running
☐	Safety	All bystanders are at a safe distance or behind the safety enclosure
☐	Safety	You know where the physical e-stop is and can reach it without moving
☐	Safety	You have a clear view of the entire cut area throughout the job

Note

The pre-cut checklist is also reproduced in Appendix C of this textbook with space for the date and job name, making it usable as a standalone laminated reference card. Students are encouraged to personalise the checklist by adding items specific to their machine setup or recurring issues.

Week 10, Day 48 — Lesson Content Overview

Chapter 20 covers Day 48 of Week 10. Days 49–50 are covered in Chapter 21 (Workshop Tool Rack Synthesis Project).

Day 48 (Create)	Advanced UGS workflow — job management and recovery. Students open UGS, connect to the machine, and run $$ in the console to verify that $20=1 (soft limits), $22=1 (homing), and $130/$131/$132 match the machine's physical travel limits. Live Feed Rate Override exercise: observe speed change at 50% vs 100% FRO during an air-cut job. Recovery drill: the instructor triggers an e-stop mid-job without warning — students apply the 10-step recovery procedure and resume from the correct line. Each student completes a full Job Setup Sheet for the drill session and carries out a post-cut evaluation: measuring three features, photographing under raking light, and filing all documents in the standard folder structure.

Chapter Summary

- UGS connects to the CNC 4040 Pro Max at 115200 baud. The port number varies by computer — identify it by unplugging and re-plugging the USB cable. Verify the GRBL version string in the UGS console on every new connection.
- GRBL soft limits ($20=1) check planned moves against the machine's travel boundaries before executing them. They require homing to be active ($22=1) and travel limits to be set ($130, $131, $132). Never disable soft limits to work around errors — investigate the root cause instead.
- Hard limits ($21) are triggered by physical limit switches at the machine's axis extremes. They halt all motion immediately and can cause position loss if triggered at high speed. They are the last line of protection after soft limits.
- Feed rate override (FRO) scales the programmed feed rate in real time without modifying the G-code. Reduce to 70–80% at corners, in hard material, or when testing new setups. Restore to 100% for normal cutting. The cutting sound is the best indicator of whether the override is helping.
- The 10-step job recovery procedure: stop spindle → note line number → raise Z → diagnose → check workpiece → re-verify zeros → re-home if needed → find restart line → start from line → monitor closely. Write the line number down immediately — it cannot be recalled after the console scrolls.
- The Job Setup Sheet records all parameters for a cutting session: material, tools, zeros, feeds, estimated and actual runtime, dimensional results, and issues. Complete it before cutting starts and finish it after evaluation. It is your audit trail and reproducibility reference.
- Post-cut evaluation: measure three critical dimensions with a calliper, photograph under raking light, document defects with labelled close-up photographs. File all files — FreeCAD, G-code, Setup Sheet, photographs — in a version-controlled folder structure. Never delete old versions.

Key Vocabulary — Chapter 20

Term	Definition
Baud rate	The speed of serial data communication between UGS and the GRBL controller, measured in bits per second. The CNC 4040 Pro Max uses 115200 baud. A mismatch produces garbled communication.
GRBL $-parameter	A firmware configuration value stored in GRBL that controls machine behaviour — axis speeds, travel limits, homing settings, and safety features. Read and written using $N=value commands in the UGS console.
Soft limits ($20)	A GRBL feature that checks planned G-code moves against configured travel boundaries ($130–$132) before executing them. Prevents the machine from moving outside its safe working volume. Requires homing to be active.
Hard limits ($21)	Physical limit switches mounted at the ends of each axis that immediately halt all motion when triggered. Act as a last-resort hardware safety backup after soft limits. Can cause position loss if triggered at high speed.
Feed Rate Override	A real-time GRBL command that scales the programmed feed rate by a percentage during a running job without modifying the G-code. Applied using the FRO slider in UGS's Machine Control panel.

Feed Hold	A UGS/GRBL command that smoothly decelerates and stops machine movement without losing position. The spindle continues running. Used to pause a job for inspection — not a substitute for e-stop.
Job recovery	The process of resuming a G-code job after a mid-cut interruption (e-stop, power loss, connection drop). Requires noting the stopped line number, diagnosing the cause, verifying position, and restarting from a known-good line.
Job Setup Sheet	A standardised documentation form completed for every CNC cutting session, recording all parameters before cutting starts and results after. The operator's audit trail for reproducibility and problem investigation.
Raking light	A technique for evaluating machined surface quality by holding a light source at a low angle across the workpiece surface, making scallop marks, tool marks, and finish defects visible that are invisible under direct overhead light.
Version control	The practice of incrementing file names (v1, v2, v3) and retaining all previous versions. Ensures any past job can be reproduced and any change between versions can be traced.

Looking Ahead — Chapter 21: Synthesis Project — Workshop Tool Rack

Chapter 21 is the synthesis project — every major skill from across the course applied to one complete, functional piece. The Workshop Tool Rack uses a parametric master Spreadsheet with 17 aliases, 7 Part Design Bodies, a French cleat Boolean geometry, 10 CAM operations with 2 tool changes, a TechDraw documentation page with embedded BOM, and 3 G-code files run in sequence.

CHAPTER 21 of 24

PART FIVE — ADVANCED SKILLS

Synthesis Project — Workshop Tool Rack

Every skill from the course applied to one complete, functional project: Project 10

Skills Synthesis

This chapter draws on every major skill developed in the course. You are not learning new tools — you are choosing which tools to apply, in what order, to solve a complex multi-component design and manufacturing problem.

Ch 5–6: Sketcher constraints + Part Design features | Ch 8: Parametric Spreadsheet | Ch 9: Linear Pattern for chisel holes | Ch 11: Multi-body design + tolerance | Ch 13: Boolean operations for cleat bevel | Ch 17: TechDraw documentation | Ch 19: Assembly verification + BOM | Ch 3 & 7: G-code and UGS workflow

Chapter Learning Objectives

By the end of this chapter, you will be able to:

- Describe the French cleat system and explain how the 45° bevel geometry is achieved on a 3-axis CNC
- Construct a fully parametric Tool Rack design driven by a single master Spreadsheet
- Design all six Tool Rack components as separate bodies in a multi-body file
- Apply Linear Pattern to the chisel hole strip
- Verify the complete assembly using the Assembly workbench
- Construct a 10-operation CAM job with correct operation ordering across multiple parts and two tool changes
- Produce a complete TechDraw documentation package including an embedded BOM
- Cut, assemble, finish, and evaluate the completed Workshop Tool Rack

The Synthesis Project — What It Means

Every project in this course taught you a new tool. This project teaches you something different: how to decide which tools to use, when, and in what order — because that is what designing and making actually requires.

The Workshop Tool Rack is a functional workshop storage piece — something you will use every time you work. It has a back panel, shelves, side panels, French cleat hanging rails, and a chisel storage strip. It is not technically complex in any single area — you have done harder things in individual chapters. What makes it the synthesis project is that it requires all of those areas simultaneously, and the decisions you make in one area affect every other area.

A change to the RackWidth parameter cascades through every component. A decision to use a 45° V-bit for the cleat bevel affects which operations need a tool change, which in turn affects the CAM job ordering, which in turn affects the setup sheet you give to the machine operator. A clearance value in the side panel joints affects whether the assembly fits together or requires recutting. Nothing is isolated.

This is what real CNC design looks like. Work through this chapter methodically. Use the spreadsheet as your single source of truth. Build the assembly before cutting. Document as you go. The physical rack at the end is the reward — but the habit of working this way is the skill.

Project 10 Overview — Workshop Tool Rack

Components	Back Panel, Top Shelf, Mid Shelf, Side Panels ×2, French Cleat Rails ×2, Chisel Rack Strip
Materials	12mm MDF (all panels and shelves), 18mm pine (cleat rails — must be solid timber for bevel strength)
Primary bit	6mm upcut end mill for all profiles, pockets, and engraving
Cleat bit	45° V-bit (chamfer bit) for the French cleat bevel face — requires one tool change
Total ops	10 CAM operations across all components — the most complex job in the course
Stock needed	Approx. 1200 × 600mm of 12mm MDF + 700 × 80mm of 18mm pine (two runs required)
Assembly	Glue and 25mm MDF screws — no mechanical fasteners needed except wall-mounting screws
Wall mount	4 × 5mm countersunk holes in back panel, 40mm timber screws into wall studs
Days	Days 49–50 of Week 10 — the final two teaching days before Week 11 certification prep

French Cleat — Design Brief

The French cleat system allows the tool rack to hang on the wall without visible hardware. Two 18mm pine strips with a 45° bevel on their top edge are screwed horizontally to the wall. The matching bevel on the back of the tool rack (also 45°, facing downward) hooks over the wall rails. The weight of the rack presses the bevels together — gravity is the fastener. To remove the rack: lift it straight up and off.

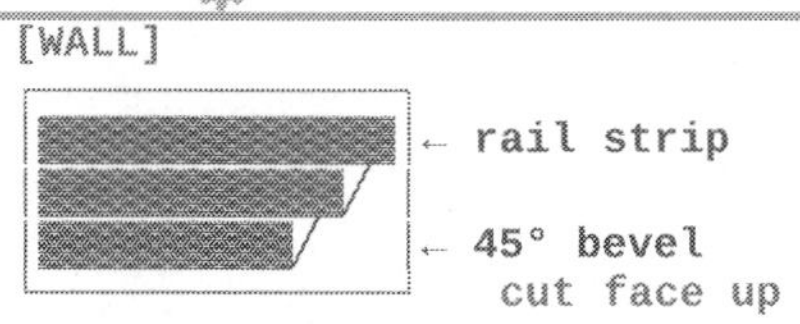

Bevel faces UP, toward the wall.

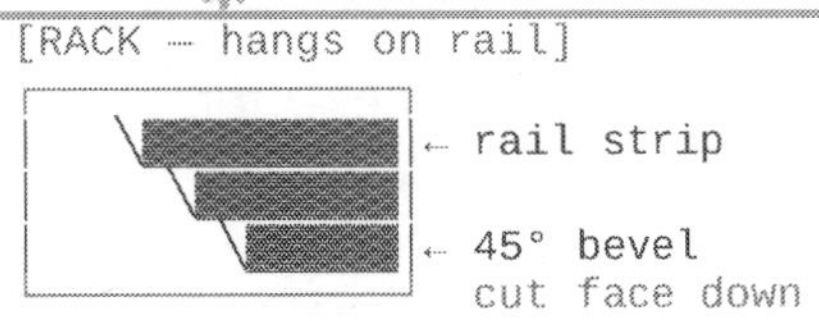

Bevel faces DOWN, away from wall.

How they interlock: The hanging piece sits on top of the wall rail. The two 45° bevel faces nest against each other — the weight of the hanging piece presses the bevels together, locking it firmly. To remove, lift the hanging piece up and off. No screws, no tools.

CNC implementation: **The 45° bevel is cut using a 45° chamfer bit (V-bit) in a Profile operation, OR a straight end mill with a 45° jig, OR by designing the rail in FreeCAD with a chamfer feature and cutting with a 3D Contour operation. The table saw + CNC hybrid method (CNC for holes and profiles, table saw for the bevel rip) is the most practical for a standard 3-axis machine.**

French Cleat on a 3-Axis CNC — The Practical Approach

A true 45° bevel rip is a table saw operation. A 3-axis CNC machine without a tilting table cannot make a true rip cut at 45°. The practical CNC approach — used in this project — is the chamfer bit method: cut the profile of the cleat strip with a standard end mill, then make a second pass along the top long edge using a 45° V-bit (chamfer bit) to remove the triangular section that creates the bevel face.

This is Operation 7 in the CAM sequence — a Profile operation along the top edge of the cleat rail stock, using a 45° V-bit, set to a depth that produces the desired bevel width. The bevel is then cleaned up with light hand sanding.

> **Tip**
> If your school workshop has a table saw, the hybrid approach is faster and produces a cleaner bevel: table saw for the 45° rip, CNC for all holes and profile cuts. Either way, the FreeCAD model shows the final shape with a chamfer feature — the CAM strategy is the variable, not the design.

The Master Spreadsheet — Single Source of Truth

Before creating any Body or sketch, build the master Spreadsheet. Every dimension of every component should flow from this spreadsheet. If a dimension is not in the spreadsheet, it cannot be updated globally — and a project this size will definitely need updates during the design process.

The spreadsheet below lists every alias required for the Tool Rack. Create each alias exactly as shown — the walkthroughs reference these names directly. Group related aliases using colour-coded group headers (these are cosmetic, not functional — select the cells and set their background colour from the Spreadsheet toolbar).

Rack Dimensions		
`RackWidth`	**600 mm**	Overall rack width — drives back panel, all shelves, and cleat rails
`RackHeight`	**450 mm**	Overall rack height — drives back panel and side panel height
Panel Thickness		
`PanelThick`	**12 mm**	Thickness of all MDF panels (back, shelves, sides, chisel strip)
`CleatThick`	**18 mm**	Thickness of French cleat rails — heavier timber for strength
Shelf Dimensions		
`TopShelfDepth`	**120 mm**	Front-to-back depth of the top shelf
`MidShelfDepth`	**100 mm**	Front-to-back depth of the middle shelf
`TopShelfY`	**380 mm**	Vertical position of the top shelf (from rack base)
`MidShelfY`	**200 mm**	Vertical position of the middle shelf (from rack base)
French Cleat		
`CleatHeight`	**60 mm**	Height (width) of each French cleat rail strip
`CleatBevelAngle`	**45 deg**	Angle of the cleat bevel — the interlocking face. Standard: 45°
`CleatY_Upper`	**310 mm**	Vertical position of the upper cleat rail on the back panel
`CleatY_Lower`	**130 mm**	Vertical position of the lower cleat rail
Chisel Rack		
`ChiselHoles`	**8**	Number of chisel/tool holes in the rack strip
`ChiselHoleDia`	**28 mm**	Diameter of each tool hole (suits 25mm handles + 3mm clearance)
`ChiselSpacing`	**44 mm**	Centre-to-centre spacing between holes: RackWidth ÷ (ChiselHoles+1)
Joinery		
`ShelfClearance`	**0.15 mm**	Clearance for shelf-to-side-panel dado joints (if used)
`MountHoleDia`	**5 mm**	Diameter of wall-mounting screw holes on the back panel

> ⚠ **Important**
>
> Create the complete spreadsheet and all aliases BEFORE creating any Body geometry. Attempting to link a sketch to a non-existent alias causes a model error that is tedious to fix after the fact. Ten minutes of spreadsheet setup saves an hour of model repair.

Component Design — Six Bodies, One File

All six Tool Rack components live in a single FreeCAD file as separate Part Design Bodies, all linked to the same master Spreadsheet. Name each Body clearly: BackPanel, TopShelf, MidShelf, SidePanel_L, SidePanel_R, CleatRail, ChiselStrip. (Left and right side panels are identical — you can use one Body and mirror it in the assembly, or simply create two identical Bodies.)

Back Panel

The back panel is the structural anchor. Sketch a rectangle at the origin: Width = =Spreadsheet.RackWidth, Height = =Spreadsheet.RackHeight. Pad to =Spreadsheet.PanelThick. Add four 5mm mounting holes (circle sketches pocketed through) at 50mm from each corner. Optionally add a shallow 3mm engraved label on the back face (your name, date, version). This is cosmetic but distinguishes your rack in a classroom setting.

Shelf Panels — Top and Mid

Top Shelf: Width = =Spreadsheet.RackWidth, Depth = =Spreadsheet.TopShelfDepth. Pad to PanelThick. No additional features needed — the shelf is a plain flat slab. Mid Shelf: identical construction using MidShelfDepth. The vertical positions (TopShelfY, MidShelfY) are used in the Assembly constraints, not in the individual part models.

Side Panels

Width = =Spreadsheet.PanelThick (the side panel is thin — its 'width' when viewed from the side equals the material thickness), Height = =Spreadsheet.RackHeight. Pad to =Spreadsheet.TopShelfDepth (or to the depth of the deepest shelf — this determines how far the side panels extend forward). Create two identical Bodies: SidePanel_L and SidePanel_R.

French Cleat Rails

This is the most geometrically interesting component. Start with a rectangle: Width = =Spreadsheet.RackWidth, Height = =Spreadsheet.CleatHeight. Pad to =Spreadsheet.CleatThick (18mm pine). Apply a Chamfer to the top long edge: 45°, size = CleatThick × tan(45°) = CleatThick (for a full-depth 45° bevel). For 18mm timber, the chamfer removes an 18mm × 18mm right triangle from the top corner — leaving the characteristic angled face.

> **Tip**
> In FreeCAD's Part Design Chamfer dialog, you can enter the chamfer size as =Spreadsheet.CleatThick to link it directly to the spreadsheet. This means if you change CleatThick (e.g., switching to 15mm pine), the chamfer updates proportionally.

Chisel Rack Strip — Linear Pattern

Width = =Spreadsheet.RackWidth × 0.7 (chisel strip is shorter than the full rack width — adjust as desired), Height = =Spreadsheet.CleatHeight. Pad to =Spreadsheet.PanelThick. Then create the chisel holes: sketch a single circle (diameter = =Spreadsheet.ChiselHoleDia) at the leftmost position. Pocket through all. Apply Part Design → Linear Pattern: feature = the hole pocket, spacing = =Spreadsheet.ChiselSpacing, occurrences = =Spreadsheet.ChiselHoles. FreeCAD creates all 8 holes at even spacing in a single operation.

> 📖 **Key Term: Linear Pattern (Part Design)**
> A Part Design feature that repeats a selected feature (such as a pocket or pad) at a defined spacing along a linear direction. The number of occurrences and spacing are set in the dialog and can both be linked to spreadsheet aliases.

Assembly Verification — Before Cutting Anything

If you built the assembly preview in Chapter 19, update the component Links to point to your finalised part files. If you are starting fresh, follow the five-step Assembly workflow from Chapter 19: Create Assembly → Insert all components as Links → Ground the Back Panel → Add Fixed joints for shelves and side panels → Solve.

For this project, the critical checks in the assembly are:

1. Shelf depth vs side panel depth — the shelf should not protrude beyond the front edge of the side panels. If it does: reduce TopShelfDepth or MidShelfDepth in the spreadsheet.
2. Side panel height matches back panel height — verify they are flush at top and bottom.
3. Cleat rail dimensions — the rail strip should sit flush against the back panel face. Check that CleatHeight and CleatThick produce a visually proportionate result.
4. Chisel strip — confirm it fits within the rack width (strip width should be ≤ RackWidth − 2 × PanelThick for clearance inside the side panels).
5. No clashes anywhere — rotate the assembly view and inspect all corners.

> 📋 **Note**
> The assembly at this stage uses Fixed joints for everything — the rack does not need to move in simulation. The goal is purely spatial verification: do all components fit in the right places at the right sizes?

CAM Strategy — Ten Operations, Two Tool Changes

This is the most complex CAM job in the course. Plan it completely before touching the machine. The operation ordering table below is your job sheet.

The Tool Rack CAM job has 10 operations across multiple parts, requiring two tool changes. The planning principles from earlier chapters all apply here simultaneously: pockets before profiles (Chapter 11), roughing + finishing for precision parts (Chapter 13), and correct zero-point management across tool changes (Chapter 3). Do not rush the CAM setup — a 20-minute error in the CAM file costs an hour of material and machine time to correct.

Op	Part	Type	Description	Bit	Why This Order
1	Back Panel	Engrave	Slot lines for shelf dadoes + mounting hole centres	6mm EM	*First — back panel anchors everything*
2	Back Panel	Pocket	Mounting holes (5mm dia, 8mm deep) — 4 holes	6mm EM	*Before profile — part must be intact*
3	Back Panel	Profile	Outer rectangle — frees the panel	6mm EM	*Profile LAST for this panel*
4	Top Shelf	Profile	Outer rectangle	6mm EM	*All shelf profiles together after back panel*
5	Mid Shelf	Profile	Outer rectangle	6mm EM	*Same stock run as Top Shelf if nesting allows*
6	Side Panels	Profile	Two side panels — nested side by side	6mm EM	*Profile only — no pockets needed*
7	Cleat Rails	Chamfer	45° bevel on top long edge — both rails	45° V-bit	*Tool change required — zero Z on flat stock*
8	Cleat Rails	Profile	Outer rectangle — frees the rail strips	6mm EM	*Tool change back — re-zero Z*
9	Chisel Strip	Pocket	All 8 chisel holes (28mm dia)	6mm EM	*Pockets before profile — rule applies*
10	Chisel Strip	Profile	Outer rectangle	6mm EM	*Profile last — critical rule*

> ⚠ **Important**
>
> Operations 1–3 (Back Panel) must complete before any other panel's profile cut, because the back panel is the widest piece — it is cut from the main MDF sheet and its profile frees the most material. Freeing large pieces first, then cutting smaller components from the remaining stock, is efficient nesting practice.
>
> Operations 7 and 8 (Cleat Rails) require a tool change. When you change from the 6mm end mill to the 45° V-bit, re-zero Z from the flat surface of the cleat stock — not from a previously cut edge. After the chamfer pass, change back to the 6mm end mill and re-zero Z again before Op 8.
>
> The Chisel Strip (Ops 9–10) must be cut last, because it is the smallest part and is most easily displaced if the stock shifts during earlier operations on the same sheet.

Nesting — Fitting All Parts on the Available Stock

The back panel (600×450mm) is the largest component and dictates the minimum MDF sheet size. Allow 10mm clearance between parts and 15mm from sheet edges. A 1200×600mm sheet accommodates the back panel plus both shelf panels side by side, with space for the side panels in any remaining area. The cleat rails are cut from a separate piece of 18mm pine.

In FreeCAD CAM, position each Body's Placement to arrange the parts within the stock area before creating any operations. This spatial arrangement is what the CAM operations follow — if a part's Placement is wrong in the model, its toolpath will also be wrong.

PROJECT 10 — SYNTHESIS WALKTHROUGH

Workshop Tool Rack

Complete design, documentation, CAM, cutting, and assembly — all skills synthesised

📋 **How to Use This Walkthrough**

This walkthrough is structured in five phases: Design, Assembly Verification, Documentation, CAM & Cutting, and Final Assembly. Work through each phase completely before starting the next. Phase completion is a natural checkpoint — if something does not look right at the end of a phase, find and fix it before proceeding.

Phase A — Design

1 **New file — ToolRack.FCStd**
File → New. Save immediately as ToolRack.FCStd in your CNC/Projects/ToolRack/ folder. This single file will contain all 7 Bodies and the master Spreadsheet.

2 **Create master Spreadsheet — all 17 aliases**
Switch to Spreadsheet workbench. Create Spreadsheet, rename 'Parameters'. Enter every alias from the spreadsheet table above. Add colour-coded group headers. Save.

3 **Create BackPanel Body**
Part Design → Body. Rename 'BackPanel'. New sketch on XY plane. Rectangle at origin: =Spreadsheet.RackWidth × =Spreadsheet.RackHeight. Fully constrain. Close. Pad: =Spreadsheet.PanelThick.

4 **Add mounting holes to back panel**
New sketch on back face. Four circles (diameter =Spreadsheet.MountHoleDia) at 50mm from each corner. Pocket → Through All.

5 **Create TopShelf and MidShelf Bodies**
New Body for each. Sketch: RackWidth × TopShelfDepth (or MidShelfDepth). Pad to PanelThick. No additional features.

6 **Create SidePanel_L and SidePanel_R Bodies**
New Body for each. Sketch: PanelThick × RackHeight. Pad to TopShelfDepth. These bodies are identical — you can copy the SidePanel_L Body's sketch and Pad into SidePanel_R.

7	**Create CleatRail Body** New Body. Sketch: RackWidth × CleatHeight. Pad to CleatThick. Apply Chamfer to the top long edge: size = =Spreadsheet.CleatThick, angle = 45°. Verify: the chamfer should remove a right-triangle section from the top corner. This is the interlocking bevel face.

8	**Create ChiselStrip Body** New Body. Sketch outer profile: (RackWidth × 0.7) × CleatHeight. Pad to PanelThick. New sketch on top face — single circle, diameter =Spreadsheet.ChiselHoleDia, positioned at X = ChiselSpacing from left edge. Pocket through all. Part Design → Linear Pattern: feature = hole pocket, direction = X, spacing = =Spreadsheet.ChiselSpacing, occurrences = =Spreadsheet.ChiselHoles. Verify: 8 holes appear evenly spaced.

9	**Visual fit check — show all bodies** Show all 7 Bodies. Use Properties → Placement to roughly position them as assembled: shelves at correct Y heights (TopShelfY, MidShelfY), side panels at the left and right edges, cleat rails on the front face. No precise assembly constraints yet — this is a quick sanity check. Does the overall geometry look right?

Phase B — Assembly Verification

10	**Create ToolRack_Assembly.FCStd** New file. Save as ToolRack_Assembly.FCStd in the same folder. Switch to Assembly workbench. Create Assembly.

11	**Insert all components and ground BackPanel** Insert Component for each of the 7 Bodies (or save each Body to its own file and insert those — either approach works in v1.1.0). Ground the BackPanel component. Add Fixed joints: TopShelf at Y=TopShelfY, MidShelf at Y=MidShelfY, SidePanel_L at the left edge, SidePanel_R at the right edge. Solve.

12	**Inspect and fix any dimensional errors** Look for clashes or gaps. Common issues: shelf depth exceeds side panel depth (reduce TopShelfDepth), cleat rails don't align flush with back panel (check CleatThick). Fix in the source part files, refresh the assembly.

Phase C — Documentation

13	**Create BOM spreadsheet in the assembly file** Switch to Spreadsheet workbench within the Assembly file. Create a BOM spreadsheet with all 7 components, their quantities, materials, stock sizes, thicknesses, bits, and G-code filenames (use v1 placeholder names at this stage — update after CAM).

14	**Create TechDraw page — A4 Landscape** Switch to TechDraw. Insert Page using A4 Landscape template. Insert Projection Group of the assembly: Front + Top + Isometric at scale 1:5 (the full rack is too large for 1:2 on A4). Position views on the left two-thirds of the page.

15 **Add dimensions — minimum 8**

Dimension: overall width (RackWidth), overall height (RackHeight), shelf depths (×2), shelf vertical positions (×2), mounting hole positions (×2). More is better — cover every critical dimension.

16 **Embed BOM and complete title block**

Insert Spreadsheet View of the BOM on the right side of the page. Complete all title block fields: ToolRack Workshop Storage, CNC-P10-001, your name, today's date, 1:5, 12mm MDF + 18mm Pine, Sheet 1 of 1, Rev A.

17 **Export PDF — print as setup sheet**

Export PDF. Print. Pin at the machine. This is your job documentation.

Phase D — CAM and Cutting

18 **Create CAM Job — back panel stock first**

CAM → Job. Insert BackPanel, TopShelf, MidShelf, SidePanel_L, SidePanel_R. Stock: 1210 × 610 × 12mm. Arrange using Placement to nest all panels within stock. Tool Library: T1 = EndMill6mm.

19 **Create Ops 1–6 (back panel and all panel profiles)**

Back panel: Engrave (shelf position lines), Pocket (mounting holes), Profile (outer). Top Shelf, Mid Shelf, Side panels: Profile only. Verify operation order: all back panel ops complete before any shelf profiles. Tabs on all profiles.

20 **Post-process first G-code file**

grbl. Save as ToolRack_Panels_12mm_v1.nc.

21 **Create second CAM Job — cleat rails (18mm pine stock)**

New CAM Job. Insert CleatRail. Stock: 700 × 80 × 18mm. Two rail instances side by side. T1 = EndMill6mm. T2 = VBit45deg (Diameter 6, Angle 45).

22 **Create Ops 7–8 (chamfer bevel then profile)**

Op 7: Profile along top long edge, T2 (V-bit), depth = 18mm × tan(45°) = 18mm, producing a full-width bevel. Op 8: Full outer profile, T1 (6mm EM). Tabs on Op 8 only — the chamfer pass does not need tabs. Post-process: ToolRack_Cleats_18mm_v1.nc.

23 **Create third CAM Job — chisel strip (12mm MDF)**

New CAM Job. Insert ChiselStrip. Stock: 450 × 90 × 12mm. Op 9: Pocket (all 8 holes). Op 10: Profile (outer). Post-process: ToolRack_ChiselStrip_12mm_v1.nc.

24 **Simulator — all three jobs**

Run CAM Simulator for each job separately. Verify: back panel engraving is at correct shelf positions, all mounting holes present, all panels have tabs, V-bit chamfer shows correctly on the cleat rails, all 8 chisel holes present on the strip. Do not proceed to the machine until all three simulations look correct.

25 **Cut Job 1 — MDF panels**

Pre-cut checklist. Clamp 1200×600mm 12mm MDF sheet. 6mm end mill. Set zero at the stock centre. Load ToolRack_Panels_12mm_v1.nc. Run. Supervise all 6 operations. Release tabs carefully — there are many pieces from this sheet.

26 **Cut Job 2 — pine cleat rails**

New stock setup: 700×80mm 18mm pine. Important: set Z zero on the flat top surface of the pine — it is thicker than the MDF. Load ToolRack_Cleats_18mm_v1.nc. T1 (6mm EM) for Op 8 is already loaded — but Op 7 needs the V-bit first. Install V-bit, set Z zero. Run Op 7 (chamfer). Tool change to 6mm EM, re-zero Z. Run Op 8 (profile). Release both cleat rails.

27 **Cut Job 3 — chisel strip**

New 12mm MDF stock piece. Load ToolRack_ChiselStrip_12mm_v1.nc. Run Ops 9–10. Release.

Phase E — Assembly and Finishing

28 **Sand all components**

All cut faces: 120 → 220 grit. Pay extra attention to the chamfered bevel faces on the cleat rails — these are the interlocking surfaces and should be smooth for reliable cleat engagement.

29 **Dry assemble before gluing**

Lay out all components and assemble without glue. Check: side panels sit flush with back panel edges, shelves seat correctly at their intended heights, chisel strip fits on the mid shelf. Confirm the cleat rail bevel engages correctly with a test against a known-flat surface.

30 **Glue and screw assembly**

Apply PVA to all mating faces. Assemble shelves to back panel first, using 25mm MDF screws from behind the back panel into the shelf edges. Then attach side panels. Clamp and allow 2 hours to cure. The cleat rails are NOT glued to the rack — they are screwed to the wall separately.

31 **Apply finish**

Danish oil, clear lacquer, or paint. Two coats minimum. Allow to dry fully before wall mounting.

32 **Wall mount**

Mark stud positions on the wall. Screw two cleat rails horizontally across at least two studs each, ensuring they are level. The lower rail at approximately 900mm from the floor is a comfortable working height. Hang the tool rack over the rails. Load your tools.

🏆 **Project 10 — Complete**

You have now completed Project 10.

This project used: a parametric master Spreadsheet (17 aliases), 7 Part Design Bodies in one file, a chamfer feature for the French cleat bevel, Linear Pattern for the chisel holes, an Assembly file for spatial verification, 3 separate CAM jobs with 10 operations and 2 tool changes, a TechDraw page with an embedded BOM, and 3 G-code files run in sequence on the CNC 4040 Pro Max.

If it is functional, it is ready for the workshop wall. If it is hanging straight and holds tools securely, every step of the workflow was correct. That is the test.

Week 10, Days 49–50 — Lesson Content Overview

Day 49 (Create)	Design and documentation. Students complete all design phases: master Spreadsheet, all 7 Bodies, Assembly verification, TechDraw documentation with BOM, PDF export. CAM setup for all three jobs. Instructor checkpoint before any student proceeds to cutting: review the CAM Simulator for all three jobs together. Any operation ordering errors, missing tabs, or dimensional problems must be fixed before Day 50.
Day 50 (Create — Project Day)	Cutting and assembly. Students cut all three jobs in sequence. Instructor supervises tool changes closely (the V-bit chamfer pass on Job 2 is the most technically demanding machine operation in the course). Assembly and finishing commenced in class; finishing coat applied as homework if time runs short. Assessment: physical tool rack, TechDraw PDF, and completed BOM submitted.

Chapter Summary

- The Workshop Tool Rack synthesis project requires all major skills from the course: parametric multi-body design, Boolean geometry, Linear Pattern, Assembly verification, multi-part CAM with correct operation ordering, TechDraw documentation, and a complete Bill of Materials.
- A master Spreadsheet with 17 aliases drives every dimension of every component. Changing RackWidth updates the back panel, both shelves, both side panels, both cleat rails, and the chisel strip simultaneously.
- The French cleat uses a 45° bevel to create a gravity-locked wall-hanging system. On a 3-axis CNC, the bevel is cut using a 45° V-bit (chamfer bit) in a Profile operation along the top edge of the 18mm pine cleat strip.
- The chisel hole strip uses Part Design Linear Pattern: one hole pocket repeated 8 times at ChiselSpacing intervals. The pattern occurrence count and spacing are both linked to spreadsheet aliases.
- The CAM job contains 10 operations across three separate G-code files (MDF panels, pine cleats, chisel strip), requiring two tool changes and three separate zero-point setups. The pockets-before-profiles rule applies throughout.
- Assembly verification (Chapter 19 workflow) must be completed before any cutting. The assembly catches dimensional errors in minutes; the machine catches them in wasted timber.
- The complete documentation package is: TechDraw drawing (minimum 8 dimensions, first-angle projection, all title block fields complete) + embedded BOM spreadsheet (7 components, all 6 BOM fields) + PDF exported setup sheet.
- The finished rack is a functional workshop tool — not a demonstration exercise. It is wall-mounted using the same French cleat rails it was designed around. If it hangs level and holds tools securely, the entire workflow was correct.

Key Vocabulary — Chapter 21

Term	Definition / Chapter Reference
Synthesis project	A project that requires the student to select and apply multiple previously-learned skills independently, without being guided to specific tools. Tests ability to make design and workflow decisions, not just follow instructions.
French cleat	A wall-hanging system using two interlocking 45° bevels — one on a fixed wall rail, one on the hanging piece. Gravity locks them together; lifting separates them. No visible hardware required.
Chamfer bit (V-bit)	A cutting bit with a conical point at a defined angle (typically 45°, 60°, or 90°). Used in this project to cut the 45° bevel face of the French cleat rail.
Master Spreadsheet	A single Spreadsheet object whose aliases drive the dimensions of all Bodies in a multi-component design. Changing one alias updates the entire project. The single source of truth for parametric design.
Linear Pattern	A Part Design feature that repeats a selected feature (pad, pocket, or other) a specified number of times at a defined spacing along one direction. Used here for the evenly-spaced chisel holes.
Nesting	Arranging multiple parts within a stock sheet to minimise material waste. In FreeCAD CAM, nesting is achieved by positioning Body Placements within the stock area before creating operations.

Job sequencing	The order in which multiple CAM jobs are run on the machine. For the Tool Rack: MDF panels first (largest stock sheet), then pine cleats (separate stock), then chisel strip. Each job requires its own stock setup and zero-point.
Tool change procedure	The sequence of steps required to swap cutting bits between operations: raise Z, stop spindle, loosen collet, swap bit, tighten collet, re-zero Z from flat stock surface, resume job. (Chapter 3 reference.)
Dry assembly	Assembling all components without glue before the final glued assembly. Confirms that all parts fit correctly, identifies any last-minute adjustments needed, and establishes the correct assembly sequence.
Assessment deliverables	For Project 10: (1) Physical tool rack, wall-mounted and functional; (2) TechDraw PDF with minimum 8 dimensions, all title block fields, embedded BOM; (3) Three G-code files submitted with version numbers.

Looking Ahead — Chapter 22: Preparing for Certification

Chapter 22 reviews everything covered in the course and prepares students for the Week 11 certification project — the Keepsake Memory Box. It includes a complete skills checklist, a practice pre-certification job, and guidance on how to approach independent design work at the Create level of Bloom's taxonomy.

CHAPTER 22 of 24

PART SIX — CERTIFICATION & REFERENCE

Preparing for Certification

Skills self-assessment, certification brief, the five golden rules, common mistakes, and the Week 11 timeline

Chapter Learning Objectives

By the end of this chapter, you will be able to:

- Complete a self-assessment of all major skills developed across the course
- Identify any skill areas needing review before the certification project
- State all seven certification project requirements
- Describe what 'independent design at the Create level' means in practice
- Apply the five golden rules to the planning of an original CNC project
- Identify the eight most common certification mistakes and how to prevent them
- Complete the 45-minute practice design sprint
- Describe the Week 11 timeline from Day 51 to Day 55

This Chapter Is Different

Every chapter until now introduced a new tool, a new technique, or a new project. This chapter introduces something harder: the expectation that you direct the work yourself.

The certification project in Chapter 24 is entirely self-directed. There is no project brief to follow, no step-by-step walkthrough, no worked example to reference. You choose what to make, you design it, you generate the G-code, and you cut it. The only constraints are the certification requirements — and those exist to ensure the work genuinely demonstrates what you have learned.

This chapter does three things to prepare you. First, it gives you a complete skills checklist so you can honestly assess where you are confident and where you need another look before Day 53. Second, it gives you the full certification requirements so there are no surprises. Third, it gives you a 45-minute practice exercise — the first genuinely independent design work of the course — so you arrive at certification having already navigated the experience of starting from a blank FreeCAD screen with no instructions.

Chapter 22 is also the moment to look back. You have covered 21 chapters, 10 projects, every major FreeCAD workbench, and every CNC workflow from G-code to finished piece. That is a significant body of knowledge. The Bloom's taxonomy table below shows you where you started and where you are now.

The Journey — Where You Started, Where You Are

At the start of Week 1, you were asked to remember: what is a CNC machine, what do the axes mean, what does G-code do. At the start of Week 11, you are being asked to create: design something original, make all the workflow decisions, and produce a finished object. The distance between those two points is the entire course.

Level	Weeks	What You Can Do	Where You Were
Remember	Wk 1–2	Name parts of the CNC machine. Recall G-code commands. Identify FreeCAD workbenches. List safety rules.	*You could describe the machine but not yet use it.*
Understand	Wk 2–3	Explain how GRBL interprets G-code. Describe what the Sketcher's degrees of freedom indicator means. Explain why dogbones are needed.	*You could explain the why behind the tools.*
Apply	Wk 3–5	Use the Sketcher to produce a fully constrained profile. Create a parametric spreadsheet. Set up a CAM job with correct feeds.	*You could complete guided projects independently.*
Analyse	Wk 5–7	Diagnose why a finger joint doesn't fit. Compare roughing vs finishing strategies. Identify which Boolean operation achieves a desired result.	*You could troubleshoot and compare options.*
Evaluate	Wk 8–9	Justify feed rate choices for a material. Assess a TechDraw drawing for completeness. Determine the best CAM strategy for a complex 3D form.	*You could make and defend design decisions.*
Create	Wk 10–11	Design an original multi-component object, produce all documentation, generate G-code, and cut a functional finished piece — independently.	*You direct the entire workflow from idea to object.*

Bloom's Revised Taxonomy (Anderson & Krathwohl, 2001). The six levels describe progressively more demanding cognitive tasks, from simple recall to independent creative production.

Skills Self-Assessment — Complete Honestly

Work through this checklist before Day 51. For each skill, mark one of three columns: Confident (you can do this without looking anything up), Review (you have done it but would benefit from re-reading the relevant chapter section), or Not Attempted (you missed this or it did not stick). Any Review or Not Attempted items become your study priorities before Day 53.

PART ONE — Foundations (Weeks 1–2, Chapters 1–7)			
Skill	**✓ Confident**	**⟳ Review**	**✗ Not Attempted**
Identify all major components of the CNC 4040 Pro Max by name	☐	☐	☐
Explain what GRBL firmware does and how UGS communicates with it	☐	☐	☐
Complete all sections of the pre-cut safety checklist from memory	☐	☐	☐
Read and explain an annotated .nc G-code file line by line	☐	☐	☐
Connect UGS to the machine, home it, and set a work zero	☐	☐	☐
Use the UGS feed rate override during a running job	☐	☐	☐
Navigate all six areas of the FreeCAD interface (Model tree, 3D view, Properties, etc.)	☐	☐	☐
Create a fully-constrained Sketcher profile — no red or yellow constraints	☐	☐	☐
Apply all 10 geometric constraints (Coincident, Parallel, Perpendicular, etc.)	☐	☐	☐
Use Pad, Pocket, Chamfer, Fillet, and ShapeString in Part Design	☐	☐	☐
Create a CAM Job with correct Stock, Tool Library entry, and at least 2 operations	☐	☐	☐
Post-process a CAM job to GRBL and load it in UGS	☐	☐	☐

PART TWO — Core Skills (Weeks 3–4, Chapters 8–11)			
Skill	**✓ Confident**	**⟳ Review**	**✗ Not Attempted**
Create a Spreadsheet with at least 5 named aliases	☐	☐	☐
Link a sketch dimension to a Spreadsheet alias using =Spreadsheet.AliasName	☐	☐	☐
Trigger a Refresh cycle and verify all linked dimensions update	☐	☐	☐
Apply a Linear Pattern to repeat a feature at defined spacing	☐	☐	☐
Apply a Polar Pattern at N occurrences over 360°	☐	☐	☐

Create a multi-body FreeCAD file with 2+ named Bodies	☐	☐	☐
Share a Spreadsheet across multiple Bodies in one file	☐	☐	☐
Apply the clearance formula to design a snug-fit slot	☐	☐	☐
Draw a dogbone fillet at an inside corner and calculate the correct diameter	☐	☐	☐
Set up a multi-part CAM job and apply the pockets-before-profiles rule	☐	☐	☐

PART THREE — Intermediate Skills (Weeks 5–7, Chapters 12–15)			
Skill	**✓ Confident**	**⟳ Review**	**✗ Not Attempted**
Install an addon via the FreeCAD Addon Manager	☐	☐	☐
Generate an FCGear InvoluteGear and configure Module, Teeth, and Height	☐	☐	☐
Calculate Pitch Diameter and gear centre distance from Module and Teeth	☐	☐	☐
Apply the Module matching rule when designing a meshing gear pair	☐	☐	☐
Use Part Boolean Cut to subtract one solid from another	☐	☐	☐
Distinguish between Part Boolean Cut and Part Design Pocket	☐	☐	☐
Configure a roughing pass (positive offset) and a finishing pass (zero offset)	☐	☐	☐
Set up a 3D Waterline and 3D Surface CAM operation	☐	☐	☐
Explain the relationship between step-over distance and surface scallop height	☐	☐	☐
Import an SVG file into FreeCAD and trace it in the Sketcher	☐	☐	☐

PART FOUR — Advanced Skills (Weeks 8–10, Chapters 16–21)			
Skill	**✓ Confident**	**⟳ Review**	**✗ Not Attempted**
Configure Adaptive Clearing and explain when it is preferred over standard pocket	☐	☐	☐
Add a Pencil Trace operation for internal fillets and corners	☐	☐	☐
Create a TechDraw drawing page from an A4 Landscape template	☐	☐	☐
Insert a Projection Group with Front, Top, and Isometric views	☐	☐	☐
Apply first-angle projection correctly (Australian standard — AS 1100)	☐	☐	☐
Add at least 6 dimension types to a TechDraw drawing	☐	☐	☐
Complete all 8 title block fields accurately	☐	☐	☐
Embed a BOM Spreadsheet into a TechDraw page	☐	☐	☐
Create an Assembly with Grounded component and at least 2 joint types	☐	☐	☐
Solve an assembly and interpret a constraint error	☐	☐	☐
Build a CNC Bill of Materials with all 6 required fields	☐	☐	☐
Plan a 10-operation multi-part CAM job with 2 tool changes	☐	☐	☐

Interpreting Your Checklist

Count your Review and Not Attempted ticks. If you have 0–3: you are well prepared. 4–8: target the relevant chapter sections for review on Day 51. 9 or more: discuss with your instructor — some targeted revision before Day 53 will make the certification project significantly less stressful.

The Certification Project — Seven Requirements

The certification project (Chapter 24) is assessed against seven non-negotiable requirements. All seven must be met for the project to be considered complete. Read them carefully — there are no partial passes.

Requirement	Detail
Independent design	The object must be your own original design — not a project from this textbook, not a copy of a class example. You may draw inspiration from anywhere, but the FreeCAD model must be created by you from scratch during the certification period.
Parametric Spreadsheet	The design must include a Spreadsheet workbench file with a minimum of 5 aliases that drive key dimensions. Dimensions typed directly into sketches without spreadsheet links do not count.
Minimum 4 workbenches	Your project must use at least 4 of the following workbenches: Sketcher, Part Design, CAM, TechDraw, Spreadsheet, FCGear, Assembly, Draft. Simply opening a workbench does not count — you must produce meaningful work in it.
Minimum 3 CAM operations	The CAM job must contain at least 3 distinct operations. Engraving, pocketing, and profiling are the typical three. A single profile cut does not meet the threshold.
TechDraw documentation	A complete TechDraw drawing page with: minimum 6 dimensions, all title block fields completed, at least 3 setup annotations, and a BOM table (minimum 2 rows). Exported to PDF and submitted.
Successful cut	The design must be successfully cut on the CNC 4040 Pro Max during the certification period. A design that is not cut, or is cut but produces an unusable result due to a CAM or design error, is not complete.
Safety compliance	All safety procedures followed throughout the certification session. Any safety violation during the session disqualifies the submission regardless of technical quality.

Note

The certification project is not assessed on aesthetic quality or complexity. A simple well-executed project that meets all seven requirements outperforms an ambitious complex project that misses one requirement. Aim for completeness and correctness, then add complexity if time allows.

What Independent Design Means — The Five Golden Rules

Independent design is not harder than following instructions. It is different. The challenge is not technical — you already have all the technical skills. The challenge is decision-making: knowing which tool to use, in what order, and why.

Every guided project in this course gave you a set of decisions already made: which workbench, which feature, which CAM operation. The certification project gives you none of those. What it gives you instead is the experience of having made all those decisions yourself over the past 10 weeks, even if you did not realise it at the time. The five golden rules below structure that experience into a reliable design process.

Rule 1 — Spreadsheet first	Before creating any geometry, build your master Spreadsheet with all key dimension aliases. Your design may be simple (5 aliases) or complex (20+). Either is fine. But start with the spreadsheet. Every dimension you type directly into a sketch instead of linking to an alias is a dimension you will have to hunt down and change manually later.
Rule 2 — Design for the machine	The CNC 4040 Pro Max has a 400×400mm work area and a maximum material thickness it can handle with your collet and bit combination. Design within those constraints from the very first sketch. A beautiful model that does not fit on the machine is worth nothing on the day.
Rule 3 — Verify before cutting	Build the assembly preview. Run the CAM Simulator. Look at both. If anything looks wrong in the simulator, it will be wrong on the machine — and machine time is more expensive than simulator time. The CAM Simulator is your last free checkpoint.
Rule 4 — Document as you go	Do not leave TechDraw documentation until after cutting. Start the drawing page once your model is at a stage you are happy with. Add dimensions progressively. This means you arrive at submission with documentation 80% complete rather than trying to produce it in the last 20 minutes of Day 55.
Rule 5 — Simple done well beats complex done poorly	An engraved coaster with a perfectly dimensioned TechDraw page, a 5-alias spreadsheet, and a clean cut scores better than an ambitious multi-body joinery assembly that has three missing dimensions and a CAM error. Scope your project honestly against the time available.

The Eight Most Common Certification Mistakes

These mistakes appear in certification projects every year. They are all preventable. Read them, recognise the patterns, and actively check for each one during your certification session.

Common Mistake	What Happens & How to Prevent It
Starting without a Spreadsheet	Designing directly in the sketch with typed numbers — then discovering a dimension needs changing. Result: hours of manual sketch editing. Fix: build the spreadsheet first, every time.
Skipping the assembly check	Cutting all parts before verifying they assemble. Result: parts that don't fit. Fix: always build the assembly preview before Job 1 runs.
Profile cut before pockets	Running a profile cut that frees a piece, then trying to pocket it. The freed piece shifts. Result: ruined workpiece. Fix: recheck the operation order table before sending any job.
Forgetting to re-zero Z after tool change	After swapping bits, running the next operation without re-zeroing Z from the stock surface. Result: cuts too deep or too shallow — often ruining the piece. Fix: re-zero Z is always the first action after every tool change.
Vague title block	Submitting a TechDraw drawing with 'Part 1', no designer name, no scale, no material. Result: the drawing conveys no actionable information. Fix: fill in all 8 title block fields before exporting PDF.
Designing beyond the machine envelope	Creating a design larger than 400×400mm without checking it fits the CNC 4040 Pro Max work area. Result: the job cannot be run. Fix: check dimensions against RackWidth/RackHeight equivalents from the very first sketch.
Under-dimensioned TechDraw	A drawing with 2 dimensions that misses all critical features. Result: the drawing cannot be used to reproduce the part. Fix: dimension every feature that defines the part — minimum 6, ideally more.
Rushing the pre-cut checklist	Skipping checklist items to save time. Result: potential safety incident or damaged workpiece. Fix: the checklist is not optional. It takes 3 minutes. A machine error can take 30 minutes to recover from.

Practice Exercise — The 45-Minute Design Sprint

This is the most important activity in Chapter 22. Set a 45-minute timer. Design, CAM-set-up (but do not cut), and document a simple original object using the skills from this course. The object should be something you would actually want — a bookmark, a phone stand, a cable tidy, a plant pot label, a decorative tile. It does not need to be complex. It needs to be yours.

🖱 **Try It Now — 45-Minute Design Sprint**

Sprint rules — read before starting the timer:

1. The object must be your own design. Not a course project, not a web template, not a traced image — your idea.

2. Minimum: one Spreadsheet alias, one Part Design feature, one CAM operation. Aim for more.

3. When the 45 minutes are up, stop — even if incomplete. The incompleteness is information.

4. After the timer: write down (a) what slowed you down most, (b) which chapter you would re-read to fix that, (c) whether your scope was too ambitious, about right, or too simple.

5. This exercise is not assessed. Its purpose is to give you a data point about your readiness before the real thing.

After the sprint, discuss with your instructor or a classmate. What did you produce? Where did you get stuck? How would you approach the same exercise differently in the 3 days before Day 53? This reflection is the most valuable outcome of the exercise.

Week 11 — Day by Day

Week 11 runs from Day 51 to Day 55 — five days covering chapters 22 through 24, ending with certification assessment. The timeline below shows what each day is for.

Day	Chapter	Phase	What Happens
Day 51	Chapter 22	Chapter 22	Certification preparation. Skills checklist self-assessment. Practice exercise (45-minute design sprint). Review any ☐ items from the checklist with the instructor. Certification project brief issued.
Day 52	Chapter 23	Chapter 23	Troubleshooting guide and machine maintenance. Independent reading of the most relevant troubleshooting sections. Machine check: clean, calibrated, and verified operational before certification begins.
Day 53	Ch 24	Design & CAM	Certification project design phase. Spreadsheet, all Bodies, Assembly verification, CAM setup for all operations. CAM Simulator check. Instructor signs off CAM before cutting.
Day 54	Ch 24	Cutting	Certification project cutting day. All operations run in correct order. No instructor assistance with CAM or design decisions — machine safety assistance only.
Day 55	Ch 24	Submit	TechDraw documentation completed. PDF exported and submitted. Finished piece presented. Certification assessment completed.

> ⚠ **Important**
>
> Day 53 is the last day you can ask your instructor design or CAM questions. From Day 54 onwards, the work must be independently executed. If your design has an unresolved problem at the end of Day 53, raise it before leaving — not at the start of Day 54 when the clock is running.

A Final Word Before the Certification Project

Ten weeks ago you learned what a CNC machine is. This week you will design something for one — independently, from scratch, with full documentation. The distance between those two points is not luck or natural talent. It is 21 chapters of deliberate practice, 10 projects, and a lot of G-code files.

The certification project is not a test of whether you can follow instructions perfectly under pressure. It is a demonstration that you have genuinely developed the skills to direct the entire CNC design and manufacturing process yourself. That is what the Bloom's Create level means — and it is what you are ready for.

Trust the process. Start with the spreadsheet. Verify before cutting. Document as you go. And if something goes wrong on the machine, troubleshoot it — that is a skill too, and Chapter 23 has you covered.

You Are More Ready Than You Think

The most common thing students say after the certification project is: 'I was more prepared than I thought I was.' The checklist, the practice sprint, and the five golden rules exist to help you realise that before Day 53 — not after.

Chapter Summary

- Bloom's taxonomy maps the course from Remember (Week 1) to Create (Week 11). The certification project requires the Create level: independent design, all workflow decisions made by the student, from idea to documented finished piece.
- The skills checklist covers 42 skills across four parts. Mark each as Confident, Review, or Not Attempted. Any Review or Not Attempted items are study priorities before Day 53.
- The certification project has seven non-negotiable requirements: independent design, a Spreadsheet with 5+ aliases, 4+ workbenches, 3+ CAM operations, complete TechDraw documentation with BOM, a successful cut, and full safety compliance.
- The five golden rules for independent CNC design: Spreadsheet first, design for the machine, verify before cutting, document as you go, and simple done well beats complex done poorly.
- The eight most common certification mistakes are all preventable: skipping the spreadsheet, skipping assembly verification, profile before pocket, forgetting Z re-zero after tool change, vague title block, designing beyond the machine envelope, under-dimensioned TechDraw, and rushing the safety checklist.
- The 45-minute design sprint on Day 51 provides a calibrated rehearsal of the certification experience. Its purpose is to reveal what slows you down and which chapter section to review — not to produce a perfect result in the time.
- Week 11 runs Days 51–55: Chapter 22 prep (Day 51), Chapter 23 troubleshooting (Day 52), Chapter 24 design and CAM (Day 53), Chapter 24 cutting (Day 54), Chapter 24 documentation and submission (Day 55).

Key Vocabulary — Chapter 22

Term	Definition
Bloom's Create level	The highest level of Bloom's Revised Taxonomy. Requires the student to produce an original work by integrating and applying knowledge independently — not by following provided instructions.
Independent design	A design produced without a provided brief, step-by-step walkthrough, or worked example to reference. The student makes all decisions: what to make, which tools to use, in what order.
Certification project	Chapter 24's independently designed and executed CNC project, assessed against seven requirements. The culminating demonstration of course competency.
Skills self-assessment	A structured checklist-based review in which a student rates their own confidence across all major course skills, identifying areas requiring revision before the certification project.

Design sprint	A time-limited design exercise (45 minutes in this context) used to calibrate readiness and identify workflow weaknesses before the assessed certification event.
Five golden rules	The five prioritised principles for independent CNC design introduced in this chapter: Spreadsheet first, design for the machine, verify before cutting, document as you go, simple done well beats complex done poorly.
Pre-certification review	The process of revisiting chapter sections corresponding to Review or Not Attempted items from the skills checklist, completed between Day 51 and Day 53.

Looking Ahead — Chapter 23: Troubleshooting Guide

Chapter 23 is a comprehensive reference chapter covering every common problem you are likely to encounter in a CNC session: machine alarms, spoilt workpieces, CAM errors, FreeCAD model problems, and UGS connection issues. Read it before Day 53 — not while something is going wrong at the machine on Day 54.

CHAPTER
23
of 24

PART SIX — CERTIFICATION & REFERENCE

Troubleshooting Guide

GRBL alarms & errors, UGS problems, FreeCAD issues, cut quality, machine maintenance, and safety incident response

📋 **How to Use This Chapter**

How to use this chapter:

This is a reference chapter — not a sequential read. When something goes wrong at the machine or in FreeCAD, find the section that matches your problem and go directly to it. Every section is organised as a symptom → cause → fix table. The GRBL Alarm and Error tables are the most commonly consulted — read those on Day 52 so you already know what the codes mean before you need them.

The Quick Reference Card at the end of this chapter is designed to be printed and kept at the machine. It summarises the five most common in-session emergencies on a single panel.

Section 1 — GRBL Alarm Codes

A GRBL alarm puts the machine into a locked state — it will not execute any movement commands until you clear the alarm. The machine displays 'ALARM:N' in the UGS status display where N is the alarm number. Always clear alarms by pressing Reset first, then re-home the machine before resuming work. Never ignore a GRBL alarm or attempt to send G-code while in alarm state.

● **Stop & Do This First**

ALARM STATE = MACHINE IS LOCKED. Do not send any G-code. Do not use the jog controls. Press RESET first. Then $H to re-home. Then re-zero your work coordinates from scratch.

Code	Name	Why It Happens	Recovery Steps
`ALARM:1`	Hard limit switch triggered	A physical hard-limit switch fired — the machine moved to the edge of its travel envelope. Cause: machine not homed correctly before job, incorrect G-code coordinates exceeding travel.	Press Reset in UGS. Re-home the machine ($ H). Re-check your work zero. Verify G-code dimensions are within machine travel (CNC 4040: 400×400×80mm).
`ALARM:2`	G-code exceeds machine travel	A G-code move would take the tool outside the machine's configured travel limits. Usually a coordinate system error.	Press Reset. Check the G-code file — look for extreme X, Y, or Z values. Verify your work zero was set correctly. Is the G-code post-processed for the correct machine?

`ALARM:3`	Reset while in motion	The Reset button was pressed while an axis was moving. GRBL cannot confirm the current position.	Press Reset, then re-home ($H). Set work zero again from scratch. Do not trust the current position after an in-motion reset.
`ALARM:4`	Probe fail — contact not expected	The probe triggered at the start of a probe cycle when it should not have been touching anything.	Check that the probe clip is not touching the bit or machine frame before probing. Inspect for debris on the probe plate. Reset and retry.
`ALARM:5`	Probe fail — no contact made	The probe cycle completed its travel distance without the bit touching the probe plate.	Ensure the probe plate is below the bit's starting height. Check probe wiring. Check that the probe plate clip is making electrical contact with the bit shank. Reset and retry.
`ALARM:6`	Homing fail — reset during homing	The Reset button was pressed during the homing cycle, or a limit switch triggered unexpectedly mid-home.	Press Reset. Manually move all axes to approximately centre of travel (hand-jog in UGS, small increments). Re-home ($H). If repeated: check limit switch wiring.
`ALARM:7`	Homing fail — door switch	A door/safety switch triggered during homing. Not all CNC 4040 configurations have this.	Check for any triggered safety switches. Press Reset. Re-home.
`ALARM:8`	Homing fail — pull-off travel	After hitting the limit switch during homing, GRBL could not pull off far enough to clear it.	Check for mechanical obstruction on the axis. Check limit switch bracket alignment. Press Reset, clear obstruction, re-home.
`ALARM:9`	Homing fail — approach failed	The axis did not reach its limit switch within the expected travel distance during homing.	Check that all axes move freely — no obstruction or binding. Check limit switch wiring and position. Press Reset. Re-home.

Section 2 — GRBL Error Codes

GRBL errors (as distinct from alarms) are responses to invalid or unprocessable G-code commands. The machine displays an error message in the UGS console but does not necessarily lock — it may pause and wait for corrective input. Errors in the middle of a running job can cause it to stop mid-cut.

Code	Name	Cause	Fix
`error:1`	Expected command letter	A character in the G-code file is not a valid G-code command.	Check the G-code file for corrupt characters. Re-post-process from FreeCAD CAM.
`error:5`	Homing not enabled	$22=0 in GRBL config. Homing is disabled in firmware.	Type $22=1 in the UGS console to enable homing, then $H to home.

`error:20`	Command in wrong state	A G-code command was sent when the machine was not in the correct mode to execute it.	Press Reset. Re-send the command from the correct machine state (e.g., ensure spindle is running before a cutting move).
`error:22`	Feed rate undefined	A cutting move was commanded with no feed rate set (no F word in the G-code).	Check CAM settings — the operation may be missing a feed rate. Re-post-process.
`error:33`	Invalid G-code ID	An unrecognised G or M code in the file. Often a post-processor mismatch.	Verify you are using the grbl post-processor in FreeCAD CAM — not LinuxCNC, Mach3, or another controller format.
`error:35`	Value out of range	A coordinate or parameter value is outside GRBL's accepted range.	Check for extreme values in the G-code file. Verify units (mm, not inches) in both FreeCAD and GRBL ($13=0 for mm).

Section 3 — UGS Connection & Operation Problems

Symptom	Likely Cause	Fix
Can't connect — no COM port listed	Controller not detected by the computer	Check USB cable is firmly seated at both ends. Try a different USB port. On Windows: Device Manager → check for unknown device. Install CH340 or CP2102 USB driver if missing. Restart UGS after connecting USB.
Connection drops mid-job	USB communication interrupted	Use a high-quality USB cable, maximum 2m. Avoid USB hubs — connect directly to computer. Disable USB power management in Windows (Device Manager → USB Root Hub → Power Management → uncheck 'Allow computer to turn off').
Soft limit alarm during job	Job coordinates exceed soft limit boundaries	Check $130 (X max travel), $131 (Y max travel), $132 (Z max travel) in GRBL config. Your job coordinates may exceed these. Re-zero from a different stock position or reduce the job extent.
Machine moves but cuts air above stock	Z zero set incorrectly — bit too high	Re-zero Z. The paper method: lower the bit until it just grips a sheet of paper when dragged. Zero from that position. Alternatively use a touch probe plate.
Machine moves in wrong direction	Axis direction inverted in GRBL config	Check $3 (direction invert mask). Consult your machine's documentation for the correct $3 value. Alternatively: in FreeCAD CAM, check that the G-code origin convention matches your machine's home position.
Feed rate override has no effect	Spindle/feed rate buttons inactive	Click inside the UGS window first — keyboard shortcuts only work when the UGS window is in focus. Use the on-screen percentage buttons rather than keyboard shortcuts if focus issues persist.
UGS shows 'ALARM' on open	Previous session ended with an alarm state	In the UGS console, type: $X to clear the alarm, then $H to re-home. Never send G-code after an alarm without clearing and re-homing first.

Section 4 — FreeCAD Model & CAM Problems

Problem	Cause	Fix
Sketch shown in red — over-constrained	Conflicting constraints applied — the sketch is over-defined.	Open the sketch. The over-constraining constraint is usually shown in red. Delete the most recently added constraint, or use Sketch → Redundant Constraints panel to identify conflicts.
Sketch shown in yellow — under-constrained	Not all sketch elements are fully constrained. The model will update unpredictably.	Open sketch. Check the DOF counter — it should read 0. Yellow elements have remaining freedom. Apply missing constraints. Common omission: forgetting to constrain the sketch origin.
Part Design feature shows error icon	A feature failed to compute — usually because the sketch it depends on has an error, or a face it references no longer exists.	Right-click the failing feature → Edit. Check the sketch it references — open and fix the sketch first. Then recompute the feature.
Spreadsheet alias 'not found' error	A sketch constraint references a Spreadsheet alias that does not exist — either misspelled or not yet created.	Open the Spreadsheet. Check the alias name exactly — aliases are case-sensitive. The sketch uses =Spreadsheet.AliasName — verify both sides match character for character.
Model does not update after Spreadsheet change	Refresh cycle not triggered, or the 3D view is stale.	Edit → Refresh Model (or press Ctrl+Shift+R in some versions). If still stale: close and reopen the file. Check that the aliases are correctly linked (yellow warnings in the Spreadsheet mean broken links).
Topological naming error — feature lost reference	A feature references an edge or face by position. After editing an earlier feature, that position changed — the reference is now wrong.	This is a known FreeCAD limitation. The solution is to use named sketches and named references where possible. For immediate fix: right-click the failing feature → Edit → re-select the correct face/edge manually.
CAM operation shows red in Model tree	The CAM operation failed to compute a valid toolpath — usually a geometry selection error.	Double-click the operation to open it. Re-select the target geometry (face, edge, or profile). Ensure the geometry is still valid in the model. Check that Final Depth is negative (below stock surface).
Assembly solver shows 'no solution'	Dimensional mismatch — the joints as defined cannot be satisfied with the current part dimensions.	Identify which joint is failing (the error message names it). Check the dimensions of the parts referenced by that joint in their source files. Fix the dimension, Refresh the assembly.

Tip

The most important FreeCAD troubleshooting habit: work in a fresh document for each new project. Files that have accumulated months of edits, failed features, and abandoned sketches become unstable. When a FreeCAD file becomes unreliable, export the good geometry to STEP format, import into a new file, and rebuild from there.

Section 5 — Cut Quality Problems

Cut quality problems are diagnosed by their visual symptom on the workpiece. The table below covers the nine most common cut quality issues encountered on the CNC 4040 Pro Max with MDF, pine, and hardwood.

Symptom	Most Likely Cause	Fix
Rough, fuzzy edges on MDF	Feed rate too high, or bit is dull. MDF fibres tear rather than cut cleanly.	Reduce feed rate by 15%. Inspect bit — a dull bit has visible wear on the cutting edge. Replace if in doubt. Ensure step-down does not exceed half the bit diameter per pass.
Burning or scorching on pine/timber	Feed rate too slow — the bit is rubbing rather than cutting, generating heat instead of chips.	Increase feed rate by 20%. Ensure spindle speed is appropriate (not too low). Clear chips with compressed air — chip re-cutting causes burning.
Chatter marks on profile walls	Bit deflecting under lateral cutting force. Too fast feed, too deep step-down, or unsupported long bit sticking out of collet.	Reduce feed by 20% and step-down by 30%. Maximise the length of bit clamped in the collet — minimise stick-out. For thin walls, consider multiple light passes.
Scalloped top surface in 3D carving	Step-over too wide for the desired surface finish — visible ridges between ball-nose passes.	Reduce step-over. For finishing passes: 0.5–1mm. Check the scallop height table in Chapter 15. A finishing pass at 0.5mm step-over removes most roughing scallops.
Cut depth inconsistent across the sheet	Material is not flat or not clamped flat against the bed. Warped MDF is very common.	Flatten the spoil board (surface-machining run). Use more clamps. For critical depth work: add a facing pass across the entire stock surface at the start of the job.
Part moves during profiling	Insufficient tabs, or clamps too far from the cut. Freed material pushed by the bit.	Add more tabs (minimum 4 per panel). Position clamps within 80mm of any profile cut path. Consider double-sided tape in addition to tabs for small parts.
Bit breaks during cut	Step-down too deep, feed too fast, or bit hitting a clamp or screw in the material.	Check stock for embedded screws or nails before cutting. Reduce step-down to half bit diameter for first attempt in new material. Reduce feed 20% in hardwood.
Ragged inside corners on pocket	Bit diameter too large — the bit cannot fit into sharp corners. Or step-down too large causing deflection.	Switch to a smaller diameter bit for tight corners. Or accept rounded corners and plan for them in the design (the FreeCAD model should reflect achievable geometry).
Z drifts upward over long job	Stepper motor losing steps — usually Z axis under load.	Reduce Z plunge rate (current recommended: 250mm/min — try 180). Check Z axis coupler is tight. Reduce step-down per pass.

Section 6 — Material and Workholding Problems

Problem	Cause	Fix
Workpiece lifts during cut	Vacuum or clamps insufficient. MDF under a profile cut tries to lift as the last web is cut.	Add tabs. Place clamps within 80mm of profile paths. Lightweight plywood: double-sided tape the entire underside to the spoil board.
Material surface waves / sags	Stock is bowed or warped — not in full contact with the spoil board.	Clamp every 150mm around the perimeter. For badly bowed sheet: run a skim facing pass (1mm depth) to reference everything to the machine plane.
Chipped edges on plywood	Cutting against grain direction on the top ply face.	Reduce feed rate 20%. Use a downcut spiral end mill for the first pass (plunges cut fibres downward rather than tearing). If unavailable: score the cut line with a knife before running the profile.
Smoke from MDF on first pass	Feed too slow or spindle too fast for the chip load. MDF generates more heat than timber at equivalent feed rates.	Increase feed. Target chip load: feed_rate ÷ (spindle_RPM × number_of_flutes). For 6mm 2-flute at 18000RPM: feed = 18000 × 2 × 0.04 = 1440mm/min (adjust down for older machines).
White powder residue on cut edges	Normal MDF dust — not a problem. However if heavier than usual: ventilation is insufficient.	Ensure workshop extraction is running during all MDF operations. Wear appropriate dust protection.

Section 7 — Safety Incident Response

A safety incident during CNC operation can range from a minor near-miss (a clamp struck by the gantry) to a serious event (bit ejection, material ejection, or personal injury). All incidents follow the same five-step response, regardless of severity.

Safety Incident Response — Five Steps	
• STOP	Press Feed Hold (or E-Stop if fitted). Turn off spindle. Wait for all movement and rotation to cease completely. Do not touch the machine or workpiece until everything is stationary.
• ASSESS	Before doing anything else: are you or anyone else injured? Is there smoke or burning? Is the bit still in one piece? Is the workpiece secure? Is the gantry in a safe position?
• ACT	If injured: seek first aid immediately. If burning: disconnect machine power. If bit broken with fragments: sweep clear before reaching near the spindle area. If workpiece loose: only remove by hand once spindle has stopped and is confirmed off.
• REPORT	Inform your instructor. Record what happened in the machine log. Do not restart a job after a safety incident without instructor clearance. All incidents — even minor ones — must be recorded.
• DIAGNOSE	Only after safety is confirmed and the incident recorded: diagnose what caused the problem. Consult the troubleshooting tables in this chapter. Fix the cause before attempting any further cutting.

Stop & Do This First

ANY personal injury — however minor — must be reported to the instructor immediately. Do not continue working. Do not attempt self-diagnosis of a cut or eye injury. Stop, report, seek first aid.

Section 8 — Machine Maintenance Checklist

The CNC 4040 Pro Max is a precision machine. The leading cause of degraded performance and unexpected alarms is accumulated MDF dust in the linear rail carriages and lead screw mechanisms. A 5-minute cleaning routine after every session prevents 90% of maintenance issues.

After every session		
☐	**Vacuum / blow out all MDF dust from rails, lead screws, and bed**	MDF dust is abrasive and clogs linear rails quickly — the leading cause of premature machine wear.
☐	**Wipe linear rails with a dry cloth**	Removes fine dust and moisture. Do not use water or solvent on linear rails.
☐	**Check collet — remove bit, inspect collet for chips or debris**	A debris-fouled collet causes runout, poor surface finish, and can cause bit slippage mid-cut.
☐	**Inspect cutting bit — check for chips, cracks, or bluntness**	Hold the bit to a bright light and rotate it. A dull bit appears slightly reflective on the cutting edge.
Weekly		
☐	**Apply a thin film of machine oil to lead screws (Z axis especially)**	Prevents rust and ensures smooth Z travel. Use light machine oil — not WD-40 (too thin) or grease (attracts dust).
☐	**Check all axis coupler grub screws — tighten if loose**	Loose couplers cause Z drift and positional error. Access with a 2mm hex key.
☐	**Run a test cut on scrap — verify depth accuracy**	A 5mm-deep pocket in MDF should be 4.9–5.1mm. If outside this range, re-calibrate steps/mm ($102 for Z).
☐	**Inspect collet for wear grooves — replace if grooved**	A worn collet cannot hold a bit with consistent runout. Replace the ER11 collet, not just the bit.
Monthly / as needed		
☐	**Check all V-slot or linear rail carriage tension**	Carriages should move smoothly without play. Adjust eccentric nuts if there is lateral play in any axis.
☐	**Verify machine geometry — run a squareness check**	Cut a 200×200mm square. Measure diagonals — they should be equal within 0.3mm. Adjust gantry if not square.
☐	**Re-surface the spoil board**	A 1–2mm facing pass across the entire bed corrects accumulated wear and re-establishes a flat Z reference for the whole cutting area.
☐	**Inspect all wiring harnesses for chafing**	Moving cables can chafe against machine structure. Damaged wiring causes intermittent alarms and communication drops.

Tip

Keep a small spray bottle of ISO 99% isopropyl alcohol and a supply of lint-free cloths at the machine. After vacuuming dust, a light wipe of the rail surfaces with IPA removes fine abrasive particles that vacuuming alone misses.

Quick Reference Card — Print and Keep at the Machine

The card below summarises the five most common in-session emergencies. Print it, laminate it if possible, and attach it to the machine or the wall beside it. When something goes wrong, consult this before anything else.

QUICK REFERENCE — Print and Keep at the Machine

ALARM Recovery

1. Spindle off 2. Wait 3. Press Reset

4. Type $H in console to re-home

5. Re-set work zero 6. Reload G-code

Z Won't Zero Correctly

Lower bit → paper grips but slides → zero Z

Always re-zero Z after every tool change

Part Moving During Cut

Stop job (Feed Hold). Spindle off. Wait.

Add clamps / double-sided tape. Add tabs.

Re-home + re-zero before restarting.

Bit Breaks

Feed Hold immediately. Spindle off. Wait.

Check stock for embedded screws/nails.

Reduce Step Down 30%. Reduce Feed 20%.

Connection Drops

Reconnect USB. Do NOT re-home.

Check current position in UGS DRO.

If position OK: resume job from correct line.

Burning / Scorching

Increase feed rate +20%. Clear chips.

Check bit is sharp. Check spindle speed.

Week 11, Day 52 — Lesson Content Overview

Activity	Detail
Independent reading (30 min)	Students read this chapter independently, focusing on the sections most relevant to problems they have encountered during the course. Instructor circulates — answer conceptual questions, but do not solve problems for students. The goal is that students can use this chapter independently by Day 54.
Fault diagnosis exercise (20 min)	Instructor presents 5 fault scenarios verbally (e.g., 'The machine shows ALARM:1 when you press cycle start. What do you do?'). Students write their diagnostic and recovery steps. Class debrief — compare answers, discuss alternatives. Emphasis: there is often more than one valid response.
Machine inspection and clean (20 min)	Students carry out the 'after every session' maintenance checklist on the CNC 4040 Pro Max: vacuum, wipe rails, inspect collet and bit, check coupler tightness. Machine is left clean and verified operational for Day 53 certification start.

Certification readiness check (20 min)	Each student reviews their Chapter 22 skills checklist. Instructor checks: (1) every student has a certification project concept, (2) the concept is feasible within the machine envelope and available time, (3) any remaining 'review' items have a plan. Day 52 ends with the machine clean and every student with a viable certification project concept.

Chapter Summary

- GRBL has 9 alarm codes (ALARM:1 through ALARM:9). All alarms lock the machine. Recovery is always: Reset → $H to re-home → re-zero work coordinates → reload G-code. Never send G-code while in alarm state.
- GRBL error codes identify invalid G-code commands. The most common are error:1 (invalid character), error:22 (undefined feed rate), and error:33 (wrong post-processor format). Fix by correcting the CAM settings and re-post-processing.
- UGS connection problems are usually caused by USB cable quality, driver issues (CH340/CP2102), or USB power management settings. Always connect USB directly to the computer — avoid hubs.
- The most common FreeCAD problems are: over-constrained sketch (delete the conflicting constraint), topological naming error (re-select the correct face manually), and broken Spreadsheet alias links (check for exact case-sensitive match).
- Cut quality problems are diagnosed symptom-first: fuzzy edges (reduce feed), burning (increase feed), chatter (reduce feed + step-down), inconsistent depth (flatten stock, add clamps), bit breakage (reduce step-down and feed, check for embedded screws).
- Safety incident response: Stop → Assess → Act → Report → Diagnose. All incidents regardless of severity must be reported to the instructor. Never restart a job after a safety incident without instructor clearance.
- Machine maintenance: vacuum and wipe rails after every session; oil lead screws and check coupler tightness weekly; check carriage tension, verify squareness, and resurface the spoil board monthly.
- The Quick Reference Card at the end of this chapter should be printed and attached to the machine. It covers ALARM recovery, Z zeroing, part movement, bit breakage, and connection drops in a scannable panel format.

Key Vocabulary — Chapter 23

Term	Definition
GRBL alarm	A GRBL firmware state that locks all machine movement until cleared. Triggered by hard limits, homing failures, probe failures, or reset-while-moving events. Cleared with the Reset command, followed by re-homing.
GRBL error	A firmware response indicating that a G-code command cannot be processed. Does not necessarily lock the machine, but halts execution of the offending command and any subsequent commands in the queue.
Hard limit	A physical switch that triggers when an axis reaches the edge of its travel range. Triggering a hard limit causes ALARM:1. Prevents the machine from mechanically damaging itself.
Soft limit	A firmware-configured coordinate boundary that triggers an alarm if a G-code move would exceed it, before the physical limit switch is reached. Configured with $20, $130, $131, $132 in GRBL.

Feed Hold	A UGS/GRBL command that pauses feed movement (the machine decelerates and stops) without resetting the controller's position knowledge. The safest way to interrupt a running job.
Chatter	Vibration of the cutting bit during a cut, leaving regular wavy marks on the machined surface. Caused by excessive feed rate, too-deep step-down, or insufficient bit clamping in the collet.
Chip load	The thickness of material removed per cutting edge per revolution. Correct chip load produces efficient cutting; too low causes rubbing and heat; too high causes deflection or breakage. Chip load = Feed Rate ÷ (RPM × Number of Flutes).
Topological naming	A FreeCAD issue where features reference faces or edges by their computational position rather than by a stable name. When earlier features are edited, positions change and the reference breaks. A known limitation of FreeCAD's geometry kernel.
Spoil board	A sacrificial flat board clamped to the CNC bed, into which the cutting bit can cut through without damaging the machine bed. Resurfaced periodically to maintain flatness.
Machine log	A written record of machine sessions, incidents, maintenance performed, and any anomalies observed. Essential for tracking machine health and for investigating repeated problems.
Runout	The amount of wobble in a spinning bit — the deviation of the bit tip from the true rotation axis. Caused by a worn collet, debris in the collet, or a bent bit shank. Excessive runout degrades surface finish and accelerates bit wear.

Looking Ahead — Chapter 24: The Certification Project — Keepsake Memory Box

Chapter 24 is the final chapter and the certification project. You will design, document, and cut an original Keepsake Memory Box — a multi-component, multi-operation project that demonstrates every major skill from the course. No walkthroughs. No hints. Just you, FreeCAD, the CNC 4040 Pro Max, and everything you have learned.

PART SIX Certification & Reference

Chapter 24 + Appendices A–J • Bloom's Level: Create (Synthesis)

📖 About Part Six

Part Six is the final section of this textbook. It contains a single teaching chapter — Chapter 24, the Certification Project — followed by Appendices A through J, which provide the reference materials, rubrics, and supplementary tables that support both the taught course and ongoing practice. Chapter 24 is different from every other chapter in this textbook: it does not teach new skills. It applies every skill in the book simultaneously, in service of an original piece of your own design. This is what the entire course has been building towards.

CHAPTER
24
of 24

PART SIX — CERTIFICATION & REFERENCE

The Certification Project

The Keepsake Memory Box — complete project brief, design guidance, assessment rubric, submission checklist, and reflection guide

This Chapter Covers Everything you need to know to plan, execute, and submit the Keepsake Memory Box certification project:

- The complete Keepsake Memory Box project brief — all nine mandatory requirements in full
- The seven assessment criteria with a four-level performance rubric
- Design guidance for each mandatory element: joinery method selection, 3D relief form selection, engraving approaches
- A detailed day-by-day certification timeline with required outputs for Days 51–55
- The complete submission checklist — every mandatory and optional item
- TechDraw minimum content checklist for certification
- Guidance on writing a reflection document that achieves distinction-level marks
- Certificate of Completion conditions and what the qualification signifies

Chapter 24 — The Certification Project

No step-by-step guide. No worked example. No known answer. You have the brief, you have the skills, and you have five days. What you make, and how well you document it, is entirely yours.

Every chapter in this textbook up to this point has given you something to follow: a sequence of steps, a worked example, a project walkthrough. Chapter 24 gives you none of those things. The certification project is assessed as independent work — that means independent design, independent decision-making, and independent problem-solving. The instructor is available for consultation but not for instruction. If you encounter a problem, you are expected to diagnose and resolve it using the skills and knowledge you have built over the preceding ten weeks.

That is not a threat. It is a description of what the Create level of Bloom's Taxonomy actually looks like in practice. Creating means producing an original, complex artefact through synthesis — combining prior knowledge in new configurations to achieve a result that did not previously exist. The Keepsake Memory Box is the evidence of that synthesis. It is the tangible answer to the question: what can you make, entirely on your own, with the skills this course has given you?

This chapter serves as your complete reference for the certification week. The project brief is reproduced in full. The assessment criteria are explained in detail, with clear descriptions of what each performance level looks like. Specific design guidance is provided for each mandatory element — not prescriptive instructions, but the considerations that will help you make good decisions independently. The submission checklist, TechDraw requirements, and reflection guide complete the chapter.

The Project Brief — Keepsake Memory Box

🏆 **Certification Project Brief — Keepsake Memory Box**

Project title: The Keepsake Memory Box

Design brief: Design and fabricate a Keepsake Memory Box. The box must be a functional, lidded container that demonstrates mastery of the core FreeCAD and CNC skills developed throughout this course.

Design freedom: Theme, proportions, decorative choices, and overall aesthetic are entirely the student's creative decision. No two certification projects should look alike.

Technical requirements: Nine mandatory technical requirements must be met. There are no exceptions to mandatory requirements. Additional recommended elements may be incorporated but are not required for a passing grade.

Assessment: Seven criteria, weighted as shown in the rubric below. Overall passing standard: 70% with a minimum of 60% on the Certification Project itself. One re-submission permitted within four weeks.

Bloom's level: Create — synthesis of all prior knowledge and skills in an original, independently-executed work.

The Nine Mandatory Requirements

#	Requirement	Description	Chapter Reference
1	Parametric Model	Complete model driven by a Spreadsheet with minimum 8 named parameter aliases. Every critical dimension must reference the spreadsheet. Demonstrate parametric update: change box height → Refresh → model updates cleanly.	8, 18
2	CNC Joinery	Box assembled using a CNC-cut joinery method: finger joints, box joints, or combination. All joints must include correct clearance tolerance values (set in Spreadsheet) and dogbone fillets at all inside corners. Assembled box holds together under its own weight without glue.	11
3	Engraved Lid	Box lid features original decorative engraving: ShapeString text, imported SVG artwork, B-spline freehand design elements, or any combination. Design must be original — not a default FreeCAD font with no customisation.	7, 10
4	3D Relief Element	At least one surface features a shallow 3D relief form created using Part Loft with minimum 3 datum plane sections. Machined using 3D CAM: Waterline roughing pass (large ball-nose) AND 3D Surface finishing pass (small ball-nose). Both passes must be visible in the CAM Simulator.	14, 15
5	Edge Treatments	All external visible edges of all box panels must have chamfer or fillet treatments applied in the FreeCAD Part Design model. Applied in the model — not as post-processing.	6, 14

6	TechDraw Drawing	A complete TechDraw engineering drawing: minimum 4 orthographic views (including at least one section view), 12 dimensions, full title block with all fields completed, Bill of Materials table, tolerance values on at least two joint dimensions.	17
7	UGS Documentation	A completed Job Setup Sheet for every cutting session. If cutting spans two days, two sheets are required. All fields must be completed including actual runtime and dimensional results.	20
8	Multi-Part CAM	All box panels must be designed to cut from a minimum of two separate stock pieces. Each stock piece must carry more than one panel — not one panel per stock piece.	11
9	Reflection Document	A project reflection document of minimum 600 words. Must address: design decisions made and reasoning, challenges encountered and how they were resolved, Bloom's Taxonomy self-assessment (which skills at which level), what you would do differently.	22, 24

Assessment Criteria — Full Rubric

The rubric below defines the four performance levels for each of the seven assessment criteria. Read every row carefully before beginning the project. The rubric is the precise instrument used to assess your work — knowing it in advance removes uncertainty and allows you to design to the standard you are aiming for.

Criterion	Distinction 85–100%	Credit 70–84%	Pass 55–69%	Not Yet Below 55%
1. Parametric Model Integrity **20%** *All critical dimensions spreadsheet-driven. Min 8 aliases.*	All critical dimensions spreadsheet-driven. 8+ named aliases correctly linked. Parametric update demonstrated cleanly — model updates with no errors or broken features on Refresh.	Most critical dimensions linked. 6–7 aliases. Minor hard-coded values in non-critical areas. Model updates substantially on Refresh.	Spreadsheet present with 4+ aliases. Some key dimensions hard-coded. Parametric update partially works.	Fewer than 4 aliases. Model has no meaningful parametric capability. Most dimensions hard-coded.

2. Joinery Quality **20%** *Finger/box joints with correct clearance and dogbones.*	All joints cut with correct clearance. Box assembles smoothly by hand (snug or press fit as designed). Dogbones present at all inside corners. No visible gaps or forced fits.	Joints mostly correct. One minor fit issue. Dogbones present. Box assembles with minimal adjustment.	Joints present. Some fit issues requiring adjustment. Dogbones partially present. Box assembles with effort.	Joinery not fully implemented. Joints do not assemble, or no dogbones, or fit is non-functional.
3. Decorative & 3D Elements **15%** *Lid engraving + 3D Loft relief with 2 CAM passes.*	Lid engraving clearly resolved — design intent evident, correct depth, clean sharp edges. 3D relief present, smooth surface, two distinct CAM passes (roughing + finishing) confirmed in Simulator and visible on piece.	Lid engraving present and readable. 3D relief present with two CAM passes but minor staircase or scallop marks visible.	Lid engraving present but shallow, unclear, or generic. 3D relief attempted — roughing only, no visible finishing pass.	Lid engraving absent or failed to machine. 3D relief element not attempted.
4. Edge Treatments **5%** *Chamfer/fillet on all external visible edges.*	Chamfer or fillet applied consistently to all external visible edges in the FreeCAD model. Treatment is appropriate to the design intent — organic forms use fillets, geometric forms use chamfers or a combination.	Edge treatments applied to most visible edges. One or two edges missed. Treatment mostly consistent.	Some edge treatments present but inconsistently applied. Visible untreated sharp corners on display faces.	No edge treatments applied in the FreeCAD model.
5. TechDraw Drawing **15%** *4+ views, 12+ dimensions, full title block, BOM, tolerances.*	4+ orthographic views including at least one section view. 12+ correctly placed and labelled dimensions. Complete title block (all fields). BOM table listing all parts. Joint tolerances shown (±0.x mm). Drawing is independently legible.	4 views, 10–11 dimensions, title block mostly complete. BOM present but incomplete. Some tolerances shown.	3 views, 8–9 dimensions, title block partially filled. No BOM or no tolerances on joints.	Fewer than 3 views. Fewer than 8 dimensions. Title block mostly empty. Not independently legible.

6. CAM & UGS Competency **15%** *All ops correct in Simulator. G-code run. Setup Sheets complete.*	All operations correct in Simulator — correct depths, strategies, tool assignments. GRBL G-code correctly post-processed and sent via UGS without instructor intervention. Job Setup Sheet completed for every session.	All operations simulate correctly. G-code sent with one minor instructor prompt (not a fundamental error). Setup Sheet mostly complete.	Most operations simulate correctly with minor errors. G-code required instructor support at one step. Setup Sheet partially complete.	CAM operations have errors in Simulator. G-code required significant instructor support. Setup Sheet not completed.
7. Project Reflection **10%** *600+ words. Critical analysis. Bloom's self-assessment. Lessons.*	600+ words. Genuine critical reflection: specific design decisions explained with clear reasoning. Challenges described with how they were resolved. Bloom's self-assessment accurate, specific, and insightful. Clear identification of what would be done differently and why.	600+ words. Good reflection with specific examples. Bloom's assessment present. Some generalisation rather than specific analysis.	400–599 words. Reflection is largely descriptive — events narrated rather than analysed. Bloom's mentioned but superficial.	Below 400 words. Purely descriptive or superficial. No critical analysis. Bloom's absent or inaccurate.

📌 **Coordinator Note on Certification**

The assessment is conducted by the subject coordinator with the physical piece, the digital submission package, and the rubric open simultaneously. The parametric update is demonstrated live. The instructor may ask questions about specific design decisions — treat this as a professional conversation, not an interrogation. Questions like 'why did you choose this clearance value?' or 'what would happen if you changed the finger count?' are opportunities to demonstrate Evaluate and Create level thinking.

Design Guidance — Making Good Independent Decisions

The following guidance is not a prescription — it is a set of design considerations that will help you make informed decisions independently. Read each section, think about how it applies to your specific design, and make the choice that best serves your project.

Box Size and Proportions

The physical size of the box affects almost every other design decision: material consumption, CAM runtime, joinery complexity, and the scale of the decorative elements. The following guidelines will help you choose proportions that are achievable in the certification week while still demonstrating the required skills.

Dimension	Recommended Range	Notes
Box width (external)	120–200 mm	Wider boxes show the finger joint pattern more impressively but require longer profile cuts. 150 mm is a good default.
Box depth (external)	80–150 mm	Shorter boxes are simpler to design. The joinery at the corners is the same regardless of depth.
Box height (sides)	50–90 mm	Taller boxes allow more finger joints per edge (minimum 3 fingers recommended). Very tall boxes require more material and longer cutting.
Lid thickness	9–12 mm	Thicker lids can accommodate a more prominent 3D relief. Thinner lids are lighter and more proportionate for smaller boxes.
Panel thickness	9–12 mm	Must match the stock material you will use. Set in the Spreadsheet as the foundational parameter.
3D relief height	15–25 mm	Keep the relief height achievable in the available machining time. 20 mm is a practical maximum for a one-day cut.

Joinery Method Selection

Three joinery methods are most practical for the certification project. The table below compares them across the key decision factors. Finger joints are the recommended default — most students completed Project 6 with finger joints and have a working model to reference.

Joinery Type	Description	Advantages	Challenges	Best For
Finger Joints (Box Joints)	Interlocking rectangular fingers of equal width cut into the panel edges. Standard CNC joinery method.	Strong, visually attractive, relatively straightforward to design. Finger count can be parametrically driven. Well-covered in Chapter 11.	Requires careful clearance design. Dogbones required at all inside corners. Finger width must be ≥ 2× tool diameter.	Recommended for most certification projects. Most students have completed this in Project 6.
Dado & Rabbet	A channel (dado) cut in one panel into which the edge of another panel fits.	Simpler CAM setup than finger joints. Good for shelf or internal divider connections.	Less visually dramatic than finger joints. Requires accurate depth — too deep or shallow and the fit fails.	Internal dividers or if the student is less confident with finger joint clearance design.
Loose Tenon	Separate wooden splines inserted into matching slots in both panels.	Clean exterior appearance — no visible joinery on outside faces.	Requires very accurate slot positioning on both mating panels. Harder to design parametrically.	Advanced option for students who want a clean exterior aesthetic and have strong parametric skills.
Dovetail (simplified)	Angled interlocking fingers — the angled geometry resists pull-out forces.	Visually impressive. Mechanically strong in one direction.	More complex sketch geometry. Requires careful angle calculation. Dogbones at acute corners are non-standard.	Optional for students who completed the optional dovetail exercises and want to demonstrate advanced geometry.

Finger Joint Design Parameters

If you choose finger joints, the following parameters must be designed correctly:

- **Finger width:** must be at least 2× the tool diameter. For a 6 mm end mill, minimum finger width = 12 mm. Recommended: 15–18 mm for a clean, readable joint on a box of this scale.
- **Finger count per edge:** odd numbers (3, 5, 7) produce joints where the finger at each corner is on the same panel — visually cleaner. For a 150 mm wide panel: 3 fingers (50 mm wide each) or 5 fingers (30 mm wide each).
- **Clearance:** start with 0.15 mm in the spreadsheet. Cut a test joint in scrap before final material. Adjust if too tight (try 0.2 mm) or too loose (try 0.1 mm).
- **Dogbone diameter:** equals the tool diameter. For a 6 mm end mill, dogbone circles are 6 mm diameter (3 mm radius). Centre of dogbone circle is at the inside corner of the notch.

3D Relief Element — Form Selection

The 3D relief element is the most time-sensitive mandatory requirement. The machining time for a 3D CAM job is significantly longer than 2.5D work — plan carefully. The table below gives guidance for selecting a Loft form that achieves the visual impact you want within the available time.

Relief Form	Datum Sections	Difficulty	Time Estimate	Design Tip
Leaf / petal	3 — base oval, mid oval, apex point	Low — ideal for first 3D CNC project	45–75 min 3D CAM	Keep the mid section at 60–70% of the base section size for a graceful taper. Do not exceed 25 mm height — deeper forms require longer roughing.
Low dome / hill	3 — large circle, medium circle, small circle	Low — very predictable Loft result	40–60 min	Simple and clean. Works well on a box lid because the dome reads clearly at any viewing angle. Diameter should be at least 80% of the lid width.
Wave / dune	4 — wide rectangle, narrow rectangle, wide rectangle, thin rectangle	Medium — start vertex alignment is critical	50–80 min	The alternating wide/narrow sections create the wave motion. All rectangles must have their start vertex at the same corner — test Loft before committing to the lid placement.
Bas-relief portrait or animal	3–5 — custom shapes traced from reference	High — freeform B-spline sections require careful shaping	60–90 min + 20 min pencil trace	Only recommended for students who are confident with B-splines and have time to iterate. Limit the relief to a simple silhouette shape rather than a complex portrait.
Geometric architectural element	3 — polygon, smaller polygon, point	Low-medium — depending on polygon complexity	40–65 min	A pyramid from a square base, or a spire from an octagon, produces strong geometric relief. Ruled Surface ON can be appropriate for strictly geometric forms.

3D Relief Placement on the Box

The 3D relief element can be placed on any surface of the box. The most common placements are:

- **Lid top face:** the most visible surface. The relief sits on top of the flat lid, visible whenever the box is displayed. Most students choose this placement. The relief footprint must fit within the lid dimensions with adequate clearance from the edges.

- **Front panel:** the relief appears on the vertical front face of the box. Dramatic when the box is positioned for display. Requires that the front panel has enough thickness (at least 25–30 mm) to accommodate the relief height.
- **Interior base:** the relief is machined into the inside of the box base — visible only when the lid is removed. An intimate, unexpected element. Requires the base to be thick enough for the relief depth without breaking through.

Engraving Design Guidance

The lid engraving is assessed on clarity, depth, and design intent — not on the specific subject matter. The design must be recognisably original and must be machined at a consistent, readable depth. The following approaches are all acceptable:

Engraving Approach	Design Considerations
ShapeString text — name, date, dedication	Choose a bold, clean font. Set the V-bit engraving depth to 0.8–1.2 mm. Single-stroke fonts engrave most cleanly. Use the Spreadsheet for text height.
Imported SVG artwork	Prepare the SVG in Inkscape: convert text to paths, simplify nodes, ungroup, save as Plain SVG. Test-import before committing to the lid design. Knotwork and botanical designs engrave particularly well.
B-spline freehand design	Draw an organic motif directly in the Sketcher using B-splines. Leaf forms, wave patterns, and abstract botanical shapes work well. Convert to a Part Shape before adding the CAM Engraving operation.
Combined text and artwork	Most visually compelling. For example: a name or date in ShapeString surrounded by an SVG decorative border. Ensure there is adequate separation between text and border to prevent visual crowding.

Certification Week Timeline — Days 51–55

The certification week is five working days. The timeline below specifies the required outputs for each day. These are not aspirational targets — they are the minimum progress required to complete and submit a full project on Day 55. Students who fall behind on Day 52 do not have time to recover. The timeline is tight by design — time management is itself a skill being assessed.

Day	Focus	Detail — Required Outputs and Key Decisions
Day 51	**Planning**	Begin with a review of the completed paper plan (from Day 50). Instructor signs off the plan before FreeCAD is opened. Create the FreeCAD file: Body, Spreadsheet with all aliases, initial parameter values. Create all datum planes. Draw and fully constrain all section sketches for the 3D Loft. Apply Part Loft and verify the form is clean (no twist). Do not proceed to Day 52 tasks unless the Loft result is acceptable. Required output: FreeCAD file with Spreadsheet, all bodies started, Loft verified.
Day 52	**FreeCAD Design**	Complete all panel sketches. Apply Pad to all panels. Create joinery geometry (finger joint sketches on mating edges with correct clearance and dogbones). Apply Pockets for all joint notches. Combine Loft form with lid panel using Boolean Union. Add chamfer/fillet edge treatments to all external visible edges. Add lid engraving elements (ShapeString, imported SVG, or B-spline). Verify the complete assembled model is correct. Required output: complete FreeCAD model — all panels, all joinery, all decorative elements.

Day 53	**CAM Setup**	Create CAM Job. Configure Stock for each stock piece. Add all tools to library. Create all operations in correct order. Run CAM Simulator for every operation — verify depths, strategies, and coverage. Post-process all G-code files with correct version naming. Load each file in UGS and verify the 3D visualiser. Record estimated runtime on Job Setup Sheet. Required output: all .nc files, all Simulator screenshots, completed pre-cut section of Job Setup Sheet.
Day 54	**Machine Cutting**	Home machine. Complete pre-cut checklist for Stock 1. Cut all panels from Stock 1. Tool changes as required — re-zero Z after each. Move to Stock 2. Repeat. After all panels cut: remove tabs, sand if required, test-assemble all joints. Record actual dimensions vs design dimensions on Job Setup Sheet. Start TechDraw: insert base views and begin dimensioning. Required output: all panels cut and test-assembled. Job Setup Sheet completed for Day 54 session. TechDraw page started.
Day 55	**Documentation**	Complete TechDraw: all 4+ views, all 12+ dimensions, complete title block, BOM, tolerances. Export as PDF. Complete Job Setup Sheet for any Day 55 work. Write project reflection (600+ words). Organise all files in version-controlled folder. Complete submission checklist — verify every mandatory item is present. Submit: physical box, digital file package (FreeCAD, G-codes, TechDraw PDF, Setup Sheets, photographs, reflection). Required output: complete submission package.

> ⚠ **Important**
> Day 51 is planning and Loft verification only. Do not begin the joinery sketches, the lid engraving, or the CAM setup on Day 51. The Loft must be verified as clean (no twist, smooth surface) before the rest of the design work begins — discovering a Loft problem on Day 53 cannot be recovered within the timeline.

FreeCAD Workflow — Certification-Specific Guidance

The following workflow guidance addresses the specific challenges that most commonly arise during the certification project. These are the places where students encounter problems — knowing them in advance allows you to avoid or quickly resolve them.

Multi-Body File Structure

A typical Keepsake Memory Box has 5–7 parts: base, front panel, back panel, two side panels, lid, and possibly an internal tray or divider. All parts must be in a single .FCStd file as separate Part Design Bodies. All Bodies must share the same Spreadsheet for their dimensional parameters.

1	**Name every body**	Right-click each Body in the Model tree → Rename. Use descriptive names: Base, FrontPanel, BackPanel, LeftSide, RightSide, Lid. Do this before drawing any sketches.
2	**Name every sketch**	Rename each sketch to describe its content: OutlineSketch, JoinerySketch, ReliefBase, ReliefMid, ReliefApex. Named sketches enable inter-body expressions and make the Model tree navigable.
3	**Name key constraints**	In the Sketcher, name dimensional constraints that may be referenced from elsewhere: Width, Height, FingerWidth, Clearance. Named constraints are referenceable in inter-body expressions.
4	**Single Spreadsheet**	All bodies share one Spreadsheet. Every alias in the Spreadsheet is available to every body. Do not create a separate Spreadsheet for each body — this makes the parametric update cycle much harder to manage.
5	**Loft in Part workbench**	The 3D Loft for the relief element is created in the Part workbench, not Part Design. After applying the Loft, use Part Boolean Union to combine it with the lid panel. This is the most reliable workflow in FreeCAD v1.1.0.
6	**Joinery geometry order**	Create the finger joint notch sketches after the panel outline sketch and Pad are complete. Sketch on the panel face, not on the XY plane. Apply Pocket for each set of notches.

Parametric Structure — Alias Checklist

The following alias list covers all the essential parameters for a typical Keepsake Memory Box. Adapt it for your specific design — you need a minimum of 8, but more is better.

Alias Name	Typical Value	Type	Controls
`BoxWidth`	150 mm	Input	External width of the box
`BoxDepth`	100 mm	Input	External depth (front-to-back)
`BoxHeight`	65 mm	Input	External height of the side panels
`PanelThick`	12 mm	Input	Material thickness — match to actual stock
`LidThick`	12 mm	Input	Lid material thickness
`FingerWidth`	16 mm	Input	Width of each finger in the joint
`Clearance`	0.15 mm	Input	Joint clearance — adjust after test cut
`InternalWidth`	BoxWidth – 2*PanelThick	Derived	Internal clear width
`InternalDepth`	BoxDepth – 2*PanelThick	Derived	Internal clear depth
`EngravingDepth`	0.8 mm	Input	V-bit engraving depth

`ReliefHeight`	20 mm	Input	Maximum height of 3D relief above panel surface
`TextHeight`	18 mm	Input	ShapeString text size

CAM Strategy — Certification Project

The certification project requires more CAM operations than any previous project. Planning the operation sequence before opening the CAM workbench prevents the most common CAM errors — incorrect operation ordering, wrong tool assignments, and missed geometry.

Operation	Tool	Target Geometry	Key Settings	Order Rule
Joinery pockets — Stock 1 panels	6 mm flat end mill	Finger joint notch faces on FrontPanel, BackPanel	Adaptive strategy if depth >8 mm. Step-down 1.5–2.0 mm. Clearance pass after main pocket.	Before profile cuts
Joinery pockets — Stock 2 panels	6 mm flat end mill	Finger joint notch faces on LeftSide, RightSide, Base	Same as above	Before profile cuts
Lid engraving — V-bit	60° V-bit	ShapeString edges, SVG Part edges, B-spline edges	Depth −0.8 mm. Feed 600–700 mm/min. Plunge 150–200 mm/min.	Before lid profile
3D Waterline roughing — lid	6–8 mm ball-nose	Full lid body surface	Step-down 2 mm. Stock-to-leave 0.3 mm. Feed 1200–1500 mm/min.	Before 3D finishing
3D Surface finishing — lid	3 mm ball-nose	Full lid body surface	Step-over 0.8–1.0 mm. Feed 800–1000 mm/min.	After roughing
Profile — all side panels	6 mm flat end mill	Outer profile edges of each panel	Full depth + 0.5 mm. Step-down 2 mm. Tabs: 4 per part. Outside cut.	After all pockets
Profile — lid	6 mm flat end mill	Outer lid profile	Full depth + 0.5 mm. Step-down 2 mm. Tabs. Outside cut.	After 3D CAM and engraving
Profile — base panel	6 mm flat end mill	Outer base profile	Full depth + 0.5 mm. Step-down 2 mm. Tabs.	After any pocket ops on base

Tip

Post-process all operations for a single stock piece into one G-code file. Do not create separate .nc files for each operation on the same stock — UGS manages the tool change pauses within a single file automatically. A separate .nc file for each operation means multiple load-and-send cycles, which increases the risk of losing XY position between operations.

TechDraw Requirements for Certification

The TechDraw engineering drawing is a mandatory submission item assessed on its independent legibility — can a competent machinist reproduce the box from the drawing alone, without access to the FreeCAD model? The checklist below defines every element required. Complete the checklist before exporting the PDF.

✓	Element	Certification Requirement
☐	**Views**	Front view (most informative face of the box)
☐	**Views**	Top view (lid or open interior)
☐	**Views**	Right-side or left-side view (showing material thickness and joint depth)
☐	**Views**	At least one section view (cutting through the box to show interior geometry — joint fingers, cavity dimensions)
☐	**Views**	Isometric or pictorial view (optional but recommended — makes the drawing immediately intelligible)
☐	**Dimensions**	Overall box width (external)
☐	**Dimensions**	Overall box height (external)
☐	**Dimensions**	Overall box depth / length (external)
☐	**Dimensions**	Material thickness (dimension one panel edge)
☐	**Dimensions**	Lid thickness
☐	**Dimensions**	Interior cavity width (internal clear dimension)
☐	**Dimensions**	Interior cavity height
☐	**Dimensions**	Finger width and depth (at least one finger joint dimensioned)
☐	**Dimensions**	Engraving depth
☐	**Dimensions**	3D relief maximum height
☐	**Dimensions**	Two further dimensions of your choice — relevant to the design intent
☐	**Title Block**	Drawing title and project name
☐	**Title Block**	Student name and date
☐	**Title Block**	Scale (e.g. 1:2 if box is large)
☐	**Title Block**	Material specification
☐	**Title Block**	Sheet number and total sheet count
☐	**Tolerances**	Joint clearance tolerance shown on at least one joint dimension (e.g. 15.00 ±0.15)
☐	**BOM**	Parts list table: Part Name / Material / Quantity / Thickness — minimum 4 rows

Section View Guidance

At least one section view is required. The most useful section for a box is a horizontal section cut through the mid-height of the assembled box — this shows the interior cavity dimensions, the joint fingers, and the panel thickness simultaneously in one view. To create a section view in TechDraw: TechDraw → Insert Section View, click the view you want to section, draw the section line through the target cut plane.

Submission Checklist

Complete this checklist before submitting. Every mandatory item must be checked. Submit the physical piece and the digital package to the instructor simultaneously — a partial submission cannot be assessed.

REPRODUCIBLE — SUBMISSION CHECKLIST

✓	Status	Item	Requirement
☐	**MANDATORY**	**Physical piece — the Keepsake Memory Box**	The finished, assembled lidded box. Must hold together under its own weight without glue. Lid must open and close. Assessed in person.
☐	**MANDATORY**	**FreeCAD model file (.FCStd)**	Final design file. All bodies named. Spreadsheet present with minimum 8 named aliases. All critical dimensions formula-driven.
☐	**MANDATORY**	**G-code file(s) (.nc)**	All .nc files used to cut the box. Named with version numbers. At minimum: one file for joinery panels, one for 3D relief surface.
☐	**MANDATORY**	**TechDraw drawing (PDF)**	Exported from FreeCAD TechDraw workbench. Minimum 4 orthographic views, 12 dimensions, complete title block, BOM, joint tolerances.
☐	**MANDATORY**	**Job Setup Sheet(s) (PDF or paper)**	One completed Job Setup Sheet per cutting session. All fields completed: material, tools, zeros, feeds, results, issues.
☐	**MANDATORY**	**Photograph(s)**	Minimum: one full assembled box, one raking-light detail of engraving, one raking-light detail of 3D relief. Labelled by filename.
☐	**MANDATORY**	**Project reflection document**	Minimum 600 words. Addresses: design decisions and reasoning, challenges encountered and resolved, Bloom's self-assessment, what you would do differently.

☐	**MANDATORY**	**Parametric demonstration record**	Either a screenshot sequence or a brief written record showing: before Refresh (original dimensions) → after parameter change → after Refresh (updated model). Confirms parametric capability.
☐	**Recommended**	**Gear or mechanical element**	A gear-shaped decorative inlay, a gear-mechanism lid, or another mechanical element integrated into the design.
☐	**Recommended**	**French cleat or keyhole mount**	A wall-mounting feature on the base of the box.
☐	**Recommended**	**Assembly workbench file**	An Assembly workbench visualisation file showing all panels in assembled position.
☐	**Recommended**	**Size variant documentation**	A second parameter set (different size) demonstrated in a TechDraw drawing or screenshot.
☐	**Recommended**	**Polar Pattern element**	At least one panel featuring a decorative element created with Polar Pattern.

Reflection Document Guide

The reflection document is 10% of the certification mark. A distinction-level reflection is not about length — it is about the depth and honesty of critical analysis. The guide below breaks the reflection into five sections with a suggested word count and example opening sentence for each. Use it as a structure, not a script.

Section	Suggested Length	What to Cover	Example Opening Sentence
Introduction	60–80 words	Briefly describe the finished project: what the box looks like, what makes it distinctive, and which mandatory requirements it addresses.	'The Keepsake Memory Box I designed is a finger-jointed pine box with a lidded compartment, a bas-relief leaf carving on the lid, and a Celtic knotwork engraving sourced from an SVG file. The box demonstrates parametric design (12 spreadsheet aliases), CNC joinery with 0.2 mm clearance, 3D Waterline and Surface CAM, and a complete TechDraw engineering drawing.'
Design decisions	150–200 words	Explain 3–5 design decisions you made during the project, and why you made them. Not what you did — why. Include at least one decision about joinery method, one about the decorative elements, and one about parametric structure.	'I chose finger joints over dado-and-rabbet because finger joints provide greater mechanical strength for a box that will be opened frequently, and because the visual rhythm of the fingers on the exterior edges was an intentional aesthetic element of the design.'

Challenges	150–200 words	Describe at least two specific challenges encountered during the project and how each was resolved. A challenge that was not resolved is worth describing too — explain why it was not resolved and what you would do differently.	'The most significant technical challenge was the Z-height mismatch on the imported SVG knotwork. The Draft Upgrade placed the converted Part objects at Z=1.47 mm, which caused the engraving CAM operation to generate rapids rather than cutting moves. I identified this through the CAM Simulator and used Draft Move to correct all objects to Z=0.'
Bloom's self-assessment	100–150 words	Identify which Bloom's Taxonomy levels your work demonstrated during the project, and give a specific example of each level. Be honest — if you found yourself working mostly at Apply level, say so.	'The parametric design and joinery geometry primarily operated at the Apply level — I was executing skills practised throughout the course. The 3D relief design decisions operated at Evaluate level — I assessed three different Loft form options (leaf, dome, and wave) before selecting the leaf because its tapered silhouette complemented the box proportions.'
What I would do differently	100–120 words	Identify 2–3 specific things you would change if you were starting the project again. These should be substantive — not 'I would plan more', but specific technical or workflow changes.	'I would design the 3D relief at 20 mm height rather than 28 mm. The 28 mm height required a Waterline roughing step-down of 1.5 mm rather than 2 mm to maintain surface quality, which increased the roughing runtime from an estimated 18 minutes to 31 minutes and nearly overran Day 54.'
Conclusion	50–80 words	A brief statement of what you are most satisfied with, and the most important single thing you learned from the project that you did not fully understand before.	'The most satisfying outcome was seeing the 3D relief carving emerge from the roughing staircase after the ball-nose finishing pass — a transformation that made the technical complexity of the week feel worthwhile. The most significant new understanding was the relationship between Loft section positioning and the final form silhouette — something I could not fully predict until I ran the Loft and saw the result.'

Bloom's Self-Assessment — Getting It Right

Every reflection must include a Bloom's Taxonomy self-assessment. The following table maps the six levels to specific certification project activities — use it to identify which level each part of your work operated at.

Bloom's Level	Certification Project Activities at This Level
Remember	Recalling constraint types, G-code commands, tool names, workbench locations
Understand	Explaining why dogbones are needed, describing what a Waterline operation does, summarising the parametric update cycle
Apply	Drawing fully constrained sketches, setting up CAM operations, running the UGS workflow from zero to finished cut
Analyse	Comparing finger joint vs dado joinery for this specific box, analysing why a CAM simulation result differs from expected, breaking down the model tree to identify a parametric error
Evaluate	Judging whether the 3D relief height is achievable in available machining time, assessing the fit quality of a test joint against the designed clearance, justifying the choice of Adaptive over ZigZag strategy
Create	Designing the box (original form, original proportions, original decorative elements), planning the complete CAM sequence independently, producing the complete submission package without external guidance

Certificate of Completion

Successful completion of this course and certification project results in the award of the Certificate of Completion in FreeCAD v1.1.0 for CNC Makers. The certificate recognises that the holder has demonstrated, through independent project work assessed against published criteria, the ability to:

- Design original parametric 3D models in FreeCAD v1.1.0 using the Part Design, Sketcher, Spreadsheet, and Part workbenches
- Generate professional-grade CAM toolpaths for a GRBL-controlled CNC router using FreeCAD's CAM workbench
- Operate the CNC 4040 Pro Max safely and competently using Universal G-code Sender
- Design and cut CNC joinery with correct tolerance and geometry
- Apply 3D CAM strategies for organic surface machining
- Produce engineering documentation using FreeCAD TechDraw
- Reflect critically on design decisions and machining outcomes in a written professional format

🏅 Certificate of Completion

FreeCAD v1.1.0 for CNC Makers: A 10-Week Foundation Course

is awarded to students who:

- Successfully complete all 10 weekly projects (Projects 1–10)
- Complete the Certification Project (the Keepsake Memory Box) with an overall mark of 70% or above
- Achieve a minimum of 60% on the Certification Project alone
- Submit all mandatory documentation components: FreeCAD file, G-code, TechDraw drawing, Job Setup Sheets, photographs, and reflection document
- Demonstrate the parametric update live to the instructor during the certification assessment
- Produce a physical piece that assembles correctly and meets the minimum technical requirements

Students who do not meet the passing standard on the first attempt are permitted one re-submission within four weeks of the original deadline. The re-submission must address all assessment feedback items identified by the instructor.

The certificate is signed by the subject coordinator and dated on the day of successful submission assessment. It represents completion of a 55-day, 10-week practical course operating at the Understand through Create levels of Bloom's Revised Taxonomy.

A Final Note

You have learned to take a blank FreeCAD document and produce a physical piece of work — something that did not exist before you designed it. That is not a small thing. The skills in this textbook — parametric design, CNC machining, precision joinery, 3D surface modelling, engineering documentation — are practised professionally by designers, fabricators, and craftspeople worldwide. You now share those skills.

The Keepsake Memory Box is not the end of that journey. It is the first piece you have made entirely on your own, without a guide. Every piece you make after this will be easier — not because the work gets simpler, but because you will approach it with the knowledge, the judgement, and the confidence that this course has built.

Make something good.

APPENDIX

A

FreeCAD v1.1.0 for CNC Makers — Student Edition

FreeCAD v1.1.0 Keyboard Shortcuts

Complete reference — all workbenches used in this course

This appendix lists the keyboard shortcuts for FreeCAD v1.1.0 relevant to this course. Shortcuts are organised by workbench and context. The most frequently used shortcuts in each section are shown first.

📋 **Note**

All shortcuts listed here are the FreeCAD v1.1.0 defaults. If a shortcut does not work as shown, it may have been reassigned on your system. To view or customise shortcuts: Tools → Customize → Keyboard tab. The default shortcut set is restored by clicking 'Reset all' in that dialog.

File & Edit — *All Workbenches*

`Ctrl + S`	Save document	`Ctrl + Shift + S`	Save As
`Ctrl + Z`	Undo last action	`Ctrl + A`	Select all objects
`Ctrl + Y`	Redo (also: Ctrl + Shift + Z)	`Del`	Delete selected object
`Ctrl + N`	New document	`Space`	Toggle visibility of selection
`Ctrl + O`	Open document	`Ctrl + Q`	Quit FreeCAD
`Ctrl + W`	Close active document	`F5`	Refresh / recalculate model

3D View Navigation — *All Workbenches — CAD Navigation Mode*

`Middle Mouse Button + Drag`	Orbit — rotate the view around the model
`Shift + Middle Mouse + Drag`	Pan — move the view horizontally and vertically
`Scroll Wheel`	Zoom in / zoom out (centres on cursor position)
`Right Mouse Button + Drag`	Tilt — tilts the view (less used; use orbit instead)
`Left Click`	Select a face, edge, or point
`Ctrl + Left Click`	Add to selection (multi-select for Chamfer, CAM, etc.)
`V then F`	Fit All — zoom to show the entire model (most useful when model is off-screen)
`V then B`	Box zoom — draw a selection box to zoom into a region

Standard Views — Numpad shortcuts:

`Numpad 0`	Default isometric / home view	`Numpad 2`	Rotate view down 10°
`Numpad 1`	Front view (looking along Y axis)	`Numpad 4`	Rotate view left 10°
`Numpad 3`	Right-side view	`Numpad 6`	Rotate view right 10°

`Numpad 7`	Top view (looking down Z axis)	`Numpad 8`	Rotate view up 10°
`Numpad 5`	Toggle orthographic / perspective projection	`Numpad 9`	Rear view

Tip

If you orbit too aggressively and lose the model, press V then F (Fit All) or Numpad 0 to reset. Both work from any workbench. This is the single most useful shortcut to memorise in Week 1.

Sketcher — Drawing Tools — *Inside Active Sketch*

Sketcher drawing tools use a two-key sequence: press G to enter drawing mode, then the letter for the tool. This is the FreeCAD v1.1.0 default — if your Sketcher has direct single-key shortcuts, those work too.

`G, L`	Line / Polyline	`G, O`	Circle by 3 points
`G, R`	Rectangle (corner to corner)	`G, S`	Slot (oblong / rounded rectangle)
`G, C`	Circle by centre and radius	`G, F`	Fillet between two lines
`G, A`	Arc by 3 points	`G, X`	External geometry reference
`G, E`	Ellipse by centre	`G, I`	Carbon copy (duplicate sketch)
`G, B`	B-spline (control point mode)	`G, T`	Trim edge (remove part of a line/arc)
`G, P`	Point	`Q`	Close sketch (exit Sketcher)

Sketcher — Constraints — *Inside Active Sketch*

Apply constraints by first selecting the target geometry (click the line, arc, or points), then pressing the shortcut. For constraints that require two elements (Equal, Symmetric, Coincident), Ctrl+click both before pressing the shortcut.

Geometric Constraints (remove DOF without entering a value):

`H`	Constrain Horizontal (line parallel to X)	`N`	Perpendicular (90° between two lines)
`V`	Constrain Vertical (line parallel to Y)	`P`	Parallel (two lines parallel to each other)
`B`	Block / Fix (lock element in place)	`C`	Coincident (merge two points)
`Q`	Equal (same length or radius as another)	`I`	Point on object (point lies on line/arc)
`S`	Symmetric about a line or point	`E`	Equal radius (arcs/circles the same size)
`T`	Tangent (smooth join between line and arc)	`X`	Constrain to X or Y axis symmetry

Dimensional Constraints (require a value to be entered):

`Shift + D`	Fix distance (between two points or line length)	`Shift + P`	Lock point to specific X,Y coordinates
`Shift + H`	Horizontal distance (X distance between two points)	`Shift + B`	Constrain internal angle of B-spline
`Shift + V`	Vertical distance (Y distance between two points)	`Shift + K`	Constrain knot of B-spline
`Shift + R`	Radius (arc or circle radius)		
`Shift + A`	Angle (between two lines, or line and axis)		

Sketcher view controls (inside active sketch):

`Ctrl + Space`	Toggle grid visibility	`Ctrl + M`	Toggle constraint markers visibility
`F`	Zoom selected element to fit view	`Ctrl + Shift + M`	Toggle redundant constraints visibility
`Esc`	Cancel current drawing tool	`Q`	Close sketch / exit Sketcher

Part Design — *3D Modelling Workbench*

Most Part Design operations are accessed from the toolbar or menu rather than keyboard shortcuts. The shortcuts below cover the most useful general-purpose actions available in Part Design.

`Ctrl + Enter`	Close Sketch and return to Part Design	`Ctrl + D`	Toggle display mode for selected object
`Shift + Esc`	Close Sketch (alternative)	`Ctrl + Shift + -`	Decrease transparency of selected object
`Space`	Toggle visibility of selected Body or feature	`Ctrl + Shift + +`	Increase transparency of selected object
`Ctrl + Shift + P`	Create Datum Plane (if toolbar not visible)	`Ctrl + G`	Group selected items (for organisation)

> 📋 **Note**
> There are no default keyboard shortcuts for Pad, Pocket, Chamfer, Fillet, or Pattern operations in Part Design. These are toolbar and menu operations. To add shortcuts for frequently used Part Design tools, go to Tools → Customize → Keyboard and search by tool name.

Spreadsheet Workbench — *Parametric Design*

`Enter`	Confirm cell entry and move to cell below	`Del`	Delete cell content
`Tab`	Confirm entry and move to cell right	`Ctrl + C`	Copy selected cells
`Shift + Enter`	Confirm entry and move to cell above	`Ctrl + V`	Paste into selected cells

`Shift + Tab`	Confirm entry and move to cell left	`Ctrl + X`	Cut selected cells
`Ctrl + Enter`	Confirm entry and stay in current cell	`Ctrl + Z`	Undo last cell edit
`F2`	Edit selected cell (enter edit mode)	`Ctrl + Shift + R`	Refresh/recalculate all expressions

CAM Workbench — *Toolpath Generation*

The CAM workbench is primarily menu and toolbar driven. The shortcuts below are the most useful general actions while working in CAM.

`Space`	Toggle visibility of selected operation	`P`	Post-process (if Jobs panel is active)
`Ctrl + A`	Select all CAM operations in the Job	`Ctrl + Shift + S`	Open CAM Simulator
`Del`	Delete selected operation	`Ctrl + Z`	Undo last operation change

Draft Workbench — *SVG Import & Object Manipulation*

`W`	Draft Wire (line with multiple points)	`M`	Draft Move
`A`	Draft Arc	`R`	Draft Rotate
`C`	Draft Circle	`S`	Snapping toggle on/off
`B`	Draft BSpline	`Shift`	Constrain movement to axis (hold while dragging)
`P`	Draft Polygon	`Ctrl`	Snap override (hold while placing point)

TechDraw Workbench — *Engineering Documentation*

`Ctrl + Shift + D`	Open TechDraw dimensioning mode	`Ctrl + Shift + A`	Insert annotation
`Shift + F`	Select mode (click to select/deselect)	`Ctrl + Shift + E`	Export page as PDF

Model Tree — Selection & Navigation — *All Workbenches*

`Left Click`	Select object in tree or 3D view	`Space`	Toggle visibility of selected item
`Ctrl + Click`	Add to selection (multi-select)	`Ctrl + G`	Group selection into a Part container
`Shift + Click`	Range select (from last selection to clicked)	`Ctrl + Shift + G`	Ungroup
`F2`	Rename selected item in tree	`Right Click`	Context menu for selected item

10 Shortcuts to Memorise First — *Master these before Week 1*

`Ctrl + S`	Save — after every significant change	`Shift + D`	Fix Distance constraint in Sketcher

`Ctrl + Z`	Undo — your most reliable safety net	`G, R / G, C`	Rectangle / Circle in Sketcher
`V then F`	Fit All — restores view when model is off-screen	`Q`	Equal constraint (also closes the sketch)
`Space`	Toggle visibility of selected element	`Numpad 0`	Isometric view — best general-purpose 3D view
`H / V`	Horizontal / Vertical constraint in Sketcher	`Del`	Delete selected object, feature, or constraint

APPENDIX
B

FreeCAD v1.1.0 for CNC Makers — Student Edition

Feeds & Speeds Reference

CNC 4040 Pro Max — timber and MDF — verified starting values for all course operations

All values in this appendix are verified starting points for the CNC 4040 Pro Max cutting the materials used in this course. They match the values introduced in the chapter tables throughout the textbook. Use them as your default starting point and adjust based on what you hear and see during cutting. No table replaces the sound of a healthy cut.

> ⚠ **Important**
> These values assume sharp, new bits of the stated diameter. Dull bits require lower feed rates and shorter tool life results in worse surface quality regardless of settings. Replace bits at the first sign of burning, chatter, or increased cutting force. MDF produces very fine carcinogenic dust — always wear your P2 dust mask.

> **Tip**
> Always run the CNC 4040 Pro Max feed rate override (FRO) at 100% for these reference values. If you need to reduce FRO below 70% to achieve a clean cut, the table values may be too aggressive for your specific material batch or bit condition.

Material Colour Key	
MDF	Medium-density fibreboard. Consistent density, no grain. Easiest to cut. Fine dust — always wear P2 mask.
Pine/Cedar	Soft timber — pine, cedar, radiata. Easy to machine but grain can tear at high feed rates. Watch for resin pockets.
Hardwood	Oak, jarrah, spotted gum, Victorian ash. Higher density requires lower feed rates. Grain direction matters for surface finish.
Dense HW	Ironbark, merbau, and similar very dense species. Reduce step-over significantly. Tool wear is noticeably higher.
Plywood/LVL	Birch ply, marine ply, LVL. Variable across plies — some passes cut with grain, some against. Use middle-range settings.

Understanding the Parameters

Feed rate (mm/min)	Horizontal cutting speed. Too fast: chatter and rough surface. Too slow: burning and poor finish. The primary variable you adjust during cutting via feed rate override.
Plunge rate (mm/min)	Vertical entry speed for each new pass. Always lower than feed rate — typically 25–35% of feed rate. Too fast causes bit breakage on entry; too slow adds unnecessary time.
Step-down (mm)	Depth of material removed per pass. More passes = less force per pass = better surface and less tool stress. Rule of thumb: step-down ≤ 0.5× tool diameter in hardwood, ≤ 0.75× in MDF.

Step-over (% or mm)	Horizontal distance between adjacent passes. Lower = more overlap = smoother floor, more passes. Higher = faster clearing, possible witness marks. 60–70% is the practical sweet spot for most work.
Spindle (RPM)	Cutting speed of the bit. Higher RPM with lower feed rate causes burning. Lower RPM with high feed rate causes chip load problems. Match RPM to material density — harder materials typically need higher RPM.

V-Bit Engraving — *60° V-Bit*

V-Bit Engraving (60°) *Following 2D profile edges at set depth — text, logos, decorative lines*

Material	Feed (mm/min)	Plunge (mm/min)	Step-Down	Step-Over	RPM	Notes
MDF	**800**	200	0.8 mm	N/A	15,000	*Standard depth. Deeper gives wider V — try 1.0mm for bolder text.*
Pine/Cedar	**700**	180	0.6 mm	N/A	15,000	*Soft wood tears at excessive depth. 0.6mm gives clean edges.*
Hardwood	**600**	160	0.5 mm	N/A	18,000	*Reduce depth if wood has strong grain running across letterforms.*
Plywood	**700**	180	0.6 mm	N/A	15,000	*Ply cross-grain layers can cause tear-out. 0.6mm is conservative and safe.*

Profile Cutting — *Outer and inner contour cuts — cut-outs and edge profiles*

6mm Flat End Mill — Profile *Full-depth cut-out of parts with tabs*

Material	Feed (mm/min)	Plunge (mm/min)	Step-Down	Step-Over	RPM	Notes
MDF	**1000**	300	1.5 mm	N/A	15,000	*Standard. Add 0.5mm to Final Depth to ensure clean cut-through.*
Pine/Cedar	**850**	240	1.2 mm	N/A	15,000	*Reduce step-down for knot-heavy sections.*

Hardwood	700	200	1.5 mm	N/A	18,000	*Hardwood tolerates 1.5mm step-down at reduced feed.*
Dense HW	550	160	1.0 mm	N/A	21,000	*Very hard species — reduce step-down and feed.*
Plywood/LVL	900	260	1.5 mm	N/A	15,000	*LVL and ply are consistent; push feed slightly.*

3mm Flat End Mill — Profile *Fine detail profiles — gear teeth, small parts*

Material	Feed (mm/min)	Plunge (mm/min)	Step-Down	Step-Over	RPM	Notes
MDF	700	200	1.0 mm	N/A	21,000	*3mm bits are fragile — never exceed 1mm step-down in MDF.*
Pine/Cedar	600	170	0.8 mm	N/A	21,000	
Hardwood	500	160	0.8 mm	N/A	24,000	*Add roughing pass (+0.25mm offset) before finishing pass.*

Pocket Clearing — ZigZag Strategy — *Parallel passes — suitable for shallow pockets in soft material*

6mm Flat End Mill — Pocket (ZigZag) *Best for pockets under 5mm depth in MDF or softwood*

Material	Feed (mm/min)	Plunge (mm/min)	Step-Down	Step-Over	RPM	Notes
MDF	1100	320	1.5 mm	65%	15,000	*Standard pocket — finish pass 0.1mm optional.*
Pine/Cedar	900	260	1.2 mm	65%	15,000	*Reduce step-down for knots or grain reversals.*
Hardwood	700	200	1.0 mm	60%	18,000	*ZigZag acceptable for shallow hardwood pockets (<5mm).*
Plywood/LVL	900	260	1.2 mm	65%	15,000	

Pocket Clearing — Adaptive Strategy — *Arc-motion clearing — required for deep pockets and hardwood*

6mm Flat End Mill — Pocket (Adaptive) *Recommended for all pockets deeper than 5mm and all hardwood*

Material	Feed (mm/min)	Plunge (mm/min)	Step-Down	Step-Over	RPM	Notes
MDF	**1100**	300	4.0 mm	65–70%	15,000	*Adaptive allows much deeper step-down in soft material.*
Pine/Cedar	**900**	250	3.0 mm	60–65%	15,000	
Hardwood	**800**	220	3.0 mm	50–60%	18,000	*Constant chip load — greatly extends bit life in hardwood.*
Dense HW	**650**	180	2.0 mm	40–50%	21,000	*Reduce step-over significantly for very hard species.*
Plywood/LVL	**950**	260	3.0 mm	55–65%	15,000	

Precision Profiling — Roughing + Finishing — *Gear teeth and toleranced parts requiring a two-pass strategy*

3mm End Mill — Roughing Pass *+0.25mm profile offset, leaves skin for finishing*

Material	Feed (mm/min)	Plunge (mm/min)	Step-Down	Step-Over	RPM	Notes
MDF	**700**	200	1.0 mm	N/A	21,000	*Extra Offset = 0.25mm*
Hardwood	**500**	150	0.8 mm	N/A	24,000	*Gear profiles — extra offset 0.25mm. Listen for chatter.*

3mm End Mill — Finishing Pass *Zero offset, full-depth single pass, slow feed for clean surface*

Material	Feed (mm/min)	Plunge (mm/min)	Step-Down	Step-Over	RPM	Notes
MDF	**400**	180	Full depth	N/A	24,000	*Single pass removes 0.25mm skin. Feed 55% of roughing.*
Hardwood	**300**	140	Full depth	N/A	24,000	*Slow feed critical for gear tooth surface quality.*

3D CAM — Waterline Roughing — *Ball-nose roughing — horizontal slice passes*

6mm Ball-Nose — Waterline Roughing *Bulk removal before surface finishing. Stock-to-leave = 0.3mm*

Material	Feed (mm/min)	Plunge (mm/min)	Step-Down	Step-Over	RPM	Notes
Pine/Cedar	**1400**	420	2.0 mm	N/A	15,000	*3D relief carving — full width horizontal passes.*
Hardwood	**1200**	400	2.0 mm	N/A	18,000	*From Ch14. Stock-to-leave 0.3mm for finishing pass.*

8mm Ball-Nose — Waterline Roughing *For larger relief panels (200mm+) where runtime reduction matters*

Material	Feed (mm/min)	Plunge (mm/min)	Step-Down	Step-Over	RPM	Notes
Pine/Cedar	**1600**	450	2.5 mm	N/A	15,000	*Faster bulk removal for large relief panels.*
Hardwood	**1400**	400	2.0 mm	N/A	18,000	*Reduce step-down if surface shows excessive staircase.*

3D CAM — Surface Finishing — *Ball-nose finishing — follows surface contours for smooth result*

3mm Ball-Nose — Surface Finishing *Step-over controls cusp height — see Ch16 for cusp height formula*

Material	Feed (mm/min)	Plunge (mm/min)	Step-Down	Step-Over	RPM	Notes
Pine/Cedar	**900**	280	N/A	0.8 mm	15,000	*0.8mm step-over ≈ 0.027mm cusp height. Display quality.*
Hardwood	**800**	250	N/A	0.8 mm	18,000	*From Ch14. Standard course finishing setting.*
Hardwood (fine)	**600**	200	N/A	0.5 mm	18,000	*For certification project — lower cusp, longer runtime.*

6mm Ball-Nose — Surface Finishing *Faster finish for larger relief areas with gentler detail*

Material	Feed (mm/min)	Plunge (mm/min)	Step-Down	Step-Over	RPM	Notes
Pine/Cedar	**1100**	330	N/A	1.2 mm	15,000	*1.2mm step-over ≈ 0.060mm cusp height. Good general finish.*
Hardwood	**900**	280	N/A	1.0 mm	18,000	

Adjusting from Reference Values

The table values are the correct starting point. Real-world cutting varies by material batch, humidity, bit wear, and machine condition. Use this table to guide deliberate adjustments rather than guessing.

Scenario	Feed/Step-Down Change	Additional Adjustment
Deeper cut needed (thicker material)	Reduce step-down to keep cutting force constant.	*Reduce feed 10–15% for extra tool wear from deeper passes.*
Cutting harder material than table	Reduce feed 20–30%. Reduce step-down 20%.	*Use Adaptive strategy for pockets. Increase RPM slightly.*
Cutting softer material than table	Increase feed 10–20%. MDF tolerates higher speeds.	*Reduce RPM slightly to avoid burning in soft materials.*
Very small detail / tight radius	Reduce feed to 70% of reference. Reduce step-down 30%.	*Use 3mm EM or smaller. Increase RPM proportionally.*
New bit, unknown quality	Start at 80% of reference values. Listen carefully.	*Increase gradually by 5% per pass if cutting sounds clean.*
Bit has been used >2 hours in hardwood	Reduce feed 15%. Dull bit requires lower chip load.	*Replace bit if burn marks or surface quality drops despite adjustment.*

Troubleshooting — What the Cut is Telling You

The sound and appearance of the cut provides real-time feedback that no simulation can replicate. A healthy cut sounds like a consistent hiss or whir — smooth, even, without vibration. Any deviation from that sound is a signal to adjust.

Symptom	Likely Cause	Adjustment
Chatter — rhythmic vibration/rattle	Feed too high, spindle too low, or bit deflecting	*Reduce feed by 20%. Check collet is tight. Check bit is not excessively long.*
Burning/scorching	Feed too low, spindle too high, or dull bit	*Increase feed by 15%. Check bit sharpness. Reduce spindle speed slightly.*
Rough, fuzzy surface (timber)	Feed too high causing tear-out, or cutting against grain	*Reduce feed 10–15%. Try a downcut spiral bit for top-face quality.*
Rough, glazed surface (MDF)	Spindle too high, bit rubbing rather than cutting	*Reduce spindle speed. Increase feed rate slightly.*
Bit breakage	Too-aggressive step-down, plunge too fast, or dull bit	*Reduce step-down to 0.5× bit diameter max. Slow plunge to 25% of feed.*
Poor dimensional accuracy	Bit deflection, loose collet, or wrong tool diameter in CAM	*Tighten collet. Verify exact tool diameter in CAM Tool Library. Add finish pass.*

Tabs breaking during profile	Feed too high on final profile pass, or tabs too thin	*Reduce feed 20% on profile. Set tab height ≥3 mm, width ≥4 mm.*
3D surface shows staircase after finishing	Step-down too large in Waterline roughing	*Reduce Waterline step-down to 1.5 mm. Ensure stock-to-leave is 0.3 mm.*
Gear teeth chipping	Feed too high for small 3mm bit in hardwood	*Reduce feed to 400–450 mm/min. Use roughing + finishing pass strategy.*

> **Tip**
> During any new cut, use UGS's Feed Rate Override (FRO) to start at 80% and increase to 100% only once the cut sounds clean. This takes 10 seconds and protects against aggressive reference values in unusual material batches.

APPENDIX

C

FreeCAD v1.1.0 for CNC Makers — Student Edition

CNC 4040 Pro Max — Technical Specifications & Maintenance

Machine specifications, collet guide, maintenance schedule, and reproducible pre-cut checklist

This appendix is the technical reference for the CNC 4040 Pro Max — the machine used throughout this course. The specifications table provides the definitive values for all machine parameters. The maintenance schedule defines the habits that keep the machine cutting accurately and extend component life. The pre-cut checklist at the end of this appendix is a reproducible form: print it, laminate it, and keep it at the machine.

Technical Specifications

Specification	Value	Specification	Value
Machine type	Moving-gantry desktop CNC router	**Controller firmware**	GRBL 1.1h
Work area (X × Y)	400 × 400 mm	**Spindle collet**	ER11 — accepts 1–7 mm shanks
Z-axis travel	Approximately 70–80 mm	**Computer connection**	USB (CH340 chip — Windows may need driver)
Frame	Aluminium extrusion — modular, rigid	**Baud rate**	115200
Work bed	T-slot aluminium extrusion. M5/M6 T-nuts for clamping	**Home position**	Back-right-top (X max, Y max, Z max)
Axes	3 (X, Y, Z) — stepper motor driven	**Supported materials**	Timber, MDF, plywood, soft plastics, aluminium (with correct feeds)

Axis Layout and Movement Directions

X axis	Left ↔ Right across the machine bed. Positive X moves the tool to the right. The X rail runs across the gantry bridge.
Y axis	Forward ↔ Backward. Positive Y moves the tool away from the operator (toward the back of the machine). The gantry moves in Y.
Z axis	Up ↕ Down. Positive Z raises the spindle (away from material). Negative Z lowers into material. Setting Z zero correctly is the single most critical setup step in any job.

Machine Components — Quick Reference

Frame	Rigid aluminium extrusion structure holding all components. Defines machine rigidity and accuracy.
Gantry	The bridge assembly carrying the Z-axis and spindle. Moves in the Y axis (forward/backward).
X-axis rail	The horizontal rail on the gantry along which the Z-carriage moves left/right.
Y-axis rails	Two parallel rails (one each side of the bed) along which the gantry travels.
Z-axis assembly	The vertical component that moves the spindle up/down. Controls cutting depth.
Spindle	Electric motor spinning the cutting bit. ER11 collet system. Maximum RPM: see machine label.
ER11 collet	Tool-holding chuck accepting 1–7 mm shanks. Must exactly match the bit's shank diameter.
Stepper motors	Three motors (one per axis) providing precise incremental movement. Controlled by GRBL.
Control board	Electronics running GRBL firmware. Interprets G-code, drives stepper motor drivers.
Limit switches	One per axis at the home position. Triggered during homing to establish machine coordinates.
T-slot table	Aluminium extrusion work surface with T-slots accepting M5/M6 T-nuts for clamping.
Drag chain	Cable management conduit protecting wiring on moving axes. Inspect weekly for chafing.
GT2 belts	Timing belts driving X and Y axes. Must be correctly tensioned — neither floppy nor over-tight.
Lead screw	Threaded rod driving Z-axis movement. Requires periodic PTFE lubrication (not thick grease).
Spoilboard	The sacrificial material layer over the T-slot table. Provides a flat cut-through surface.
E-stop button	Physical emergency stop. Cuts power to all motors and spindle immediately when pressed.

ER11 Collet Size Reference

The ER11 collet system uses separate collet inserts for each bit shank diameter. Always use the correct collet size — a collet that is even 0.5 mm too large will not hold the bit securely, will cause runout (vibration), and risks bit ejection during cutting.

Collet Size	Shank Dia.	Bits That Fit	Used In This Course
1/8" (3.175 mm)	**3.175 mm**	Engraving V-bits (60°, 90°), 1/8" upcut/downcut spirals	*Projects 1–5, 7 — engraving and fine detail*
6 mm	**6 mm**	6 mm flat end mills (2-flute), 6 mm ball-nose bits	*Projects 1–11 — profiling, pocket clearing, 3D roughing*

3 mm	**3 mm**	3 mm flat end mills (2-flute), 3 mm ball-nose bits	*Projects 4, 7, 8, 9 — fine profiles, gear teeth, 3D finishing*
4 mm	**4 mm**	4 mm flat end mills (occasional use)	*Not used in standard course — note for future reference*
3 mm → 6 mm	**ER11 range**	ER11 accepts 1–7 mm shanks (with matching collet size)	*Always match the collet to the shank diameter exactly — do not use the wrong size*

> **Safety**
> Never use a 6 mm collet with a 3.175 mm (1/8") bit, or vice versa. An undersized shank in an oversized collet is not gripped securely. At cutting speeds, an improperly seated bit can eject from the collet — a serious safety hazard. Always match collet to shank exactly.

> **Tip**
> To change a collet: loosen the collet nut while holding the spindle shaft with the second spanner. The collet will drop out with the nut. Insert the new collet (it clicks into the nut), thread the nut onto the spindle, insert the bit to at least 2/3 of the collet depth, and tighten firmly with both spanners. Never tighten with only one spanner — you risk damaging the spindle shaft threads.

Workspace Requirements

Clearance	Minimum 600 mm clearance on all four sides of the machine for safe access during operation and maintenance.
Dust extraction	CNC routing produces significant fine dust — especially from MDF (carcinogenic). A dust extraction hose connected to the spindle dust shoe is strongly recommended. At minimum, use a P2 dust mask for every session.
Lighting	Good overhead lighting over the full cutting area. You must be able to clearly see the workpiece and spindle during any cut.
Power	Stable mains power. Avoid power boards shared with other high-draw equipment. Power fluctuations during a job will cause GRBL to reset and lose position.
Ventilation	Adequate ventilation to prevent fine dust accumulation in the air. Open windows or use a ventilation fan pointed away from the operator.
Cable management	USB cable and power cord secured so they cannot be snagged or pulled during a job. The most common non-technical cause of job interruptions is a snagged cable.
Computer placement	Computer screen visible from the machine position without turning away from the cut. You must be able to monitor UGS while watching the spindle.
Emergency access	The physical e-stop must be reachable from your normal operating position without stepping toward the machine. Never stand directly over the spindle during cutting.

Maintenance Schedule

Regular maintenance is not optional — it directly affects cut quality and dimensional accuracy. A machine with dirty or under-lubricated rails will exhibit inconsistent positioning, increased chatter, and accelerated component wear. The schedule below defines the minimum maintenance intervals.

Frequency	Task	Action Required
After every session	Clean	Remove all chips and dust from the T-slot table, X/Y/Z rails, and work area using a soft brush and vacuum. MDF dust is especially fine and abrasive — never blow it with compressed air into the rails.
After every session	Inspect	Check the cutting area and spoilboard for loose clamps, embedded fasteners, or damage that could cause problems in the next session.
Weekly	Lubricate	Apply a drop of light machine oil (3-in-1 or equivalent) to each linear rail. Run the axis back and forth the full length after oiling to distribute. Wipe off excess with a cloth.
Weekly	Belt tension	Gently press each GT2 belt at its midpoint. It should feel firm, not floppy or drum-tight. A loose belt causes axis backlash; an over-tight belt causes bearing wear. Adjust by loosening the motor mount screws if needed.
Weekly	Lead screw	Check the Z-axis lead screw for visible debris. Apply a small amount of dry PTFE lubricant to the lead screw thread. Do not use thick grease — it collects wood dust and clogs.
Monthly	V-wheels	Inspect the V-wheels on the gantry and Z-carriage. They should roll smoothly without lateral wobble. Tighten the eccentric nut (the off-centre nut) on any loose V-wheel. Overtightening causes flat spots — adjust until the wheel rolls smoothly under light finger pressure.
Monthly	Wiring	Inspect all cable chains and wiring runs for chafing, kinking, or wear. Pay particular attention to the drag chain on the Y axis — it flexes with every job.
Monthly	Fasteners	Check all structural bolts and motor mount screws with a hex key. Vibration can loosen fasteners over time, particularly on the gantry cross-beam.
As needed	Collets	Inspect ER11 collets for visible wear, scoring, or deformation of the inner bore. A worn collet grips the bit unevenly and causes runout (vibration). Replace if bore shows wear or if the bit shifts during tightening.
As needed	Spindle brush	On brushed DC spindles: inspect the carbon brushes for wear. Signs of worn brushes include excessive sparking, reduced spindle speed at normal voltage, or spindle noise. Replace brushes in pairs.
As needed	Spoilboard	Resurface the spoilboard when it becomes significantly scarred or when cut-through depths are no longer consistent. Run a surfacing job at 0.5–1.0 mm depth across the entire spoilboard surface.

📋 **Lubrication Reference**

Lubrication summary:

Rails (X, Y, Z linear rails): light machine oil — 3-in-1 or equivalent — weekly.

Z-axis lead screw: dry PTFE spray lubricant — never thick grease — monthly.

Bearings and V-wheels: dry or light oil only — do not use thick grease near timber dust.

Collet threads: clean and dry — a drop of anti-seize compound on the collet nut thread is acceptable if corrosion is a problem in your environment.

Reproducible Pre-Cut Checklist

Print this checklist on card stock and laminate it for use at the machine. Use a whiteboard marker to tick items during setup — wipe clean after each session. Complete it before sending any G-code job, without exception. It corresponds to the checklist introduced in Chapter 20.

COMPLETE PRE-CUT CHECKLIST *Job Name: ______________ Date: ____________ Operator: ____________*		
☐	**Machine**	Machine is homed and position is verified
☐		Soft limits are enabled ($20=1 confirmed in GRBL settings)
☐		Spoilboard is flat, clean, and free of debris in the cut path
☐	**Material**	Workpiece is correctly positioned and securely clamped
☐		Material dimensions measured and match the CAM stock settings
☐		No knots, voids, or defects in the planned cut path
☐	**Tool**	Correct tool is installed and fully seated in the collet
☐		Collet nut is tight — use spanners, not finger-tight
☐		No visible damage or wear on the cutting edges
☐	**Zeros**	X/Y zero is set at the correct reference point on the workpiece
☐		Z zero is set correctly — paper method or touch plate confirmed
☐	**G-code**	Correct .nc file is loaded in UGS — check filename and version
☐		3D Visualiser in UGS shows expected toolpaths — all operations visible
☐		Estimated runtime has been noted on the Job Setup Sheet
☐	**Safety**	PPE on: safety glasses, hearing protection, dust mask or extraction running
☐		All bystanders are at a safe distance or behind the safety enclosure
☐		You know where the physical e-stop is and can reach it without moving
☐		You have a clear view of the entire cut area throughout the job
		All checks complete. Clear to send the job.

> 📋 **Note**
> A digital version of this checklist is included in the companion files as JobSetupSheet_Blank.pdf. The full Job Setup Sheet (also in the companion files) includes this checklist plus fields for recording cutting session parameters, results, and dimensional measurements.

APPENDIX

D

FreeCAD v1.1.0 for CNC Makers — Student Edition

GRBL Settings Reference

Complete $-parameter guide for the CNC 4040 Pro Max — GRBL 1.1h

GRBL's $-parameters control every aspect of how the CNC 4040 Pro Max behaves — from motor timing to travel limits to homing speed. This appendix is the complete reference for all parameters relevant to this machine. It is not a configuration guide. It is a reference: you use it to verify that your machine is correctly set up, to understand what a parameter does when you see it mentioned in UGS, and to diagnose unexpected machine behaviour.

> ⚠ **Important**
> Most GRBL parameters should never be changed by an operator. The values in the CNC 4040 Pro Max Value column are the correct factory settings for this machine. Changing motor step values ($100–$102), direction masks ($2–$3), or pulse timing ($0–$1) without precise knowledge of your hardware can permanently damage the machine or corrupt calibration. The Change? column in every table marks what you can safely interact with.

Change Indicator Legend

NO	Do not change without specific instruction. Changing these can cause machine damage or loss of calibration.
VERIFY	Check that the current value matches the expected value. Do not change unless the value is wrong.
CAUTION	Can be adjusted by an experienced operator for specific purposes. Changes affect machine performance.
SET	Must be set correctly for the machine to operate within its physical limits. Verify after any firmware reset.

Reading and Writing Parameters in UGS

Read all settings	Type $$ in the UGS console and press Enter. GRBL responds with every current parameter value. Take a screenshot at the start of every new module and save it in your project folder.
Read single setting	Type $N (e.g. $20) and press Enter. GRBL responds with the current value for that parameter only.
Write a value	Type $N=value (e.g. $20=1) and press Enter. GRBL confirms with 'ok' if the value was accepted. Values take effect immediately — no restart required.
Restore defaults	Type $RST=$ and press Enter. This restores all parameters to GRBL factory defaults. Note: this will erase machine-specific calibration values ($100–$102, $130–$132). Only use this as a last resort.

💡 **Tip**

Before making any $-parameter change, write down the current value first. If the change produces unexpected behaviour, you can restore the original value immediately. The $$ readout is your before-snapshot.

Group 1 — Motor Pulse & Timing — $0–$6 DO NOT CHANGE

These parameters define the electrical signals sent to the stepper motor drivers. Changing them without matching them to your specific driver hardware will cause missed steps, motor faults, or physical motor damage. They are set correctly at the factory.

Param	Name	4040 Value	Range	What It Controls	Change?
$0	**Step pulse time**	10	*2–20 µs*	Duration of each electrical pulse sent to the motor driver. Matched to the driver specification.	NO
$1	**Step idle delay**	25	*0–254 ms*	Time the stepper motors remain energised after the last move. Prevents losing position from motor drift.	NO
$2	**Step port invert**	0	*0–7 (mask)*	Inverts step signal polarity for specific axes. Hardware-specific — matched to motor driver.	NO
$3	**Direction invert**	0	*0–7 (mask)*	Inverts direction signal for specific axes. Determines which direction is 'positive' for each axis.	NO
$4	**Step enable invert**	0	*0 or 1*	Inverts the enable signal to the motor driver. Hardware-specific.	NO
$5	**Limit pin invert**	0	*0 or 1*	Inverts limit switch signal logic. Set based on whether limit switches are normally-open or normally-closed.	NO
$6	**Probe pin invert**	0	*0 or 1*	Inverts the probe contact signal logic.	NO

Group 2 — System Behaviour — $10–$13 Verify only

Param	Name	4040 Value	Range	What It Controls	Change?
$10	**Status report**	1	*0–3*	Controls what data UGS requests from GRBL in its status poll. 1 = machine position + work position (default). Do not change.	VER
$11	**Junction deviation**	0.010	*0.001–0.100 mm*	Controls how aggressively GRBL rounds corners at direction changes. Lower = slower, smoother corners. Higher = faster but may cause vibration.	VER
$12	**Arc tolerance**	0.002	*0.001–0.100 mm*	Controls the accuracy of circular arc approximation. Default is fine for all course operations.	VER
$13	**Report in inches**	0	*0 or 1*	0 = report in millimetres (correct for this course). 1 = report in inches. Must be 0. A value of 1 causes all UGS coordinate displays to show in inches.	VER

Group 3 — Limits & Homing — $20–$27 Verify and understand these

These are the parameters most likely to need verification after a firmware update or factory reset. They directly affect machine safety. All values here were confirmed from Ch20.

Param	Name	4040 Value	Range	What It Controls	Change?
$20	Soft limits	1	*0 or 1*	Enables GRBL to check planned moves against $130/$131/$132 before executing them. 1 = enabled (required). Must have $22=1 and machine homed to function.	SET
$21	Hard limits	0 or 1	*0 or 1*	Enables physical limit switch monitoring. If a switch triggers, GRBL halts immediately. 1 = enabled when limit switches are installed and wired.	SET
$22	Homing cycle	1	*0 or 1*	Enables the homing cycle. 1 = machine must home before operations. Required for soft limits to function. Never set to 0.	SET
$23	Homing dir invert	0	*0–7 (mask)*	Sets which direction each axis moves to find its home switch. Default 0 = all axes home toward minimum position (back-left-up).	VER
$24	Homing feed	25	*1–1000 mm/min*	Slow final approach speed to the limit switch. Slower = more precise home position.	VER
$25	Homing seek	500	*1–5000 mm/min*	Fast initial approach speed to find the limit switch. 500 mm/min is safe for the 4040 Pro Max.	VER
$26	Homing debounce	250	*1–1000 ms*	Delay after limit switch contact before GRBL confirms the switch has triggered. Prevents false triggers from vibration.	VER
$27	Homing pull-off	1.0	*0.1–100 mm*	Distance the machine backs away from the limit switch after homing. Creates a small clearance so the switch is not continuously triggered.	VER

Group 4 — Steps per mm — $100–$102 DO NOT CHANGE

These are the machine's fundamental calibration values. They define how many motor steps equal one millimetre of movement on each axis. They are determined by the hardware: motor step angle, microstepping setting, pulley teeth, and belt pitch. Changing them without recalibrating against a known physical measurement will cause all dimensions to be wrong.

Param	Name	4040 Value	Range	What It Controls	Change?
$100	X steps/mm	800	*depends on hardware*	Motor steps required to move X axis 1 mm. Determined by stepper specification and driver microstepping. Typical: 800 for 1/8 microstepping with GT2 belt.	NO
$101	Y steps/mm	800	*depends on hardware*	Same as $100 for Y axis. Must match actual Y drive hardware.	NO

$102	**Z steps/mm**	**800**	*depends on hardware*	Same as $100 for Z axis. Z may differ from X/Y if a different lead screw pitch is used.	**NO**

> 📋 **Note**
>
> If your machine consistently cuts undersized or oversized parts by a fixed percentage, the steps/mm value may be wrong. This is a calibration issue requiring measurement against a known distance and formula: new_value = current_value × (expected_distance ÷ actual_distance). Do not guess — measure with a precision calliper.

Group 5 — Maximum Rates — $110–$112 Adjust with caution

These limit the maximum speed each axis will attempt. The values below are the CNC 4040 Pro Max defaults. They can be reduced if the machine exhibits missed steps at high rapid speeds, but should never be increased beyond the rated values for the stepper motors and drivers.

Param	Name	4040 Value	Range	What It Controls	Change?
$110	**X max rate**	**3000**	*100–5000 mm/min*	Maximum speed the X axis will travel at any movement — including rapids. 3000 mm/min is the safe limit for the 4040 Pro Max.	**CAU**
$111	**Y max rate**	**3000**	*100–5000 mm/min*	Maximum Y axis speed. Should match $110 for a symmetric gantry machine.	**CAU**
$112	**Z max rate**	**500**	*100–2000 mm/min*	Maximum Z axis speed. Deliberately lower than X/Y to protect the Z axis lead screw and prevent violent plunges.	**CAU**

Group 6 — Acceleration — $120–$122 Adjust with caution

Acceleration affects how quickly each axis reaches its maximum speed. Higher acceleration = snappier motion but more mechanical stress. Lower acceleration = smoother motion but longer ramp-up time at the start of each move.

Param	Name	4040 Value	Range	What It Controls	Change?
$120	**X acceleration**	**200**	*1–2000 mm/s²*	How quickly the X axis accelerates to full speed. 200 mm/s² is a conservative, reliable value for the 4040 Pro Max.	**CAU**
$121	**Y acceleration**	**200**	*1–2000 mm/s²*	Y axis acceleration. Matches X for symmetric performance.	**CAU**
$122	**Z acceleration**	**100**	*1–2000 mm/s²*	Z axis acceleration. Lower than X/Y to reduce mechanical shock on the Z axis lead screw.	**CAU**

Group 7 — Travel Limits — $130–$132 Must be set correctly

These define the physical working envelope of the machine. GRBL uses these values with soft limits ($20=1) to prevent moves that would drive the spindle into the frame. Verify these match the machine's actual physical travel after any firmware reset.

Param	Name	4040 Value	Range	What It Controls	Change?
`$130`	**X max travel**	350	*1–1000 mm*	Maximum X axis travel in mm. Must not exceed the physical X axis movement range. GRBL will reject any move that would go beyond this.	SET
`$131`	**Y max travel**	350	*1–1000 mm*	Maximum Y axis travel. Must match physical Y axis range.	SET
`$132`	**Z max travel**	80	*1–500 mm*	Maximum Z axis travel. Should match the spindle's physical vertical range of movement.	SET

> ⚠ **Important**
> If $130/$131/$132 are set larger than the actual physical travel, soft limits will not protect the machine from crashing into the frame. After any factory reset ($RST=$), verify these three values immediately before running any job.

GRBL Alarm Codes — Quick Reference

When GRBL enters an alarm state, UGS displays 'ALARM:N' in the console and the machine halts. Type $X to clear the alarm (unlock GRBL). Always diagnose the cause before clearing — clearing an alarm without fixing the root cause will cause the same alarm again.

Code	Name	Cause and Response
`ALARM:1`	**Hard limit triggered**	A limit switch was physically triggered — the machine moved into its end-stop. Clear with $X, then re-home. Check why the machine reached the limit.
`ALARM:2`	**Soft limit violated**	A G-code move would exceed the configured travel limits. Clear with $X. Check CAM coordinates, verify correct zero point, check $130/$131/$132 values.
`ALARM:3`	**Reset during motion**	Machine was reset while axes were moving. GRBL has lost position. Re-home required before any further operation. Never reset during active cutting.
`ALARM:4`	**Probe fail — contact**	Probe sequence did not detect contact within the expected distance. Check probe wiring and probe tool Z position.
`ALARM:5`	**Probe fail — lost**	Probe sequence detected initial contact but contact was then lost. Probe signal may be intermittent — check wiring.
`ALARM:6`	**Homing fail — reset**	Homing cycle was reset or interrupted. Re-attempt homing with $H.
`ALARM:7`	**Homing fail — door**	Safety door was opened during homing. Close door, clear alarm, re-home.
`ALARM:8`	**Homing fail — cycle**	Homing failed to complete within the expected distance. Check that limit switches are wired and triggering correctly ($21=1 required).
`ALARM:9`	**Homing fail — switch**	No limit switch detected during homing search. Check limit switch wiring, connection to control board, and $22=1 setting.

GRBL Error Codes — Common Causes

Error codes appear when GRBL receives a G-code command it cannot process. Unlike alarms, errors do not halt the machine — GRBL skips the line and continues. However, a skipped line means a missing cut or movement. Errors in a G-code file must be investigated.

Code	Name	Likely Cause
`error:1`	**Expected command letter**	G-code line starts with an unexpected character. Check the .nc file for corruption or wrong post-processor.
`error:2`	**Bad number format**	A numeric value in the G-code is malformed. Usually caused by wrong post-processor settings in FreeCAD CAM.
`error:20`	**Unsupported command**	GRBL does not support the G-code command sent. Check post-processor is set to 'grbl' not another machine type.
`error:22`	**Invalid target**	A motion command requests a position outside soft limits. Check zero point and verify CAM Final Depth values.
`error:33`	**Gcode command not in range**	A G-code value is outside the valid range for that command. Common cause: negative spindle speed S value.
`error:34`	**Value out of range**	A parameter value exceeds GRBL's valid range. Common in $-commands — check the valid range column in this appendix.

> **Tip**
> The most common cause of GRBL errors in this course is using the wrong post-processor in FreeCAD CAM. Always select the 'grbl' post-processor, not 'linuxcnc', 'fanuc', or any other machine type. Each post-processor generates different G-code dialects — GRBL only speaks its own.

APPENDIX
E

FreeCAD v1.1.0 for CNC Makers — Student Edition

Bloom's Taxonomy Reference

Six cognitive levels, CNC-specific action verbs, and course progression map

Bloom's Revised Taxonomy (Anderson & Krathwohl, 2001) is the educational framework that structures this entire course. You will encounter Bloom's level indicators in every chapter, in every project brief, and in every reflection document. This appendix is your complete reference — the six levels in full, the action verbs associated with each level in a CNC and FreeCAD context, the course progression map, and guidance on using Bloom's in your project reflections.

The levels build on each other. You cannot reliably Apply something you do not first Understand. You cannot Evaluate something you have not yet Applied. The course ascends through all six levels across 10 weeks, finishing with the certification project — an independently-executed Create-level task.

The Six Levels — Complete Reference

Level	Definition	Key Action Verbs (CNC context)	Example Activities in This Course
1 Remember	Recall and recognise information. The foundation level — before you can do anything with knowledge, you must first have it.	*Name, List, Identify, Recall, Label, State, Define, Recognise, Match, Repeat, Locate*	Name the FreeCAD workbenches and their functions. Identify G-code command types (G0, G1, G2/G3, M3, M5). Label the six areas of the FreeCAD interface. Recall the GRBL $-parameters for soft limits.
2 Understand	Explain and interpret information in your own words. You understand something when you can describe it to someone else without reading the definition.	*Explain, Describe, Summarise, Interpret, Classify, Compare, Paraphrase, Illustrate, Discuss*	Explain why dogbones are required at inside corners. Describe what a Waterline CAM operation does. Summarise the parametric update cycle. Interpret a CAM Simulator result and describe what it shows.
3 Apply	Use knowledge to carry out a task or solve a familiar problem. You are no longer explaining — you are doing.	*Use, Execute, Implement, Operate, Demonstrate, Build, Sketch, Construct, Set up, Run, Produce*	Draw a fully constrained sketch to given dimensions. Set up a Pocket Shape CAM operation with Adaptive strategy. Run the UGS workflow from zero point to finished cut. Build a parametric model using a Spreadsheet.

4 **Analyse**	Break information down, compare, find patterns, differentiate. You begin making decisions based on evidence, not instruction.	*Examine, Compare, Differentiate, Investigate, Distinguish, Deconstruct, Break down, Troubleshoot, Test*	Compare ZigZag vs Adaptive pocket clearing for a given scenario. Diagnose why a CAM simulation result differs from expected. Analyse a Part Loft result for start vertex alignment issues. Break down a complex design into its constituent bodies and features.
5 **Evaluate**	Make judgements, critique quality, justify decisions. You are not just describing or using — you are forming and defending reasoned positions.	*Justify, Assess, Judge, Critique, Argue, Recommend, Select, Decide, Defend, Rate, Rank*	Justify the choice of Adaptive over ZigZag for a specific hardwood pocket. Assess whether a 3D relief height is achievable in the available CAM time. Evaluate the fit quality of a test joint against the designed clearance. Critique your own project output against design intent.
6 **Create**	Produce original work by synthesising knowledge and skills. This is the highest level — not imitating, not following, but generating something new.	*Design, Plan, Construct, Compose, Develop, Generate, Produce, Invent, Build, Formulate*	Design the Keepsake Memory Box with original proportions, decorative elements, and joinery. Plan a complete multi-operation CAM sequence from a new design brief. Produce a TechDraw engineering drawing independently. Formulate a parametric Spreadsheet structure for a new design.

> **Tip**
>
> Whenever a task feels difficult, ask yourself: which Bloom's level is this asking me to work at? If it feels harder than previous tasks in the same chapter, you are probably being stretched to a new level. Difficulty is not a sign that something is wrong — it is a sign that learning is happening.

Course Progression — Parts, Weeks, and Bloom's Levels

The course is structured so that every successive Part requires deeper cognitive engagement than the previous one. The table below maps each Part and its chapters to the expected Bloom's level range and the kinds of cognitive activities that characterise it.

Part	Chapters	Module	Weeks	Bloom's Level	Key Cognitive Activities
Part One	**Ch 1–3**	Before You Begin	Pre-course	**Remember**	*Names and functions of tools, interface elements, G-code command types, safety procedures*
Part Two	**Ch 4–7**	Foundations of FreeCAD	Weeks 1–2	**Remember → Understand**	*Explaining what constraints do, describing the parametric update cycle, summarising a CAM operation*

Part Three	Ch 8–10	Core Skills	Weeks 3–5	Understand → Apply	*Building parametric models, setting up CAM operations, running complete workflows from brief to cut*
Part Four	Ch 11–14	Intermediate Skills	Weeks 5–7	Apply → Analyse	*Comparing joinery methods, troubleshooting Loft twist, evaluating import failures, designing gear pairs*
Part Five	Ch 15–21	Advanced Skills	Weeks 8–10	Analyse → Evaluate → Create	*Justifying CAM strategy choices, critiquing surface quality, designing complete synthesis projects*
Part Six	Ch 22–24	Certification	Week 11	Evaluate → Create	*Designing the Keepsake Memory Box entirely independently — no step-by-step guide*

All 11 Projects — Bloom's Level by Week

Every project in the course is designed to operate primarily at a specific Bloom's level. Projects in later weeks do not abandon lower-level skills — they build on them. A Create-level project still requires you to Remember commands, Understand principles, and Apply techniques. The Bloom's level describes the highest cognitive demand of the project.

Proj	Days	Project Title	Bloom's Level
1	1–5	Name Badge	Remember
2	6–10	Engraved Timber Coaster	Understand
3	11–15	Parametric Luggage Tag	Understand → Apply
4	16–20	Decorative Patterned Tile	Apply
5	21–25	Artistic Knotwork Panel	Apply → Analyse
6	26–30	Two-Panel Finger Joint Box	Analyse
7	31–35	Spur Gear Pair Display	Analyse → Evaluate
8	36–40	Organic Relief Carving Panel	Evaluate
9	41–45	Decorative Wall Clock Face	Evaluate → Create
10	46–50	Workshop Tool Rack	Create
11	51–55	Keepsake Memory Box (Certification)	Create (Synthesis)

Assessment Language — How Bloom's Shows Up in Criteria

Assessment criteria are written using action verbs that reflect the expected Bloom's level for each week. The table below shows how the language of assessment shifts across the course — from lower-order verbs in the early weeks to higher-order verbs as the course progresses.

Level	Weeks	Bloom's Stage	Assessment Verbs Used	Example Assessment Language
●	1–4	Remember → Understand	*Name, identify, label, describe, explain, recall*	Name the constraint type. Identify the error. Label this G-code move.
●	5–7	Apply → Analyse	*Use, demonstrate, compare, investigate, deconstruct*	Demonstrate the complete CAM workflow. Compare these two joint options.
●	8–9	Evaluate	*Justify, assess, critique, recommend, defend*	Justify your choice of strategy. Assess the quality of the finished surface.
●	10–11	Create	*Design, plan, produce, generate, formulate, construct*	Design an original parametric model. Plan the complete CAM sequence independently.

Using Bloom's Taxonomy in Your Project Reflections

Every project includes a reflection document. The strongest reflections identify which Bloom's levels your work demonstrated and give specific evidence for each claim. Vague self-assessments ('I think I was at the Apply level') score poorly. Specific, evidenced self-assessments score at distinction.

✗ **Weak reflection**	"I think the parametric design was at the Apply level."
☑ **Strong reflection**	"The parametric Spreadsheet design operated at the Evaluate level. I assessed three different alias structures before selecting a hierarchy with 12 cells, justifying the choice because it minimised circular reference risk while keeping all critical dimensions formula-driven. This required me to judge between valid approaches rather than follow a procedure."

The strong reflection names the level, provides a specific example of what was done, and explains the reasoning behind a decision. The weak reflection states a level without evidence. Use the verbs table at the top of this appendix to identify the language that matches the level you are claiming.

Five questions for a strong Bloom's self-assessment:

1. What did I produce, and what level of thinking did it require? (Remember → Create)
1. What decisions did I make independently, and what was the basis for each decision?
2. Where did I get stuck, and which Bloom's level was I trying to operate at when that happened?
3. What would I do differently — and what does that tell me about which level I am now at?
4. Which specific skill or concept did I move from Understand to Apply (or Apply to Analyse) during this project?

Quick Reference — Six Levels at a Glance

Cut out or photograph this strip and keep it at your workstation for reference during project days and reflection writing.

1 Remember	2 Understand	3 Apply	4 Analyse	5 Evaluate	6 Create
Name *Identify* *Recall*	*Explain* *Describe* *Summarise*	*Use Build* *Demonstrate*	*Compare* *Investigate* *Deconstruct*	*Justify* *Assess* *Defend*	*Design Plan* *Produce*

APPENDIX

F

FreeCAD v1.1.0 for CNC Makers — Student Edition

Assessment Rubrics — All 11 Projects

Formative and summative rubrics for Projects 1–10 and the Certification Project

This appendix contains the complete assessment rubrics for all 11 projects in this course. Read the rubric for each project before you begin — understanding what Distinction-level work looks like in advance allows you to design to that standard rather than discovering it after the fact.

Course Assessment Weighting Summary
Overall passing standard: 70% overall with at least 60% on the Certification Project. One re-submission permitted per project within one week of grade return.

Projects 1–4	**5% each**	20% total	*Remember / Understand*	Name Badge, Coaster, Luggage Tag, Patterned Tile
Projects 5–8	**7.5% each**	30% total	*Apply / Analyse / Evaluate*	Knotwork Panel, Finger Joint Box, Gear Pair, Relief Carving
Projects 9–10	**10% each**	20% total	*Evaluate / Create*	Wall Clock Face, Workshop Tool Rack
Project 11	**30%**	30% total	*Create (Synthesis)*	Keepsake Memory Box — Certification Project

P1 Personalised Name Badge

Week 1 • Days 1–5 • Bloom's: Remember • Course weighting: 5%

Criterion	Distinction 85–100%	Credit 70–84%	Pass 55–69%	Not Yet <55%
Model Integrity *20%*	Sketch fully constrained (all green). Rectangle dimensions match brief exactly. Body, Sketch, Pad, and ShapeString present and named in model tree.	Sketch fully constrained. Dimensions within ±0.5 mm of brief. Model tree mostly clean.	Sketch mostly constrained with one or two under-constrained elements. Dimensions approximately correct.	Sketch not fully constrained. Model fails to build or has red/yellow features.
ShapeString Text *20%*	Name clearly rendered, correct font, height as specified. ShapeString correctly positioned on model face at Z=0. All letter edges selected for Engraving operation.	Name rendered correctly. Font and height appropriate. Minor positioning offset. Most letter edges selected.	Name present but font or height not as specified, or Z position slightly off causing engraving path issues.	ShapeString absent, incorrect text, or text not at correct Z position — engraving path not generated.

CAM Operations *20%*	Both operations (Engraving + Profile) present, verified in Simulator. Correct depths, feeds, and tool assignments. Tabs correctly placed, no yellow rapids through material.	Both operations present and simulated correctly. Minor tab placement issue or feed rate slightly off reference.	Both operations present with one minor Simulator error (e.g. slight depth deviation or missing tab). Can proceed with adjustment.	Operations missing, incorrect tool assigned, or Simulator shows significant errors that would produce an incorrect cut.
Machine Output *20%*	G-code runs successfully without operator intervention. Physical badge matches design dimensions within ±0.3 mm. Engraving is clean and consistent.	G-code runs with one minor self-corrected issue. Badge within ±0.5 mm. Engraving mostly clean.	G-code required one instructor prompt. Badge has minor dimensional deviation or shallow engraving.	G-code failed to run, or required significant instructor support, or physical piece not achieved.
Reflection *20%*	200+ words. Specific design decisions explained with reasoning. Bloom's Remember level accurately identified and evidenced. Notes what would be done differently.	150+ words. Good reflection with specific examples. Bloom's level mentioned.	100+ words. Largely descriptive. Bloom's mentioned without evidence.	Under 100 words. Purely descriptive or Bloom's absent.

P2 Engraved Timber Coaster

Week 2 • Days 6–10 • Bloom's: Understand • Course weighting: 5%

Criterion	Distinction 85–100%	Credit 70–84%	Pass 55–69%	Not Yet <55%
Model Integrity *20%*	Circular sketch fully constrained with fixed centre and correct diameter. Border ring pocket geometry correct. ShapeString correctly placed at Z=0.	Sketch fully constrained. Dimensions within ±0.5 mm. Minor tree organisation issues.	Sketch mostly constrained. Circular geometry approximately correct. One under-constrained element.	Sketch not fully constrained or circular geometry incorrect. Model fails or has errors.
Pocket & Engraving Design *20%*	Border ring pocket clearly defined with correct inner/outer radii and depth. Engraving centred on coaster face. Design shows deliberate aesthetic choices in font and placement.	Border ring pocket correct. Engraving present and readable. Minor centring offset.	Border ring pocket present with minor depth or dimension error. Engraving present but not well centred.	Border ring pocket missing or incorrect. Engraving absent or fails to generate a toolpath.

CAM Operations *20%*	Three operations (Engraving, Pocket, Profile) in correct order. Simulator confirms complete coverage. All feeds/depths match reference values. Tabs correctly applied.	Three operations correct and simulated. Minor order issue or slightly off reference feed rate.	Operations present with one simulation error or incorrect depth. Correctable with adjustment.	Operations missing, in wrong order, or Simulator shows significant errors.
Machine Output *20%*	G-code runs successfully. Coaster within ±0.3 mm. Border ring machined to correct depth. Engraving clean and legible.	G-code runs with one minor self-corrected issue. Coaster within ±0.5 mm. Engraving readable.	G-code required one instructor prompt. Minor dimensional issue or engraving depth inconsistent.	G-code failed or required significant support. Physical piece not achieved.
Reflection *20%*	200+ words. Explains understanding of Engraving vs Pocket operations. Describes what the three-operation sequence achieves. Bloom's Understand level evidenced.	150+ words. Good explanation of operations. Bloom's mentioned.	100+ words. Describes events rather than understanding. Bloom's superficial.	Under 100 words. No explanation of skill development.

P3 Parametric Luggage Tag

Week 3 • Days 11–15 • Bloom's: Understand → Apply • Course weighting: 5%

Criterion	Distinction 85–100%	Credit 70–84%	Pass 55–69%	Not Yet <55%
Model Integrity *20%*	Sketch fully constrained with all critical dimensions driven by Spreadsheet aliases. Model tree clean. Pad and all features correctly ordered.	Sketch constrained. Most dimensions formula-driven. One hard-coded value in non-critical position.	Sketch constrained. Some key dimensions hard-coded. Spreadsheet present with 3+ aliases.	Sketch not fully constrained or Spreadsheet missing. Model fails or has no parametric capability.
Parametric Design *20%*	Minimum 6 named aliases in Spreadsheet. All critical dimensions (width, height, thickness, hole diameter, string slot width) formula-driven. Parametric update demonstrated cleanly — change one value, model updates without errors.	6+ aliases. Most critical dimensions formula-driven. Parametric update works substantially.	4–5 aliases. Some key dimensions hard-coded. Parametric update partially works.	Fewer than 4 aliases. Model has no meaningful parametric capability.

CAM Operations *20%*	Profile, slot, and engraving operations correct. All depths, tools, and feeds verified in Simulator. Tab placement appropriate for small part.	Operations correct. Minor feed rate deviation or tab placement slightly off.	Operations present with one Simulator error. Correctable.	Operations missing or significant Simulator errors.
Machine Output *20%*	G-code runs successfully. Luggage tag within ±0.3 mm. String slot correct depth and width. Engraving clean.	G-code runs with one self-corrected issue. Tag within ±0.5 mm.	G-code required one instructor prompt. Minor dimensional issue.	G-code failed or piece not achieved.
Reflection *20%*	200+ words. Explains what parametric design means in practice. Describes the experience of changing an alias and observing the model update. Bloom's transition from Understand to Apply evidenced.	150+ words. Good explanation of parametric design. Bloom's mentioned.	100+ words. Describes events. Limited explanation of why parametric design matters.	Under 100 words. No engagement with parametric concept.

P4 Decorative Patterned Tile

Week 4 • Days 16–20 • Bloom's: Apply • Course weighting: 5%

Criterion	Distinction 85–100%	Credit 70–84%	Pass 55–69%	Not Yet <55%
Model Integrity *20%*	Sketch fully constrained. Pattern element (Linear or Polar) correctly applied and visible in model tree. Border pocket sketch correct. All features build cleanly.	Sketch constrained. Pattern applied correctly. Minor tree organisation issue.	Sketch mostly constrained. Pattern present but count or spacing slightly off brief.	Sketch not constrained or Pattern feature missing/incorrect.
Pattern Design *20%*	At least one Linear or Polar Pattern element demonstrably drives a repeated feature. Pattern count and spacing set by design intent, not arbitrary. Border pocket clearly defines edge of design area.	Pattern element present and functional. Spacing and count approximately match design intent.	Pattern element present with minor spacing error or single pattern instance only.	No Pattern operation used, or pattern produces incorrect geometry.

CAM Strategy *20%*	Pocket operation uses Adaptive strategy correctly. Step-over in reference range (60–70%). All pocket depths correct. Profile operation with tabs last. Simulator confirms complete coverage.	Adaptive strategy used. Step-over within 10% of reference. Simulator correct.	ZigZag strategy used where Adaptive is specified, or step-over outside reference range. Simulator mostly correct.	Pocket strategy incorrect or Simulator shows significant errors.
Machine Output *20%*	G-code runs successfully. Tile within ±0.3 mm. Pattern elements clearly visible and evenly spaced on physical piece.	G-code runs with one self-corrected issue. Pattern visible. Tile within ±0.5 mm.	G-code required one instructor prompt. Pattern present but minor spacing inconsistency.	G-code failed or piece not achieved.
Reflection *20%*	200+ words. Explains the Apply level experience — using Pattern tools to produce a specific outcome, not just following steps. Describes decisions made about pattern count and spacing. Bloom's evidenced.	150+ words. Describes pattern design process. Bloom's mentioned.	100+ words. Describes events. Minimal design reasoning.	Under 100 words. No engagement with design decisions.

P5 Artistic Knotwork Panel

Week 5 • Days 21–25 • **Bloom's: Apply → Analyse** • Course weighting: 7.5%

Criterion	Distinction 85–100%	Credit 70–84%	Pass 55–69%	Not Yet <55%
Model Integrity *20%*	Native sketch fully constrained. Imported SVG correctly scaled and Z-positioned at model face. All imported paths visible and connected as Part shapes. No geometry gaps.	Sketch constrained. SVG imported with minor scaling deviation. Paths mostly correct.	Sketch mostly constrained. SVG imported but Z-height issue present — corrected with guidance.	Sketch not constrained, or SVG import failed, or paths at incorrect Z causing toolpath errors.
SVG Import Quality *20%*	Imported artwork visually matches source SVG without distortion. All paths Draft Upgraded successfully. No broken or missing path segments. Artwork correctly positioned relative to panel border.	Artwork imported correctly. Minor segment gap or slight position offset.	Artwork imported with one broken segment or Z-height error corrected with guidance.	Artwork import failed or produces incorrect toolpaths in CAM Simulator.

CAM Operations *20%*	Three operations in correct order: V-bit engraving of artwork, Pocket border, Profile. Z-height of imported paths confirmed in Simulator (green cutting moves, no yellow rapids across surface). Tabs correct.	Three operations correct. Minor operation order issue or one small Simulator anomaly.	Operations present with one Z-height error identified and corrected. Simulator mostly correct.	Operations missing or Z-height issue not corrected — Simulator shows rapid moves across artwork.
Machine Output *20%*	G-code runs successfully with tool change. Engraved artwork matches design intent. All paths machined — no missed sections. Panel within ±0.3 mm.	G-code runs with one self-corrected issue. Most artwork paths machined. Panel within ±0.5 mm.	G-code required one instructor prompt. Minor missed section or depth inconsistency.	G-code failed or artwork significantly incomplete.
Reflection *15%*	250+ words. Analyses SVG import process — specifically the challenges of Z-height alignment and Draft Upgrade. Identifies skills at Apply vs Analyse level. What would be done differently with SVG preparation.	200+ words. Good analysis of import process. Bloom's transition evidenced.	150+ words. Describes import events. Some analytical comments.	Under 150 words. Purely descriptive.

P6 Two-Panel Finger Joint Box

Week 6 • Days 26–30 • Bloom's: Analyse • Course weighting: 7.5%

Criterion	Distinction 85–100%	Credit 70–84%	Pass 55–69%	Not Yet <55%
Model Integrity *20%*	Multi-body file with both panels as separate Bodies. All finger joint notch sketches fully constrained on panel faces. Dogbones present at every inside corner. Clearance value in Spreadsheet.	Two Bodies present. Sketches constrained. Dogbones mostly present. Clearance in Spreadsheet.	Two Bodies present. Some notch sketches under-constrained. Dogbones partially present.	Single-body file, or notch geometry incorrect, or no dogbones.

Joinery Design *25%*	Finger joints designed with correct clearance (0.15–0.20 mm). Dogbones correctly sized (radius = tool radius). Box assembles by hand — snug fit. No visible gaps or forced fit.	Joints designed with correct clearance. Box assembles with minimal adjustment. Dogbones present.	Joints present. Fit requires moderate adjustment. Dogbones partially correct.	Joints do not assemble, clearance not designed, or dogbones missing.
CAM Quality *20%*	All joinery pocket operations before profile. Nesting of both panels on stock verified in Simulator. Correct depths and feeds for hardwood/MDF. Tabs on both profiles.	Operations in correct order. Nesting correct. Minor feed rate deviation.	Operations mostly correct. One ordering issue (e.g. profile before pocket) corrected.	Operations in wrong order or nesting incorrect.
Machine Output *15%*	G-code runs successfully. Both panels cut correctly. Assembled box holds together under its own weight without glue.	Both panels cut. Box assembles with minor tab-removal adjustment needed.	Both panels cut. Box assembles with effort — fit slightly tight or loose.	Piece not cut or box does not assemble.
Reflection *20%*	250+ words. Analyses the joinery design decisions — clearance value choice, finger count, dogbone sizing. Describes what 'analyse' means in practice through this project. Bloom's Analyse level evidenced.	200+ words. Good analysis of joinery decisions. Bloom's mentioned.	150+ words. Describes process. Some analytical comments on fit.	Under 150 words. No analysis of design decisions.

P7 Spur Gear Pair Display

Week 7 • Days 31–35 • **Bloom's: Analyse → Evaluate** • Course weighting: 7.5%

Criterion	Distinction 85–100%	Credit 70–84%	Pass 55–69%	Not Yet <55%
Model Integrity *20%*	Both gears generated with matching Module values. Gear ratio correct to brief. Centre bores correctly subtracted via Boolean Cut. All geometry clean — no overlapping surfaces. Model tree clearly named.	Both gears with matching Module. Ratio correct. Boolean operations applied. Minor tree naming issue.	Both gears with matching Module. One Boolean operation incomplete or minor ratio deviation.	Module values do not match (gears cannot mesh) or Boolean operations missing.

Gear Design *25%*	Gear ratio demonstrates understanding of Module × Teeth = pitch diameter relationship. Centre bore correctly sized for shaft. Correct teeth count for specified ratio. Gears visually mesh when assembled on display frame.	Gear ratio correct. Centre bore present. Gears appear to mesh.	Gear ratio approximately correct. Centre bore present but slightly wrong diameter.	Gear ratio incorrect or gears do not mesh geometrically.
CAM Precision *20%*	Roughing pass uses +0.25 mm Extra Offset. Finishing pass uses zero offset. Feeds for roughing and finishing match Ch13 reference values for material. Both passes in Simulator — separate toolpaths visible.	Roughing + finishing strategy applied. Extra offset approximately correct. Simulator correct.	Single pass only (no roughing/finishing strategy), or roughing offset missing.	Profile operations missing or Simulator shows significant errors.
Machine Output *15%*	G-code runs successfully. Gear profiles clean — teeth smooth, no chatter marks. Gears assembled on frame and rotate smoothly when meshed by hand.	Gear profiles clean. Gears mesh with minor resistance.	Gear profiles present but minor surface roughness. Gears mesh with moderate resistance.	Gear profiles not achieved or gears do not rotate when meshed.
Reflection *20%*	250+ words. Evaluates the roughing/finishing strategy — why it was needed, what it achieved. Analyses gear ratio decision. Bloom's Evaluate level evidenced with specific design judgements.	200+ words. Evaluates strategy with specific observations. Bloom's mentioned.	150+ words. Describes process. Minimal evaluation of strategy choice.	Under 150 words. No evaluation of decisions made.

P8 Organic Relief Carving Panel

Week 8 • Days 36–40 • **Bloom's: Evaluate** • Course weighting: 7.5%

Criterion	Distinction 85–100%	Credit 70–84%	Pass 55–69%	Not Yet <55%
Model Integrity *20%*	Datum planes named. All section sketches fully constrained on correct planes. Part Loft result clean — no twist, smooth surface. Boolean Union with base panel correct. Chamfer/Fillet applied to all external edges.	Datum planes present. Sketches constrained. Loft result smooth with minor surface imperfection. Edge treatments applied.	Datum planes present. Loft result has minor twist or surface issue. Chamfer/Fillet partially applied.	Loft result has significant twist or fails to build. Datum planes or edge treatments missing.
3D Loft Form *20%*	3+ datum plane sections clearly demonstrate considered design intent. Loft form matches stated design brief. Height appropriate (15–25 mm). Smooth transitions between sections. Organic or geometric form clearly resolved.	3+ sections. Loft form matches brief with minor deviation. Form reasonably smooth.	3 sections minimum but form partially resolved. Height outside recommended range.	Fewer than 3 sections, or form does not match brief, or Loft fails.
3D CAM Quality *20%*	Waterline roughing verified in Simulator — horizontal passes at correct step-down (2 mm), stock-to-leave 0.3 mm. 3D Surface finishing verified — step-over ≤1.2 mm. Both passes visible and distinct in Simulator.	Both operations present and simulated. Step-down or step-over slightly outside reference range.	Both operations present. One parameter significantly outside reference range. Correctable.	Only one 3D operation (roughing or finishing only), or Simulator shows significant errors.
Machine Output *20%*	Two tool changes managed correctly in UGS (Z re-zeroed after each). Physical relief matches simulation. Finishing pass eliminates staircase from roughing. Surface quality appropriate for display.	Two tool changes completed. Relief visible. Minor staircase remaining after finishing pass.	Two tool changes completed with one instructor prompt. Relief present but finishing pass incomplete.	Only one operation run, or piece not achieved.
Reflection *20%*	250+ words. Evaluates step-down and step-over choices — justifies values selected against CAM runtime. Identifies Evaluate-level decisions made. Connects Loft form design to final machined result.	200+ words. Evaluates CAM strategy choices. Bloom's mentioned.	150+ words. Describes 3D CAM process. Minimal evaluation.	Under 150 words. No evaluation.

P9 Decorative Wall Clock Face

Week 9 • Days 41–45 • Bloom's: Evaluate → Create • Course weighting: 10%

Criterion	Distinction 85–100%	Credit 70–84%	Pass 55–69%	Not Yet <55%
Model Integrity *20%*	All sketches fully constrained. Polar Pattern produces 12 correctly-spaced hour markers. Clock face proportions match brief. Model tree named and clean.	Sketch constrained. 12 markers present and approximately evenly spaced. Tree mostly clean.	Sketch mostly constrained. Polar Pattern present but count or spacing slightly off.	Sketch not constrained or Polar Pattern absent/incorrect.
Design & Pattern *15%*	Polar Pattern produces visually even 12-marker arrangement. Hour marker proportions appropriate to clock face diameter. Engraving design (text, SVG, or B-spline) resolved — depth and scale appropriate.	Markers evenly spaced. Engraving present. Minor proportion issue.	Markers present but spacing slightly uneven. Engraving present but shallow or poorly scaled.	Polar Pattern absent or engraving not present.
CAM Operations *20%*	All operations in correct order (engraving before pocket before profile). Adaptive pocket strategy for hour marker recesses. Feeds match reference. Simulator confirms complete and correct toolpaths.	Operations in correct order. Adaptive strategy used. Simulator correct.	Operations mostly correct. One ordering issue or incorrect pocket strategy.	Operations in wrong order or significant Simulator errors.
TechDraw Drawing *25%*	Minimum 3 orthographic views including one section view. 10+ correctly placed and labelled dimensions. Title block fully completed (all fields). Drawing is independently legible.	3 views including section. 8–9 dimensions. Title block mostly complete.	2–3 views. 6–7 dimensions. Title block partially filled.	Fewer than 2 views, fewer than 6 dimensions, or title block mostly empty.
Machine Output *10%*	G-code runs successfully. Clock face within ±0.3 mm. Hour markers evenly machined. Engraving clean.	G-code runs. Clock face within ±0.5 mm. Minor marker depth variation.	G-code required one instructor prompt. Minor dimensional issue.	G-code failed or piece not achieved.

Reflection *10%*	250+ words. Evaluates the relationship between parametric design and TechDraw. Identifies Create-level thinking in the design decisions. Specific judgements about dimension placement and drawing legibility.	200+ words. Evaluates drawing quality. Bloom's transition evidenced.	150+ words. Describes drawing process.	Under 150 words.

P10 Workshop Tool Rack

Week 10 • Days 46–50 • Bloom's: Create • Course weighting: 10%

Criterion	Distinction 85–100%	Credit 70–84%	Pass 55–69%	Not Yet <55%
Parametric Model *20%*	5+ Part Design Bodies all sharing one Spreadsheet with 12+ named aliases. All critical dimensions formula-driven. Parametric update demonstrated — change rack width, all panels update cleanly. French cleat Boolean Cut correct.	5 Bodies. 10+ aliases. Most dimensions formula-driven. Update mostly works.	5 Bodies. 7–9 aliases. Some hard-coded values. Update partially works.	Fewer than 5 Bodies, or fewer than 7 aliases, or no meaningful parametric capability.
Design & Synthesis *20%*	All required elements present: 5 panels, French cleat, hook brackets. Design shows considered proportions. Assembly logic (how parts relate) evident in model structure. Demonstrates synthesis of skills across course.	All required elements present. Proportions appropriate. Assembly logic evident.	All required elements present. Some proportion issues. Assembly logic partially evident.	Required elements missing or design not functional as a tool rack.
CAM Operations *20%*	6 operations in correct sequence across 2 stock pieces. All operations verified in Simulator. French cleat slot operation correct depth and angle. Multi-part nesting efficient. Feeds match reference for material.	6 operations mostly correct. Nesting on 2 stocks. Simulator correct.	Operations present with one sequencing or nesting error. Correctable.	Operations missing, in wrong order, or significant Simulator errors.

TechDraw + BOM *20%*	4+ orthographic views including section view. 12+ dimensions. Complete title block. BOM table with all 5 parts listed (name, material, qty, thickness). Joint tolerances shown on at least one dimension.	4 views. 10–11 dimensions. BOM present. Title block mostly complete.	3 views. 8–9 dimensions. BOM incomplete. Some title block fields empty.	Fewer than 3 views, fewer than 8 dimensions, or no BOM.
Machine Output *10%*	G-code runs on both stock pieces without instructor intervention. French cleat slot machined correctly. All panels cut and test-assembled. Tool rack is structurally sound.	Both stocks cut with one self-corrected issue. Assembly mostly correct.	Both stocks cut with one instructor prompt. Minor assembly issue.	Piece not fully cut or assembly fails.
Reflection *10%*	300+ words. Identifies Create-level synthesis — specific moments in the project where prior skills combined in new ways. Evaluates CAM strategy decisions. Honest assessment of what could be improved.	250+ words. Identifies synthesis. Evaluates decisions. Bloom's evidenced.	200+ words. Describes process. Some synthesis identified.	Under 200 words or no synthesis identified.

P11 Keepsake Memory Box — Certification Project

Week 11 • Days 51–55 • **Bloom's: Create (Synthesis)** • Course weighting: 30%

Criterion	Distinction 85–100%	Credit 70–84%	Pass 55–69%	Not Yet <55%
Parametric Model Integrity *20%*	All critical dimensions Spreadsheet-driven. Min 8 named aliases correctly linked. Parametric update demonstrated live — model updates cleanly with no errors on Refresh.	All critical dimensions linked. 8+ aliases. Model updates substantially on Refresh.	Spreadsheet present with 4+ aliases. Some key dimensions hard-coded. Update partially works.	Fewer than 4 aliases. No meaningful parametric capability. Most dimensions hard-coded.
Joinery Quality *20%*	All joints cut with correct clearance. Box assembles smoothly by hand. Dogbones present at all inside corners. No visible gaps or forced fits.	Joints mostly correct. One minor fit issue. Dogbones present. Box assembles with minimal adjustment.	Joints present. Some fit issues requiring adjustment. Dogbones partially present.	Joinery not fully implemented. Joints do not assemble, or no dogbones, or fit is non-functional.

Decorative & 3D Elements *15%*	Lid engraving — design intent evident, correct depth, clean sharp edges. 3D relief present, smooth surface, two distinct CAM passes (roughing + finishing) confirmed in Simulator and visible on piece.	Lid engraving present and readable. 3D relief with two CAM passes but minor staircase visible.	Lid engraving present but shallow, unclear, or generic. 3D relief attempted — roughing only.	Lid engraving absent or failed to machine. 3D relief element not attempted.
Edge Treatments *5%*	Chamfer or fillet applied consistently to all external visible edges in the FreeCAD model. Treatment appropriate to design intent.	Edge treatments applied to most visible edges. One or two edges missed.	Some edge treatments present but inconsistently applied.	No edge treatments applied in the FreeCAD model.
TechDraw Drawing *15%*	4+ orthographic views including at least one section view. 12+ correctly placed dimensions. Complete title block (all fields). BOM table listing all parts. Joint tolerances shown.	4 views. 10–11 dimensions. Title block mostly complete. BOM present but incomplete.	3 views. 8–9 dimensions. Title block partially filled. No BOM or no tolerances.	Fewer than 3 views. Fewer than 8 dimensions. Not independently legible.
CAM & UGS Competency *15%*	All operations correct in Simulator. GRBL G-code correctly post-processed and sent via UGS without instructor intervention. Job Setup Sheet completed for every cutting session.	All operations simulate correctly. G-code sent with one minor instructor prompt. Setup Sheet mostly complete.	Most operations simulate correctly with minor errors. G-code required instructor support at one step.	CAM operations have errors in Simulator. G-code required significant instructor support.
Project Reflection *10%*	600+ words. Genuine critical reflection: specific design decisions explained with clear reasoning. Challenges described with resolutions. Bloom's self-assessment accurate, specific, and insightful.	600+ words. Good reflection with specific examples. Bloom's assessment present.	400–599 words. Largely descriptive. Bloom's mentioned but superficial.	Below 400 words. Purely descriptive or superficial. No critical analysis.

APPENDIX
H

FreeCAD v1.1.0 for CNC Makers — Student Edition

Further Reading & Online Resources

FreeCAD documentation, community forums, SVG sources, CNC communities, and where to go next

The resources listed in this appendix extend well beyond what any single textbook can cover. FreeCAD is a complex, actively-developed application with a large and genuinely helpful community. The CNC machining world has decades of accumulated knowledge across forums, videos, and maker communities. This appendix points you toward the most reliable and relevant sources across each area of the course.

All URLs were verified at the time of publication. If a URL no longer works, search for the resource name directly — established projects rarely disappear, but they occasionally move domains.

FreeCAD — Official Documentation and Community

Resource	URL	What It Provides
FreeCAD Official Site	`freecad.org`	Download page for FreeCAD v1.1.0. Also lists release notes, donation page, and links to all community channels.
FreeCAD Documentation	`wiki.freecad.org`	The primary reference for all workbenches, tools, and features. Covers FreeCAD from first principles to advanced workflows. Essential reading for topics beyond this course.
FreeCAD Forum	`forum.freecad.org`	The official community forum. The single most useful resource for troubleshooting problems not covered in this textbook. Highly active, generally welcoming to beginners.
FreeCAD GitHub	`github.com/FreeCAD/FreeCAD`	Source code repository. Also the place to report genuine software bugs (as opposed to user questions, which go to the forum). Check open issues before reporting — your bug may already be known.
FreeCAD Addon Repository	`github.com/FreeCAD/FreeCAD-addons`	The curated list of all addons available through the Addon Manager. Browse this to discover addons not covered in Chapter 12.
FreeCAD Reddit	`reddit.com/r/FreeCAD`	Active community for tips, project showcases, and questions. Good for quick answers and inspiration. Less technical depth than the official forum.

💡 **Tip**

The FreeCAD Forum is the best place to get help with problems. Search before posting — most common problems have been discussed before. When posting, include your FreeCAD version, operating system, a screenshot of the issue, and a brief description of what you were trying to do. A clear question receives a fast, useful answer.

Universal G-code Sender (UGS)

Resource	URL	What It Provides
UGS GitHub	`github.com/winder/Universal-G-Code-Sender`	Official source for UGS downloads, release notes, and bug reporting. Always download UGS from this source, not from third-party sites.
UGS Wiki	`github.com/winder/Universal-G-Code-Sender/wiki`	Documentation for UGS features, keyboard shortcuts, and configuration. Covers the platform-specific version (UGS Platform) as well as the classic version used in this course.
UGS Issues / Forum	`github.com/winder/Universal-G-Code-Sender/issues`	Bug reports and feature requests. Search here if UGS is behaving unexpectedly — version-specific issues are often documented with workarounds.

GRBL Firmware

Resource	URL	What It Provides
GRBL GitHub	`github.com/gnea/grbl`	Official GRBL firmware repository. The complete $-parameter reference, configuration guide, and G-code support table are all here. Essential reading alongside Appendix D.
GRBL Wiki	`github.com/gnea/grbl/wiki`	The definitive reference for all GRBL $-parameters, alarm codes, error codes, and G-code support. More detailed than Appendix D for edge cases and advanced configuration.
GRBL Configuration	`github.com/gnea/grbl/wiki/Grbl-v1.1-Configuration`	Specific to GRBL 1.1 — the version on the CNC 4040 Pro Max. Explains every $-parameter, its valid range, and its effect on machine behaviour.

Inkscape and SVG Artwork Sources

Used in Chapter 10 (Importing Artwork). Inkscape is required for preparing SVG files for import into FreeCAD's Draft workbench.

Resource	URL	What It Provides
Inkscape	`inkscape.org`	Free, open-source vector graphics editor. Required for SVG preparation: converting text to paths, simplifying nodes, ungrouping, and saving as Plain SVG. Download version 1.3 or later.
Inkscape Documentation	`inkscape-manuals.readthedocs.io`	Comprehensive Inkscape reference. The most relevant sections for this course: the Node editor, Path operations (Simplify, Object to Path), and SVG export settings.
SVGRepo	`svgrepo.com`	Large library of free, licence-clear vector graphics. Excellent source for CNC engraving artwork — Celtic knotwork, botanical, geometric, and decorative designs. Check individual file licences.
Vecteezy	`vecteezy.com`	Vector art library with free and premium tiers. Many designs are suitable for CNC engraving after Inkscape preparation. The free tier requires attribution for commercial use.
Open Clipart	`openclipart.org`	Public domain vector artwork — no attribution required, no licence restrictions. Quality varies, but excellent for finding simple outlines suitable for engraving.
The Noun Project	`thenounproject.com`	Icon-style vector graphics. Clean outlines, simple geometry — ideal for CNC engraving without complex node simplification. Free with attribution; paid licence available.

📋 **Note**

Always verify the licence of any SVG file before using it in work that may be distributed or sold. 'Free' does not always mean 'unrestricted'. SVGRepo and Open Clipart are the safest choices for unrestricted use.

CNC Machining Communities and Forums

Resource	URL	What It Provides
r/hobbycnc	`reddit.com/r/hobbycnc`	Active community for desktop and hobby CNC routing. Good for project inspiration, troubleshooting, and feeds-and-speeds discussion for small routers.
r/CNC	`reddit.com/r/CNC`	Broader CNC community including professional and industrial machines. More technical depth; useful for understanding principles that apply at any scale.

CNC Zone	cnczone.com	Long-running CNC forum covering routers, mills, lathes, and plasma cutters. Extensive archives of troubleshooting discussions. Use the search function before posting.
Maker Forums	makerforums.info	Community covering 3D printing, CNC routing, and laser cutting. Friendly to beginners. Good cross-disciplinary discussion between CNC and other digital fabrication tools.

Feeds, Speeds, and Tooling References

Appendix B of this textbook provides the course-specific feeds and speeds reference for the CNC 4040 Pro Max with timber and MDF. The resources below provide broader context and calculators for different materials and machine types.

Resource	URL	What It Provides
Carbide 3D Calculator	carbide3d.com/carbide-create/carbide-create-how-to/speeds-and-feeds	Free feeds and speeds calculator covering common materials. Designed for desktop CNC routers similar to the 4040 Pro Max. A useful sanity check against the course reference values.
G-Wizard Calculator	cnccookbook.com/gwizard-cnc-speeds-feeds-calculator	Paid feeds and speeds calculator with extensive material database. Industry-standard tool. The free trial is useful for understanding how chip load and surface speed interact.
CNC Cookbook	cnccookbook.com	Extensive articles on machining fundamentals — chip load, surface speed, runout, tool deflection, and more. The theory behind the numbers in Appendix B.

Going Beyond This Course — What to Learn Next

Completing this course gives you a solid foundation in parametric CAD, 2.5D and 3D CNC machining, and professional workflow. The table below maps the most natural directions for further development — and notes which skills from this course transfer directly to each one.

Direction	What to Learn Next
Sheet metal and flat-pack furniture	SheetMetal workbench (Appendix A lists it). Master flat-pattern DXF export, bend calculations, and kerf compensation for laser cutting. The same parametric Spreadsheet skills from this course apply directly.
Mechanical assembly and simulation	Assembly4 workbench. Build constrained multi-body assemblies with joints, verify spatial fit, and simulate kinematic motion. Directly extends the multi-body skills from Chapters 11 and 19.
3D printing and additive manufacturing	FreeCAD Part Design exports .STL directly. Your Sketcher, Part Loft, and parametric skills transfer completely. The main addition is learning slicer software (Cura, PrusaSlicer) for print preparation.

CNC aluminium and harder materials	The FreeCAD skills are identical. The differences are in feeds, speeds, and tooling. Research single-flute end mills, chip evacuation, and cutting fluid before attempting aluminium on a desktop CNC.
Laser cutting	FreeCAD produces DXF files for laser cutting directly from 2D sketches. Your constraint and parametric skills transfer completely. Kerf compensation is the main new concept.
CNC router for signage and production	Scale your skills from hobby to professional use. Research Vectric VCarve or Aspire for signage-specific CAM. Your understanding of toolpaths, feeds, and G-code from this course makes the learning curve manageable.

A Note on Community

The FreeCAD and CNC machining communities are genuinely open and helpful. Post your project photos on Reddit or the FreeCAD Forum when you complete the certification project. Ask questions when you get stuck. Share what you learn. The communities that built the tools you have been using throughout this course are maintained by people who do exactly that.

APPENDIX

I

FreeCAD v1.1.0 for CNC Makers — Student Edition

Companion Digital Files Index

Complete listing of all .FCStd, .nc, .svg, and .pdf files included with this course

The companion digital files are provided to support your learning throughout this course. They are available from your course coordinator or the course's digital learning platform. This appendix lists every file, its format, and how to use it.

> ⚠ **Important**
> The companion files are reference tools — not shortcuts. Using a provided .FCStd file instead of building your own is like copying the answers to an exercise: you produce a correct result but learn nothing. Always build your own design files for each project. Open the companion files after completing your own version to compare approaches.

File Type Overview

Format	Type	Count	What It Is and How to Use It
`.FCStd`	**FreeCAD Project File**	**11 files**	One complete reference design per weekly project. Open in FreeCAD to inspect the model, parametric structure, and CAM setup.
`.nc`	**G-code File**	**11 files**	One post-processed G-code file per project, ready for UGS. Compare your output against this to identify CAM errors.
`.svg`	**SVG Artwork File**	**5 files**	Pre-prepared SVG artwork for use in the Week 5 importing project. Verified to import correctly into FreeCAD's Draft workbench.
`.pdf`	**Job Setup Sheet**	**2 files**	One blank template and one completed example from Project 5. Print and use as a reference when setting up your own sheets.
`.pdf`	**Assessment Rubrics**	**1 file**	Complete rubrics for all 11 projects in a single reference document. Read before beginning each project.

FreeCAD Project Files (.FCStd) — 11 Files

Each file is a complete reference design for one project. Open these in FreeCAD v1.1.0 to inspect the parametric structure, model tree organisation, Spreadsheet aliases, and CAM setup. The most useful application is comparing your own Model tree structure with the reference — not to copy the design, but to see how the reference was organised and ask yourself why.

Filename	Days	Proj	Project Name	Key Skills Demonstrated
Proj01_NameBadge.FCStd	1–5	1	**Personalised Name Badge**	Sketcher, ShapeString, basic CAM Profile and Engraving. Single-body, single-material-thickness model.
Proj02_TimberCoaster.FCStd	6–10	2	**Engraved Timber Coaster**	ShapeString text layout, Pocket engraving, TrueType font selection, UGS workflow.
Proj03_LuggageTag.FCStd	11–15	3	**Parametric Luggage Tag**	Spreadsheet with 6 aliases, parametric update demonstration, mounting hole, string slot.
Proj04_PatternedTile.FCStd	16–20	4	**Decorative Patterned Tile**	Linear and Polar Pattern operations, array-driven pocket geometry, multi-operation CAM.
Proj05_ArtworkPanel.FCStd	21–25	5	**Artistic Knotwork Panel**	Draft Import, SVG scaling, Draft Upgrade, Z-height verification, combined native + imported CAM.
Proj06_FingerJointBox.FCStd	26–30	6	**Two-Panel Finger Joint Box**	Multi-body file, finger joint geometry with clearance and dogbones, multi-part CAM nesting.
Proj07_GearPairDisplay.FCStd	31–35	7	**Spur Gear Pair Display**	FCGear InvoluteGear, gear ratio, Part Boolean operations, precision gear-profile CAM.
Proj08_ReliefCarvingPanel.FCStd	36–40	8	**Organic Relief Carving Panel**	Datum planes, Part Loft (3 sections), Boolean Union, 3D Waterline roughing, Surface finishing.
Proj09_WallClockFace.FCStd	41–45	9	**Decorative Wall Clock Face**	Polar Pattern hour markers, Adaptive pocket clearing for recesses, TechDraw engineering drawing.
Proj10_WorkshopToolRack.FCStd	46–50	10	**Workshop Tool Rack**	Full synthesis: 5 bodies, 17 Spreadsheet aliases, French cleat Boolean, 6 CAM operations, TechDraw + BOM.
Proj11_KeepsakeMemoryBox.FCStd	51–55	11	**Keepsake Memory Box (Sample)**	Certification-level example only. Full parametric, joinery, engraving, 3D relief, TechDraw. Instructor reference.

> **Note**
>
> Proj11_KeepsakeMemoryBox.FCStd is the instructor reference example for the certification project. It is included in the digital files package for coordinator use. Students should not open this file before completing their own certification project — doing so would compromise the independent work requirement.

G-code Files (.nc) — 12 Files

Each G-code file has been post-processed through the GRBL post-processor and verified to run correctly on the CNC 4040 Pro Max. Use these files to understand what a correct post-processed output looks like, and to compare against your own exports if your G-code is producing unexpected results in UGS.

Filename	Proj	Tool(s)	Operations
Proj01_NameBadge_3mm_v1.nc	1	3 mm flat EM	Profile cut + ShapeString engraving
Proj02_TimberCoaster_12mm_v1.nc	2	3 mm flat EM	Pocket engraving, border profile
Proj03_LuggageTag_6mm_v1.nc	3	3 mm flat EM	Profile, slot, engraving — single file
Proj04_PatternedTile_12mm_v1.nc	4	3 mm flat EM	Multiple pocket patterns, profile
Proj05_ArtworkPanel_15mm_v1.nc	5	60° V-bit + 6 mm EM	V-bit engraving, Adaptive pocket, profile — 2 tool changes
Proj06_FingerJointBox_12mm_v1.nc	6	6 mm flat EM	Joinery pockets + profile — all panels
Proj07_GearPairDisplay_18mm_v1.nc	7	3 mm + 6 mm EM	Gear profile roughing + finishing, bore pocket, frame profile
Proj08_ReliefCarving_35mm_v1.nc	8	6 mm + 3 mm BN	3D Waterline roughing + Surface finishing — 2 tool changes
Proj09_ClockFace_12mm_v1.nc	9	60° V-bit + 6 mm EM	Engraving + Adaptive hour recesses + profile
Proj10_ToolRack_Stock1_v1.nc	10	6 mm flat EM	Stock 1: cleat slot + panel profiles
Proj10_ToolRack_Stock2_v1.nc	10	6 mm flat EM	Stock 2: hook brackets + smaller panels
Proj11_MemoryBox_Ref_v1.nc	11	Multiple	Instructor reference only — not for student use

> 📋 **Note**
> Project 10 (Workshop Tool Rack) has two separate G-code files because the design requires two stock pieces. Each file contains all operations for one stock piece. This multi-file approach is covered in Chapter 21. All other projects use a single G-code file per project.

SVG Artwork Files (.svg) — 5 Files

These SVG files are prepared for use in the Week 5 importing project (Chapter 10). Each has been processed in Inkscape: text converted to paths, nodes simplified, objects ungrouped, and saved as Plain SVG. They will import into FreeCAD's Draft workbench without further preparation.

Filename	Description	Notes for Use
SVG_01_CelticKnotwork.svg	**Celtic knotwork interleaved border**	Week 5 — 3 closed paths. Complex organic form, ideal for the border recess design.
SVG_02_LeafBotanical.svg	**Botanical leaf and stem outline**	Week 5 — 2 open and 1 closed path. Flows naturally as a V-bit engraving on the panel surface.
SVG_03_AnimalSilhouette.svg	**Simple animal silhouette (rabbit)**	Week 5 — 1 closed path, simplified to under 40 nodes. Good introductory example.
SVG_04_DecorativeBorder.svg	**Geometric rectangular border frame**	Week 5 — 4 paths. Straight-edged design — imports cleanly without simplification needed.
SVG_05_WavePattern.svg	**Repeating wave decorative motif**	Week 5 — 5 paths. More complex; requires node simplification before import. Used in class exercises.

Supporting Documents (.pdf) — 3 Files

Filename	Description	How to Use It
`JobSetupSheet_Blank.pdf`	**Blank Job Setup Sheet template**	Print one per cutting session. All fields blank. Use as your session documentation worksheet.
`JobSetupSheet_Example_Proj05.pdf`	**Completed example — Project 5**	Fully completed example from the Artwork Panel session. Use as a reference for your own completions.
`AssessmentRubrics_AllProjects.pdf`	**All 11 project rubrics**	Complete assessment rubrics for Projects 1–11. Read the rubric before beginning each project.

```
Recommended folder structure for your companion files:
FreeCAD_CNC_Course/
  Companion_Files/
    FCStd/          ← All 11 .FCStd reference files
    GCode/          ← All 12 .nc G-code files
    SVG/            ← All 5 .svg artwork files
    PDFs/           ← Job Setup Sheets + Assessment Rubrics
  My_Projects/
    Proj01_NameBadge/
    Proj02_Coaster/
    ...
```

Keeping your own project files in a separate My_Projects folder (organised per project) prevents any risk of overwriting a companion reference file. It also makes it easy to compare your finished design with the reference — both are accessible from one parent folder.

APPENDIX
I

FreeCAD v1.1.0 for CNC Makers — Student Edition

4th Axis Integration

A-axis GRBL setup, indexed workflow, G-code reference, workholding, and safety for the CNC 4040 Pro Max

The 4th axis (A-axis) adds rotational movement to the CNC 4040 Pro Max, enabling machining on cylindrical workpieces, engraving on curved surfaces, and multi-face indexing without re-clamping. This appendix covers everything needed to begin using the 4th axis.

This is supplementary content. All 24 course chapters cover 3-axis machining. The skills you have built are the foundation — most transfers directly to 4th axis work.

What Transfers from This Course — and What Is New

Status	Type	Topic	Detail
✓ Direct		**FreeCAD design (Part Design, Sketcher, Parametric, Assembly)**	Unchanged. Design rotary work as a normal 3D Body.
✓ Direct		**UGS connection, homing, FRO, e-stop, Job Setup Sheet**	Identical workflow. A-axis appears as 4th DRO coordinate.
✓ Direct		**CAM 2.5D operations (engraving, profile, pocket)**	Work on cylindrical surfaces. Stock setup accounts for cylinder.
✓ Direct		**G-code fundamentals (G0,G1,M3,M5,M6, feeds)**	A-axis adds one new word — A — to existing vocabulary.
⚠ New		**$103, $133, $143 — A-axis GRBL parameters**	Three new parameters. Must be set and verified before use.
⚠ New		**A-axis zeroing via G92 (no limit switch homing)**	Manual reference procedure each session. No auto-home.
⚠ New		**G-code A word, indexed vs continuous rotation**	Covered in this appendix.
⚠ New		**FreeCAD CAM indexed workflow**	Manual combination of per-face G-code files.
● New		**Rotary workholding safety rules**	Additional rules — covered in this appendix.

Hardware Overview

The 4th axis kit consists of a rotary headstock (stepper motor and chuck), a tailstock for workpiece support, and T-slot mounting hardware. The stepper motor connects to the A-axis output on the machine's 4-axis control board.

Headstock	Motor-driven chuck assembly. Holds and rotates the workpiece under GRBL control. Check your unit's documentation for gear ratio — needed for the $103 calculation.
3-jaw chuck	Self-centring for round stock. All three jaws move simultaneously. Fast and approximately concentric for cylindrical workpieces up to the chuck's rated capacity.
Tailstock	Non-driven support at the far end of the workpiece. Use whenever workpiece length exceeds 3× diameter. Prevents vibration and flex during machining.
Mounting alignment	The headstock rotation axis must be parallel to the machine X axis. Misalignment produces taper across the workpiece length. Verify with a dial indicator on a test cylinder while moving X.
Control board	The CNC 4040 Pro Max board must have a 4-axis stepper output header. Check the board silkscreen. If absent, a separate stepper driver and power supply are required.

GRBL A-Axis Parameters ($103, $133, $143)

Three $-parameters configure the A-axis. These supplement the parameters in Appendix D. The standard $0–$132 range covers 3-axis only. A-axis parameters use $103, $133, and $143.

Param	Name	Typical	Range	What It Controls	Change?
$103	**A steps/degree**	**~35–36**	*See formula*	How many motor steps equal one degree of rotation. The critical calibration value — wrong steps/degree means wrong angles.	**SET**
$133	**A max rate**	**360°/min**	*100–3600*	Maximum A-axis speed. Start conservatively — 360°/min is a safe default. Increase only after confirming no missed steps.	**CAU**
$143	**A acceleration**	**50°/s²**	*10–1000*	Acceleration rate. Keep low to prevent chuck jaw loosening from rotational shock. Increase gradually once workholding is confirmed.	**CAU**

Calculating $103 — Steps per Degree

The value for $103 depends on your specific hardware. Check your 4th axis documentation for motor steps/revolution, driver microstepping setting, and headstock gear ratio:

$103 = (Motor steps/rev × Microstepping × Gear ratio) ÷ 360

```
Example:  200 steps  ×  16 (1/16 micro)  ×  4 (4:1 gear)  ÷  360  =  35.56 steps/°
```

Check documentation for: motor steps/rev (200 or 400), microstepping (8/16/32), gear ratio

> ⚠ **Important**
>
> Do not guess $103. Verify by commanding G0 A90, measuring actual rotation with a protractor, and adjusting: new_$103 = current × (90 ÷ actual_degrees). Retest until all four quadrant checks (A90, A180, A270, A360) pass.

Verify $103	G0 A90, measure actual angle, adjust until 90.00° confirmed.
Verify $133	Command G1 A360 F360. If motor skips (audible clicks, position loss), reduce $133 by 20%.
Verify $143	Command rapid full rotation. If chuck jaws show movement during start/stop, reduce $143.
Save configuration	After all three values confirmed: screenshot $$ and save alongside your X/Y/Z configuration reference.

UGS Workflow with the 4th Axis

The UGS interface does not change significantly. The DRO shows A as a fourth coordinate. All other controls (homing, FRO, e-stop, job recovery) work identically to Chapter 20.

> 📖 **Key Term: G92 — Set Position**
>
> A G-code command that defines the current position as a specified value without moving. G92 A0 tells GRBL the current A-axis position is zero degrees — the standard session-start zeroing procedure. The A-axis has no limit switch homing.

A DRO display	Shows current A-axis position in degrees from the last G92 A0. Does not reset at power-on. Set G92 A0 explicitly at the start of every session.
A-axis jogging	If A jog buttons don't appear in Machine Control, jog via console: $J=G91A10F200 jogs 10° incrementally. Return to G90 after any G91 jog session.
Setting A zero	Rotate chuck to reference orientation by hand (spindle off). Type G92 A0. A coordinate resets to 0.000.
No soft limits	GRBL does not apply soft limits to A-axis. Take care with cable routing — continuous rotation in one direction can wrap wiring.

G-code A-Axis Reference

The A word specifies angular position in degrees. In G90 (absolute) mode, A values are from G92 A0. In G91 (incremental) mode, A values are relative to current position. Always return to G90 after any G91 block.

G-code	Action
`G0 A90`	Rapid rotate to A=90° from current zero
`G1 A180 F300`	Controlled rotation to A=180° at 300°/min
`G1 A360 F200`	Full rotation — returns to start (G90 absolute)
`G91 G1 A45 F200`	Incremental 45° rotation from current position
`G90`	Return to absolute mode — always use after G91 blocks
`G0 A0`	Rapid return to A-axis zero
`G92 A0`	Set current position as A=0 — the session reference command
`G0 Z20`	Raise spindle 20mm clear — always do this before any A rotation

> **Tip**
> Always raise Z clear before any A-axis rotation. Type G0 Z[safe_height] before G0 A[angle]. A rotation command while the spindle is near the workpiece will drag the bit through material.

Indexed motion	A-axis rotates to position, stops. X/Y/Z cutting happens at that position. Z raises, A rotates to next position. Standard workflow for the 4040 Pro Max.
Continuous motion	A rotates simultaneously with X/Y/Z — helical cuts, wrapped toolpaths. Requires specialised CAM software. FreeCAD v1.1.0 does not generate these natively.

FreeCAD CAM — Indexed Workflow

FreeCAD v1.1.0 does not generate A-axis rotation commands automatically. For indexed work: create separate CAM Jobs for each face, post-process to separate .nc files, then manually combine with A-axis rotation commands between them.

Step 1 — Design	Model the complete workpiece as a single Part Design Body. All faces to be machined are present in one model.
Step 2 — One Job per face	Create a CAM Job for each face (Face_A0, Face_A90, etc.). Each Job targets different geometry on that face.
Step 3 — Post-process	Post-process each Job to separate GRBL .nc files: Part_A0.nc, Part_A90.nc, etc. Verify each in UGS's 3D Visualiser.
Step 4 — Combine	Edit a master .nc file in a text editor. Structure: [Face_A0 content] → G0 Z[safe] G0 A90 → [Face_A90 content] → G0 Z[safe] G0 A180 → etc.
Step 5 — Verify	Load combined file in UGS. Step through to confirm rotation commands appear at correct positions between face operations.

Note

For continuous 4-axis toolpaths (wrapped text, helical fluting, 4-axis relief), consider Vectric VCarve Pro or Fusion 360 CAM — both export GRBL-compatible G-code. Export your FreeCAD model as STEP or STL for import.

Indexed Workflow — Step by Step

Indexed A-axis Workflow — Step by Step		
1	Home X/Y/Z	Home machine ($H) as normal. A-axis has no limit switch — zero is set manually.
2	Mount workpiece	Secure in chuck. Tighten all jaws with chuck key. Verify no wobble. Engage tailstock if length > 3× diameter.
3	Set X/Y/Z zero	Z zero = top of cylinder. X/Y zero = chuck spindle axis. Use paper method or touch plate as normal.
4	Set A zero	Rotate chuck to Face 1 reference orientation. Type G92 A0 in UGS console. A coordinate resets to 0.
5	Test rotation	G1 A360 F180 — watch full slow rotation. Verify clearance from spindle, gantry, and fixturing before cutting.
6	Run Face 1	Send Face 1 G-code. All operations at A=0 position.
7	Raise Z, rotate	G0 Z[safe]. G0 A90 (or required angle). Wait for rotation. Verify visually.
8	Run remaining faces	Repeat for A90, A180, A270 as required.
9	Return and remove	G0 A0. Stop spindle. Remove workpiece from chuck.

Workholding for Rotary Work

Round stock	3-jaw self-centring chuck. Tighten all jaws with chuck key. Test rotation by hand — no wobble. Use tailstock if length > 3× diameter.
Square/irregular stock	3-jaw chuck does not grip square stock securely. Use a 4-jaw independent chuck or a square fixture plate on the chuck face.
Tailstock	Advance until firm contact, then clamp. No end-play. For hardwood, a small centre-drill hole in the stock end prevents grain splitting.
Overhang limit	Unsupported overhang beyond chuck jaws should not exceed 3× stock diameter. Always use tailstock beyond this limit.
Checking run-out	After mounting: rotate by hand and watch a reference point. Visible wobble = re-chuck. Runout over 0.5 mm produces periodic toolmarks across the diameter.

Safety — Additional Rules for Rotary Work

All safety fundamentals from Chapter 2 apply. The following additions are specific to the 4th axis.

> **Safety**
> Never reach toward rotating stock while the spindle or A-axis motor is energised. A 50 mm cylinder at 60 RPM has a surface speed of approximately 9 m/min — faster than it visually appears. Do not judge safety by perceived rotation speed.

Chuck key rule	Remove the chuck key before ANY A-axis movement. A key left in the chuck during rotation is a serious ejection hazard. Hang it on a fixed hook away from the machine.
Jaw tightness check	Verify all jaws are fully tightened with the chuck key before every rotation. A loosened jaw will eject at speed.
Tailstock clamped	If tailstock is engaged, verify it is clamped before any A-axis movement. Unclamped tailstock can slide, releasing the workpiece end.
Inspect stock	Do not mount cracked, split, or poorly-dried timber. Defective stock can fracture under rotational and cutting forces.
Stand to the side	Operate from the side of the machine — not in front of or behind the chuck. Ejected stock travels radially outward from the chuck face.
Test rotation first	G1 A360 F180 before starting spindle. Watch the full rotation. Verify clearance from spindle, gantry, and all fixturing throughout.
Checklist additions	Add to pre-cut checklist (Appendix C): (1) Chuck jaws tight with chuck key. (2) Tailstock clamped if engaged. (3) Test rotation complete and clearance confirmed.

What You Can Make with the 4th Axis

Engraved cylinders	Text or pattern engraving on curved surfaces. Indexed work — rotate to each face, engrave, rotate.
Multi-sided objects	Hexagonal, octagonal, or polygon cross-sections. Each flat face machined at the corresponding indexed A position.
Knobs & shafts	Control knobs, decorative handles, gear shafts with flats or keyways. Flat features milled at specific A positions.
Box legs & spindles	Table legs with fluting or decorative grooves. Combine X-axis travel with A-axis indexing for repeating patterns.
Stamp rollers	Cylindrical stamps with relief pattern around the circumference. Indexed face panels or continuous-axis CAM software.

Chess pieces (advanced)	Complex curved profiles using continuous 4-axis CAM in Vectric VCarve Pro or Fusion 360. FreeCAD STEP exports cleanly to both.

APPENDIX J

FreeCAD v1.1.0 for CNC Makers — Student Edition

Glossary

142 terms defined — from Alias to ZigZag

This glossary contains 142 terms used throughout the FreeCAD v1.1.0 for CNC Makers course. Terms are arranged alphabetically. Where a term is defined more fully in a specific chapter, the chapter reference is noted in the definition. Use the chapter-end vocabulary tables as the first reference for terms introduced in context; use this glossary to locate and cross-reference terms across the full course.

A	
A-axis	The fourth axis in a CNC machine, providing rotational movement. Controlled by GRBL with parameters $103 (steps/degree), $133 (max rate), and $143 (acceleration). Used for machining cylindrical workpieces and multi-face indexed work.
Adaptive (strategy)	A CAM pocket-clearing strategy that maintains constant chip load by varying the toolpath arc. More efficient than ZigZag for pockets deeper than 5 mm and for hardwood. Requires a step-over setting expressed as a percentage of tool diameter.
Addon	A community-developed extension to FreeCAD, installed via the Addon Manager. Adds workbenches, tools, or macros not included in the core FreeCAD installation. FCGear is the primary addon used in this course.
Addon Manager	The FreeCAD tool (Tools → Addon Manager) for browsing, installing, updating, and disabling community addons. Limits the list to addons hosted on GitHub with public source code.
Air cut	Running a G-code job with the spindle raised well above the material surface. Used to verify toolpath motion safely before cutting real material. Also called a dry run.
Alarm state	GRBL's locked state when an unexpected event occurs — hard limit trigger, soft limit violation, or homing failure. The machine halts all movement. Cleared with $X in the UGS console.
Alias	A user-assigned name given to a FreeCAD Spreadsheet cell. Aliases allow sketch constraints and feature properties to reference cells by descriptive name (=Spreadsheet.Width) rather than grid location (=Spreadsheet.B2). Aliases are case-sensitive and cannot contain spaces.
Analyse	The fourth level of Bloom's Taxonomy. Involves breaking information down, comparing options, finding patterns, and differentiating. In this course: comparing CAM strategies, diagnosing model errors, analysing joinery fit quality.
Attachment plane	The plane on which a Sketcher sketch is drawn. Selected when creating a new Sketch. The XY plane is the most common choice for flat CNC profiles.

B	
Backlash	The small amount of play in a mechanical drive system (belts, lead screws, gears) where reversing direction does not immediately produce equivalent movement. Reduces dimensional accuracy and can contribute to chatter at direction changes.

Ball-nose end mill	A cutting bit with a hemispherical tip, used for 3D surface machining. Leaves a scalloped surface whose quality is determined by step-over. Used in Waterline roughing and 3D Surface finishing operations.
Baud rate	The speed of serial data communication between UGS and the GRBL controller, measured in bits per second. The CNC 4040 Pro Max communicates at 115200 baud. Using a different baud rate causes garbled communication or no response.
Bill of Materials (BOM)	A structured list of all parts required to assemble a design, including part names, quantities, materials, and thicknesses. Inserted in TechDraw engineering drawings as a table.
Bloom's Taxonomy	A hierarchical framework for classifying learning objectives and cognitive skills, developed by Benjamin Bloom (1956) and revised by Anderson and Krathwohl (2001). Six levels: Remember, Understand, Apply, Analyse, Evaluate, Create. The course is structured to ascend through all six levels.
Body	In FreeCAD Part Design, a container holding all the features that form one solid part. A single FreeCAD document can contain multiple Bodies for multi-body designs.
Boolean operation	A solid geometry operation combining or subtracting 3D bodies. Types: Part Fuse (union — combines two bodies), Part Cut (subtracts one body from another), Part Common (retains only the intersecting volume).

C	
CAD	Computer-Aided Design. The use of software to create precise 2D drawings or 3D models of objects before manufacturing.
CAM	Computer-Aided Manufacturing. The process of using software (FreeCAD's CAM workbench in this course) to generate machine toolpaths and G-code from a 3D model.
Cell alias	See Alias.
CH340	A USB-to-serial chip used on the CNC 4040 Pro Max's control board. Windows computers may require a CH340 driver installation before UGS can establish USB communication.
Chamfer	An angled flat surface applied to an edge, replacing a sharp 90° corner with a bevel at a specified angle and distance. Applied in FreeCAD Part Design using the Chamfer tool.
Chip load	The thickness of material each cutting edge removes per revolution of the bit. Determined by feed rate, spindle speed, and flute count. Correct chip load produces clean cutting. Too low causes rubbing and burning; too high causes vibration or breakage.
Circular reference	An expression error in FreeCAD where a formula references a value that itself depends on the original — creating an unresolvable loop. FreeCAD displays an error and the model cannot update.
Clearance	The small gap added to a mating dimension to allow two parts to assemble without force. In finger joint design, typically 0.15–0.20 mm per side. Set as a named alias in the Spreadsheet.
CNC	Computer Numerical Control. A machining process where the cutting tool is guided by a computer following numerical G-code instructions, producing repeatable, precise results.
CNC 4040 Pro Max	The desktop CNC router used throughout this course. Work area: 400 × 400 mm. Z-axis travel: ~70–80 mm. Controller: GRBL 1.1h. Spindle: ER11 collet. Connected via USB.
COM port	The Windows name for a serial communication port. On macOS and Linux, equivalent ports are named /dev/ttyUSB0 or /dev/ttyACM0. Identified in UGS by unplugging and re-plugging the USB cable.

Compound formula	A FreeCAD expression that combines multiple spreadsheet aliases using arithmetic operators: e.g. =Spreadsheet.Width – 2*Spreadsheet.BorderInset calculates a derived value from two base parameters.
Constraint	In FreeCAD's Sketcher, a rule that restricts the position, size, or relationship of sketch geometry. Types: geometric (defining relationships) and dimensional (defining exact values). Together, constraints reduce degrees of freedom to zero.
Construction geometry	In FreeCAD Sketcher, reference lines and curves that define layout positions but are not included in the profile used for Pad or Pocket operations. Displayed as blue dashed lines.
Control board	The electronics in the CNC 4040 Pro Max that run the GRBL firmware, interpret G-code commands, and drive the stepper motor drivers for each axis.
Create	The sixth and highest level of Bloom's Taxonomy. Involves producing original work by synthesising knowledge and skills in new configurations. In this course: designing original parametric models, planning complete CAM sequences, producing the certification project independently.

D	
Datum plane	A reference plane inserted in the FreeCAD model tree, positioned relative to existing geometry at a defined distance or angle. Used to create Sketches at non-standard positions and as the section planes for Part Loft operations.
Degrees of freedom (DOF)	The number of independent ways a sketch element can still move or change. Each constraint applied removes one or more DOF. A fully constrained sketch has 0 DOF remaining. Displayed numerically in the status bar during Sketch editing.
Depth per pass	See Step-down.
Dimension	A measured annotation on an engineering drawing indicating a distance, diameter, radius, or angle. Placed using TechDraw's Insert Dimension tools. The collection of dimensions must be sufficient for an independent machinist to reproduce the design.
Dogbone	A small circular pocket at each inside corner of a profile or slot, compensating for the radius of the cutting bit. Without dogbones, mating parts cannot seat fully at inside corners because the bit leaves a curved residue. Dogbone diameter equals the tool diameter.
Draft Upgrade	A FreeCAD Draft workbench operation that converts imported SVG wire objects into Part Shape objects suitable for selection as base geometry in CAM Engraving operations.
Draft workbench	A FreeCAD workbench providing 2D/3D drawing tools, array operations, and SVG import. Used in this course primarily for importing SVG artwork files (File → Import or via the Draft menu).
Driven constraint	A sketch constraint whose value is determined by a formula rather than a typed number. Displayed with a small = indicator. Changes automatically when the referenced spreadsheet value changes after Refresh.

E	
E-stop	Emergency stop. A physical button (or software command) that immediately halts all machine movement and spindle rotation. Must be accessible from the operator's normal standing position at all times during machine operation.

End mill	A cylindrical cutting tool with cutting edges on its end face and sides. Used for profile cuts, pocket clearing, and slot cutting. Specified by diameter (3 mm, 6 mm in this course), flute count (typically 2), and overall length.
Engineering drawing	A standardised 2D technical document produced in TechDraw showing orthographic views, dimensions, tolerances, title block, and BOM. Must be independently legible — a competent machinist should be able to reproduce the design from the drawing alone.
ER11 collet	The precision tool-holding chuck system on the CNC 4040 Pro Max spindle. Each collet size accepts one specific shank diameter range (1–7 mm total). The collet is tightened and loosened using two spanners — never finger-tight.
Evaluate	The fifth level of Bloom's Taxonomy. Involves making judgements, justifying decisions, critiquing quality, and defending positions. In this course: justifying CAM strategy choices, assessing joint fit against designed clearance, evaluating surface finish against design intent.
Expression	In FreeCAD, a mathematical formula entered in a property or constraint field that calculates its value from other model parameters or spreadsheet aliases. Expressions are created using the Expressions Editor (blue = button).

F	
.FCStd	FreeCAD's native file format. A compressed ZIP archive containing the full model geometry, feature tree, CAM data, spreadsheets, and all project settings. Version numbers should be included in filenames (Project_v1.FCStd, Project_v2.FCStd).
Feature	In FreeCAD Part Design, a single modelling operation applied to a Body — Pad, Pocket, Chamfer, Fillet, Linear Pattern, Polar Pattern, etc. Features are listed in the Model tree in creation order and applied sequentially.
Feed hold	A UGS/GRBL command that smoothly decelerates and stops machine movement without losing position. The spindle continues running. Triggered via the Feed Hold button in UGS. Not a substitute for e-stop.
Feed rate	The horizontal cutting speed in mm/min. How fast the tool moves sideways through material during a controlled cutting move (G1). One of the four fundamental CAM settings alongside plunge rate, step-down, and spindle speed.
Feed rate override (FRO)	A real-time GRBL command that scales the programmed feed rate by a percentage without modifying the G-code. Applied using the FRO slider in UGS's Machine Control panel. Range: 10–200% of programmed rate.
Fillet	A rounded surface applied to an edge, replacing a sharp 90° corner with a smooth curve of defined radius. Applied in FreeCAD Part Design using the Fillet tool. Edge treatment for organic or flowing forms.
Finger joint	A CNC joinery method where interlocking rectangular fingers are cut into mating panel edges. Requires correct clearance (typically 0.15–0.20 mm), dogbones at all inside corners, and a minimum finger width of 2× the tool diameter.
Firmware	Software permanently embedded in a hardware device's microcontroller, as distinct from software installed on a general-purpose computer. GRBL is the firmware on the CNC 4040 Pro Max's control board.
FreeCAD	A free, open-source, parametric 3D CAD application. Version 1.1.0 is used throughout this course. Available from freecad.org.
Fully constrained	A Sketcher state where all elements have zero degrees of freedom — no element can move without changing or removing a constraint. Displayed as all-green geometry. Required before closing a Sketch.

G	
G0	Rapid positioning move in G-code. Moves the machine to specified coordinates at maximum speed without cutting material. Used for positioning between cutting operations.
G1	Controlled linear cutting move in G-code. Moves the machine to specified coordinates at the programmed feed rate (F value), cutting as it travels.
G2 / G3	Clockwise (G2) and counter-clockwise (G3) arc cutting moves in G-code. Used for curved profile cuts and circular operations. Require centre coordinates (I, J) or radius (R) as parameters.
G92	A G-code command that sets the current position as a specified coordinate value without physically moving the machine. Used to define A-axis zero each session: G92 A0.
Gantry	The bridge-like assembly in the CNC 4040 Pro Max that carries the Z-axis and spindle across the machine bed. The gantry travels in the Y axis (forward and backward).
Gear ratio	The relationship between the speeds of two meshing gears, determined by tooth count: ratio = driven_teeth ÷ driving_teeth. A 3:1 ratio means the output rotates once for every three input rotations.
G-code	The programming language used to control CNC machines. A sequence of movement, spindle, and auxiliary instructions (G0, G1, M3, M5, S, F, etc.) that the GRBL firmware interprets into motor movements.
GRBL	Open-source CNC controller firmware running on the CNC 4040 Pro Max's control board. Interprets G-code commands into precise stepper motor pulses. Version 1.1h is standard for this machine. Pronounced 'gerbil'.
Green light rule	The course principle that every sketch element must be fully constrained (all green) before closing the Sketch. Never close a Sketch with white, yellow, or red elements.

H	
Hard limits	Physical limit switches at the extreme ends of each axis travel. When triggered, GRBL immediately halts all motion — including spindle. Acts as emergency backup after soft limits. Enabled with $21=1. Can cause position loss if triggered at high speed.
Homing	The machine start-up procedure where each axis moves at controlled speed until it triggers its limit switch, establishing a known machine reference position. Triggered with $H in UGS. Required before soft limits ($20=1) can function.

I	
Idle	GRBL's normal ready status when the machine is connected, not executing movement, and no alarm is active. The expected state before sending a job and after a job completes.
Inter-body expression	A FreeCAD expression in one Body that references a dimension, alias, or constraint from another Body in the same document. Enables multi-body models to be parametrically linked without manual value matching.

J	
Job	In FreeCAD CAM, the container object that links toolpaths to a 3D model Body and holds all Stock, Tool, and Operation settings. Every CAM workflow begins with creating a Job.

Job recovery	The procedure for resuming a G-code job after a mid-cut interruption. Steps: stop spindle, note line number, raise Z, diagnose cause, check workpiece, re-verify zeros, re-home if needed, restart from line, monitor closely.
Job Setup Sheet	A standardised documentation form completed for every CNC cutting session, recording material, tools, zeros, feeds, estimated and actual runtime, dimensional results, and issues encountered. The operator's audit trail for reproducibility.

K	
Kerf	The width of material removed by a cutting tool. For end mills, kerf equals the tool diameter. The FreeCAD CAM workbench automatically offsets profile toolpaths by half the tool diameter to account for kerf.

L	
Limit switch	A sensor mounted at the end of each axis travel. Triggered during homing to establish machine coordinates. Can also trigger a hard limit alarm if the machine approaches the physical end of travel during operation.
Linear Pattern	A FreeCAD Part Design feature that repeats a feature or Body at equal intervals in one or two linear directions. Defined by occurrence count and spacing distance.
Loft	A FreeCAD Part workbench feature that creates a 3D solid by interpolating a smooth surface between multiple 2D section profiles placed on separate datum planes. At least three datum sections are recommended for well-resolved organic forms.

M	
M3	G-code command to start the spindle rotating. Format: M3 S[RPM], e.g. M3 S18000 starts the spindle at 18,000 RPM.
M5	G-code command to stop the spindle motor.
M6	G-code command for a tool change pause. The machine stops movement and waits for the operator to change the cutting bit and re-zero Z before resuming.
Machine coordinates	Position values measured from the machine's home position (established by homing). Distinct from work coordinates, which are measured from the operator-set zero point for a specific job.
Model tree	The hierarchical panel on the left side of the FreeCAD interface showing every object, Body, and Feature in the current document. Features are listed in creation order and applied sequentially top-to-bottom.
Module (gear)	The metric unit of gear tooth size. Module = pitch_diameter ÷ number_of_teeth. Two gears can only mesh if they have identical Module values. Minimum practical Module for timber gears: 3.

N	
.nc	A common file extension for G-code files. Contains machine instructions exported from FreeCAD CAM via the Post Processor and sent to the CNC machine via UGS.
Node	In Inkscape and SVG editing, a control point defining the path of a curve or line. Simplifying nodes (Path → Simplify) reduces SVG complexity and improves FreeCAD import reliability.

O	
Open-source	Software whose source code is publicly available and free to use, modify, and distribute. FreeCAD, GRBL, and UGS are all open-source projects.
Origin	The point (0, 0, 0) in the FreeCAD model coordinate system. Fixing sketch geometry to the origin using constraints is a best practice for stable, well-defined parametric models.
Orthographic view	A 2D projection of a 3D model showing one face without perspective distortion. Standard engineering drawings include front, top, and right-side orthographic views as a minimum.

P	
P2 dust mask	A respirator rated to filter particles at the P2 level (95% filtration of 0.3 µm particles). Required for all CNC routing. Standard surgical masks are insufficient — MDF dust contains carcinogenic formaldehyde resins.
Pad	A FreeCAD Part Design feature that extrudes a 2D Sketch profile into a 3D solid in the normal direction of the sketch plane. Creates or adds material to a Body.
Parameter	A named variable in a FreeCAD model or Spreadsheet that controls a dimension or property. Changing the parameter value updates all geometry linked to it after a model Refresh.
Parametric design	A design approach in which model dimensions are controlled by named variables (parameters) rather than fixed numbers. Changing a single parameter value automatically propagates to all linked dimensions throughout the model.
Part Cut	A FreeCAD Part workbench Boolean operation that subtracts the volume of one solid (the tool) from another solid (the base), removing the overlapping region.
Part Fuse	A FreeCAD Part workbench Boolean operation that combines two separate solid bodies into a single unified solid.
Part Loft	See Loft.
Pitch diameter	The effective working diameter of a gear, calculated as Module × number_of_teeth. Two gears mesh correctly when their pitch circles are tangent to each other.
Plain SVG	The Inkscape export format (File → Save As → Plain SVG) that removes Inkscape-specific metadata, producing a clean SVG file for reliable import into FreeCAD's Draft workbench.
Plunge rate	The vertical entry speed in mm/min for a cutting tool descending into material at the start of each pass. Always slower than feed rate — typically 20–35% of feed rate. Too fast causes bit breakage on entry.
Pocket	In FreeCAD Part Design, a feature that removes material from an existing solid by cutting a Sketch profile into it. Also the name of the FreeCAD CAM Pocket Shape operation that clears an area of material.
Pocket Shape (CAM)	A FreeCAD CAM operation that clears material from a defined area using parallel passes (ZigZag strategy) or arc-motion passes (Adaptive strategy). Produces a flat-floored recess.
Polar Pattern	A FreeCAD Part Design feature that repeats a feature or Body at equal angular intervals around a specified axis. Used for arrangements in a circle, such as clock face hour markers.
Post-processor	Software that converts FreeCAD CAM's generic toolpath data into G-code formatted for a specific CNC controller. The 'grbl' post-processor is used in this course. Selected in CAM → Post Process.

Pre-cut checklist	A standardised list of machine, material, tool, zero point, G-code, and safety checks completed before sending any job. Reproduced as a reproducible form in Appendix C.
Profile (CAM)	A CAM operation that cuts along the outer or inner boundary of a profile at progressive depths, automatically offsetting by half the tool diameter. Always the last operation in any job that produces a cut-out part.
Properties panel	The panel in FreeCAD showing the properties of the selected feature. The Data tab contains functional values (dimensions, settings); the View tab contains display settings (colour, transparency).

R	
Rapid move	A machine movement at the machine's maximum speed (G0 command). Moves the tool between positions without cutting. Rapid moves are shown in yellow in the CAM Simulator — they must always stay above the material surface.
Reflection	A written document submitted with each project in which the student explains design decisions made, challenges encountered and resolved, and a Bloom's Taxonomy self-assessment of the level at which they operated.
Remember	The first and foundational level of Bloom's Taxonomy. Involves recalling and recognising information. In this course: naming FreeCAD interface elements, recalling G-code command functions, identifying constraint types.
Refresh	The FreeCAD operation (Ctrl+Shift+R or Edit → Refresh model) that recalculates all spreadsheet-driven dimensions after a parameter change. Required after every spreadsheet edit before the model updates.

S	
Scallop height	The height of ridges left on a 3D machined surface between adjacent ball-nose tool passes. Smaller step-over = lower scallop height = smoother surface = longer machining time. Controlled by the step-over setting.
Section view	An engineering drawing view showing internal geometry as if the object had been cut by a plane. Required in TechDraw drawings from Project 9 onwards. Created with TechDraw → Insert Section View.
ShapeString	A FreeCAD Part workbench tool that creates a 3D solid representation of text from a chosen font and character height. The resulting letter edges are used as base geometry for CAM Engraving operations.
Sketch	A 2D drawing created in FreeCAD's Sketcher workbench, attached to a plane or face. The foundation for all Part Design features — every Pad, Pocket, and Pattern starts with a fully constrained Sketch.
Sketcher	The FreeCAD workbench for drawing and constraining 2D geometry. Accessed automatically when a new Sketch is created in Part Design, and closed automatically when the Sketch is finished.
Soft limits	A GRBL feature that checks planned G-code moves against travel limits ($130/$131/$132) before executing them. If a move would exceed the limits, GRBL refuses it before it happens. Enabled with $20=1. Requires homing ($22=1) to function.
Spoilboard	A sacrificial layer of material (typically MDF) placed over the T-slot table work surface. Provides a flat reference surface for cut-through operations and protects the aluminium table from end mill contact.
Spreadsheet (workbench)	A FreeCAD workbench providing a spreadsheet embedded inside the .FCStd file. Used for parametric design: cells hold named aliases that drive model dimensions via formula expressions.

Staircase	The stepped surface appearance produced by Waterline roughing, caused by the discrete horizontal slices of the operation. Removed by the subsequent 3D Surface finishing pass.
Step-down	The depth of material removed in each cutting pass, in millimetres. Also called depth per pass. Smaller step-down = more passes = less force per pass = better surface quality = longer machining time. Maximum guide: ≤0.5× tool diameter in hardwood.
Step-over	The horizontal distance between adjacent cutting passes, expressed as a percentage of tool diameter or as an absolute distance. Lower step-over = more passes = smoother floor = longer runtime. Typical value: 60–70% for MDF, 50–60% for hardwood.
Stock	In FreeCAD CAM, a virtual block representing the physical raw material on the machine bed. Must match actual material dimensions exactly — the Stock Z height defines the material's top surface from which all depth calculations are made.
Stock-to-leave	An offset applied during 3D roughing operations that preserves a thin skin of material on all surfaces for removal by the finishing pass. Typical value: 0.3 mm. Prevents over-cutting during roughing.
SVG	Scalable Vector Graphics. An XML-based vector image format used in this course for CNC artwork files. Prepared in Inkscape and imported into FreeCAD via the Draft workbench.

T	
T-slot table	The aluminium extrusion work surface of the CNC 4040 Pro Max. T-shaped channels accept M5/M6 T-nuts for flexible clamping arrangements. Keep clean of chips and debris after every session.
Tab (holding tab)	A small bridge of uncut material that holds a cut-out part attached to the surrounding stock until manually removed. Prevents the part from shifting or lifting during the final profile pass. Minimum: 4 tabs, 4 mm wide × 3 mm tall.
TechDraw	The FreeCAD workbench for generating 2D engineering drawings from 3D models. Produces orthographic projections, section views, dimensions, tolerances, title blocks, and Bills of Materials. Required for Projects 9, 10, and the Certification Project.
Title block	The standardised information panel in the corner of an engineering drawing containing the drawing title, designer name, date, scale, material specification, and sheet number.
Tolerance	An acceptable range of dimensional variation. Written as ±value (e.g. 15.00 ±0.15 mm). Joint dimensions are toleranced in TechDraw drawings to specify the designed clearance explicitly.
Tool Library	In FreeCAD CAM, a database of cutting tool definitions stored within a Job. Each definition specifies the tool type, diameter, flute count, and cut length. Build once, reuse across all projects.

U	
Understand	The second level of Bloom's Taxonomy. Involves explaining and interpreting information in your own words. In this course: explaining why dogbones are needed, describing what a Waterline operation does, summarising the parametric update cycle.
UGS	Universal G-code Sender. The free, open-source desktop application used in this course to send G-code files from a computer to the CNC 4040 Pro Max via USB. Connects at 115200 baud.

V	
V-bit	A CNC cutting tool with a V-shaped profile. The effective cutting width varies with depth — deeper cuts produce wider V-grooves. Used for engraving text and decorative lines. Common angles: 60° and 90°.
Vector graphic	A digital image defined by mathematical path descriptions (lines, arcs, and curves) rather than pixels. Can be scaled to any size without quality loss. SVG is the standard vector format used in this course.
Version control	The practice of incrementing filenames (v1, v2, v3) and retaining all earlier versions. Enables tracing changes, reproducing any past state, and identifying which version of a design was cut on a specific date.

W	
Waterline (3D CAM)	A 3D CAM roughing strategy that removes material in horizontal slices at a fixed Z step-down interval. Produces a staircase pattern on curved surfaces. The staircase is removed by the subsequent 3D Surface finishing pass.
Workbench	A collection of tools, menus, and toolbar buttons in FreeCAD activated together for a specific type of task. Switching workbenches changes the available tools without affecting the document.
Work coordinates	Position values measured from the operator-set zero point for a specific job. Distinct from machine coordinates. Set at the start of each session using the Zero button in UGS after jogging the spindle to the reference position.
Work envelope	The three-dimensional space within which the cutting tool can reach. For the CNC 4040 Pro Max: 400 × 400 mm (X × Y) by approximately 70–80 mm (Z). Designs that exceed the work envelope cannot be cut in a single setup.

X	
X axis	The horizontal axis of the CNC machine, running left-right across the bed. Positive X moves the tool to the right. The X rail runs along the gantry bridge.

Y	
Y axis	The axis running forward-backward along the machine bed. Positive Y moves the tool away from the operator (toward the back of the machine). The gantry travels in the Y direction.

Z	
Z axis	The vertical axis. Positive Z moves the spindle upward (away from material). Negative Z moves the spindle into material. Setting Z zero correctly is the single most critical machine setup step in any cutting session.
Zero point	The operator-defined origin for a specific cutting job, set at the start of every session by jogging the spindle to the reference position and zeroing X, Y, and Z in UGS. All G-code coordinates in the job are measured from this point.
ZigZag (strategy)	A CAM pocket-clearing strategy using back-and-forth parallel passes. Simple and predictable. Best for shallow pockets (under 5 mm) and MDF. For deeper pockets or harder materials, the Adaptive strategy is preferred.

APPENDIX
K

FreeCAD v1.1.0 for CNC Makers — Student Edition

Index

364 entries — chapters, appendices, and key terms

This index covers all key terms, concepts, G-code commands, FreeCAD features, and tools introduced across the 24 chapters and 10 appendices of this textbook. References are by chapter (**Ch.N** in teal) and appendix (*App.X*). Page numbers will be inserted during final typeset layout.

Key: Ch.N = Chapter N *App.A–J* = Appendix A–J App.A Keyboard Shortcuts App.B Feeds & Speeds App.C Machine Specs App.D GRBL Settings App.E Bloom's App.F Rubrics App.I 4th Axis App.J Glossary

Symbols & Numbers

A

B

C

D

E

F

G

H

I

P

R

S

T

U

V

W

X

Made in the USA
Coppell, TX
02 November 2026

10699189R00251